T0215476

Introductory Physics

Physics describes how motion works in everyday life. Clothes washers and rolling pins are undergoing rotational motion. A flying bird uses forces. Tossing a set of keys involves equations that describe motion (kinematics). Two people bumping into each other while cooking in a kitchen involves linear momentum.

This textbook covers topics related to units, kinematics, forces, energy, momentum, circular and rotational motion, Newton's general equation for gravity, and simple harmonic motion (things that go back and forth). A math review is also included, with a focus on algebra and trigonometry.

The goal of this textbook is to present a clear introduction to these topics, in small pieces, with examples that readers can relate to. Each topic comes with a short summary, a fully solved example, and practice problems. Full solutions are included for over 400 problems.

This book is a very useful study guide for students in introductory physics courses, including high school and college students in an algebra-based introductory physics course and even students in an introductory calculus-level course. It can also be used as a standalone textbook in courses where derivations are not emphasized.

Michael Antosh teaches physics at the University of Rhode Island, USA.

Introductory Physics

Summaries, Examples, and Practice Problems

Michael Antosh

CRC Press
Taylor & Francis Group
Boca Raton London New York

CRC Press is an imprint of the
Taylor & Francis Group, an **informa** business

First edition published 2023
by CRC Press
6000 Broken Sound Parkway NW, Suite 300, Boca Raton, FL 33487-2742

and by CRC Press
4 Park Square, Milton Park, Abingdon, Oxon, OX14 4RN

ISBN: 978-0-367-43685-8 (hbk)
ISBN: 978-0-367-43423-6 (pbk)
ISBN: 978-1-003-00504-9 (ebk)

DOI: 10.1201/9781003005049

Typeset in Minion
by Deanta Global Publishing Services, Chennai, India

Contents

Dear Reader:

The goal of this textbook is to have topics laid out clearly, in small pieces, with examples that apply to everyday life.

If you could use some additional information on math, it might help to go through the math review chapter first – that's Chapter 13.

Physics can be a very difficult subject, but it can also be very doable and enjoyable.

Good luck!

Acknowledgements

Thank you to my wife and to my mother, for your support and for helping me make examples by answering questions like "what situation in everyday life has something that spins faster and faster?"

Thank you to my father for support, lots of advice, and inspiration.

Thank you to Leon N Cooper for teaching me to teach Newton's second law before the first law, since the first law is just a special case of the second law.

I have had the opportunity to teach introductory physics topics using four different textbooks, and each textbook has influenced this book.

"College Physics" by Michael Tammaro (ISBN 978-1119361053) contains similar:

Sections: a math review chapter, a section on significant figures (and keeping to three significant figures in the textbook), and forces that aren't named and come from objects making contact.

Problems: more than one force adding up to the net force in the centripetal direction (and calculating those forces), kinematics problems where you find all unknown variables, force problems where more than two forces have vertical components (and you find the normal force for example), and problems with the largest possible value of position/velocity/acceleration in simple harmonic motion.

Explanations: when to use radian units, the number of known variables needed to solve kinematics problems, describing the angle in Equation 6.2.1, and the importance of two and three-dimensional motion.

Equations: this textbook's Equation 2.6.3 and most equations for Sections 6.9 and 6.10.

Applications: there may be similar problems with forces on cars, forces on circling airplane pilots, and spinning fans.

"Physics for Scientists and Engineers" by Tipler and Mosca (ISBN 978-1429201247) contains similar:

Sections: rolling without slipping, rope and pulley problems.

Problems: mathematical style vector math, static friction problems with maximum and non-maximum amounts of force, ramp problems asking for acceleration, two-dimensional kinematics problems that are not projectile motion, problems where the change in mechanical energy is computed, and rope and pulley problems. In particular, some of the rope and pulley problems contain the same setup as in this textbook.

Equations: Equation 7.6.1 in this textbook is a slightly modified version of an equation that I have only seen in Tipler and Mosca's text; the torque vector component equations are also very similar or the same.

Explanations: I used very similar explanations for dimensions behaving separately (Section 3.2), describing projectile motion as a special case of two-dimensional kinematics, maximum height velocity being between upward and downward, stating that forces cause motion, using a dot for free body diagrams, the derivation for conservation of mechanical energy (including all work being conservative or non-conservative), potential energy depending on position, collision forces being large enough to assume linear momentum conservation, the connection between linear momentum conservation and center of mass, the use of centripetal and tangential directions (including as applied to centripetal force), larger shapes' moment of inertia coming from adding point masses, conservation of mechanical energy with Newton's law of gravity, the equations for rolling without slipping, and the relationships of force and acceleration as well as mechanical energy for simple harmonic motion.

Applications: there may be similar problems with energy and projectiles (soccer balls), forces doing opposite but equal work, mechanical energy problems with vertical springs, airplane pilots or toy cars making a vertical loop, rotating bowling balls, and tools.

"College Physics" by Urone, Hinrichs, Dirks, and Sharma (ISBN 978-1-938168-00-0) has similar:

Sections: similar placement of uniform circular motion in the text.

Problems: problems with kinematics and forces combined.

Equations: simple harmonic motion equations without ω.

Explanations: I used very similar explanations for different time points used in larger kinematics problems, naming problem types with forces, that frequency is the number of repeats per second in simple harmonic motion, and I made similar use of the terms prefix (Section 1.3) and quantity (throughout).

Applications: there may be similar problems with cars on ramps, three or more vertical forces alone, multiple problems on different physics concepts but with the same situation, spring and gravity-based mechanical energy problems, work or force problems with four forces in different directions, problems with carts (or similar), problems that use north/east/south/west for directions, and motion of the moon.

"Physics: Structure and Meaning" by Cooper (ISBN 978-0874515923) may contain problems where the acceleration changes but can be broken into pieces with constant acceleration, similar to the problem outlined in Section 2.8.

Units and Significant Figures

1.1 INTRODUCTION: UNITS HELP TELL US HOW MUCH

Units are part of describing how much of something there is. Units give us two types of information:

- What type of quantity you are talking about – for example, "seconds" indicate time, and "meters" indicate distance.

- The size of each number you are talking about – for example, 5 feet versus 5 centimeters.

In this chapter, we'll learn how to convert between different sets of units and how to use units to double-check our answers.

1.2 UNIT CONVERSIONS

There are two major unit systems used in this book:

- SI units. SI stands for "système international". It is used in most countries, and physics is almost always done using these units. Kilometers and kilograms are two examples.

- Imperial units. Some countries use these units in everyday life, including the United States and United Kingdom. Feet and pounds are two examples.

DOI: 10.1201/9781003005049-1

For the motion problems that we will do, every single unit will involve some combination of the following four quantities: length, time, mass, and angle, as shown in Table 1.1.

There are many combinations of the SI units in physics. One particularly long example is power, which is kg·m²/s³ and has been given the name Watts.

For each quantity, the units can be converted between SI and imperial units, and also between related units like seconds and hours. Here is a list of common conversion factors:

$$\text{Length} : 1\,\text{m} = 3.281\,\text{ft} \tag{1.2.1}$$

$$\text{Length} : 1609\,\text{m} = 1\,\text{mi} \tag{1.2.2}$$

$$\text{Length} : 5280\,\text{ft} = 1\,\text{mi} \tag{1.2.3}$$

$$\text{Length} : 5280\,\text{ft} = 1609\,\text{m} \tag{1.2.4}$$

$$\text{Length} : 12\,\text{in} = 1\,\text{ft} \tag{1.2.5}$$

$$\text{Time} : 60\,\text{min} = 1\,\text{hr} \tag{1.2.6}$$

$$\text{Time} : 60\,\text{s} = 1\,\text{min} \tag{1.2.7}$$

$$\text{Angle} : 2 \cdot \pi \text{ radians} = 360 \text{ degrees (note} : \pi \approx 3.14) \tag{1.2.8}$$

Sometimes, it's useful to convert between two types of units. Maybe you solve for an answer in SI units (like meters/second), and you want to do a sense check by converting it to something you see in everyday life (like miles/hour).

To do unit conversions, you use a conversion like a fraction and multiply it by the number you wish to convert. The top and bottom (numerator and denominator) of the fraction are chosen such that the old units cancel out.

TABLE 1.1   Units for Some Commonly Used Quantities, with Abbreviations

Quantity	SI Unit	Imperial Unit
Length	Meters (m)	Feet (ft), miles (mi)
Time	Seconds (s), minutes (min), hours (hr)	Seconds (s), minutes (min), hours (hr)
Mass	Kilograms (kg)	Slugs
Angle	Radians (rad) or degrees (°)	Radians (rad) or degrees (°)

Example 1.2.1: Conversion from Miles to Meters

You travel 0.500 miles to school. How far is that in meters?

Solution:

This one only requires one conversion factor – a conversion between meters and miles. Above, we can see that the conversion is:

$$\text{Length} : 1\ \text{mi} = 1609\ \text{m}$$

We know that we need to make this into a fraction. But is it (1 mi/1609 m), or is it (1609 m/1 mi)? You start with an answer in miles, so the conversion that would change this answer to meters would be (meters/mile). If you do it that way, your units become:

$$\text{miles} \cdot \frac{\text{meters}}{\text{miles}} = \text{meters}$$

If you do the conversion the other way, you get:

$$\text{miles} \cdot \frac{\text{miles}}{\text{meters}} = \frac{\text{miles}^2}{\text{meters}}$$

The second choice is definitely not the units we want! The first choice works.

Multiply our original amount (0.500 meters) by a conversion factor that is meters/miles

$$0.500\ \text{miles} \cdot \frac{1609\ \text{meters}}{1\ \text{miles}} = 805\ \text{meters}$$

Note: in this first example, "meters" and "miles" are written out. It is more common to abbreviate units – see Table 1.1 for abbreviations. Miles are abbreviated "mi" and meters are abbreviated "m". In that case, the solution would look like:

$$0.500\ \text{mi} \cdot \frac{1609\ \text{m}}{1\ \text{mi}} = 805\ \text{m}$$

Example 1.2.2: Converting from Miles/ Hour (mi/hr) to Meters/Second (m/s)

You are in a bus, on the highway, traveling at 60.0 mi/hr. What is this speed in SI units, m/s?

Solution:

This problem will involve two conversions – one for the length quantity in the units (mi to m) and one for the time quantity in the units (1/hr to 1/s). Further, we don't have a direct conversion from hours to seconds, so we'll look for a middle step. It turns out that we can convert from hours to minutes, and then from minutes to seconds. Based on that information, we'll use three conversion factors from the list of factors above:

$$\text{Length} : 1609 \text{ m} = 1 \text{ mi} \tag{1.2.2}$$

$$\text{Time} : 60 \text{ min} = 1 \text{ hr} \tag{1.2.6}$$

$$\text{Time} : 60 \text{ s} = 1 \text{ min} \tag{1.2.7}$$

We can make each factor a fraction. And we'll choose the numerator (top) and denominator (bottom) of that fraction such that the old unit cancels each time.

Let's do this in steps. First, we'll convert the length, using the same strategy as example problem 1.2.1 above:

$$60.0 \frac{\text{mi}}{\text{hr}} \cdot \frac{1609 \text{ m}}{1 \text{ mi}} = \left(60.0 \cdot 1609\right) \frac{\text{m}}{\text{hr}}$$

We didn't do the math out, because we don't need to until the last step. We now have units of m/hr (meters/hour), which is correct for the length but incorrect for the time. We will now convert hours to minutes:

$$60.0 \frac{\text{mi}}{\text{hr}} \cdot \frac{1609 \text{ m}}{1 \text{ mi}} \cdot \frac{1 \text{ hr}}{60 \text{ min}} = \frac{\left(60.0 \cdot 1609\right)}{60} \frac{\text{m}}{\text{min}}$$

In the conversion from 1/hr to 1/min, we put the hr in the numerator (top) of the fraction so that it would cancel out the existing hr in the denominator (bottom) of the original units. We now have units of meters/minute. To finish, we convert minutes to seconds:

$$60.0 \frac{\text{mi}}{\text{hr}} \cdot \frac{1609\,\text{m}}{1\,\text{mi}} \cdot \frac{1\,\text{hr}}{60\,\text{min}} \cdot \frac{1\,\text{min}}{60\,\text{s}} = \frac{(60.0 \cdot 1609)}{(60 \cdot 60)} \frac{\text{m}}{\text{s}}$$

This is the same process as going from hours to minutes. We can now calculate the final number:

$$60.0 \frac{\text{mi}}{\text{hr}} \cdot \frac{1609\,\text{m}}{1\,\text{mi}} \cdot \frac{1\,\text{hr}}{60\,\text{min}} \cdot \frac{1\,\text{min}}{60\,\text{s}} = 26.8\,\text{m/s}$$

Problem to Try Yourself

Note: there will be many "problem to try yourself" problems in this textbook. You'll get the most out of them if you try them before looking at the solution.

Convert 17.0 m/s to mi/hr.

Solution:

$$\frac{17.0\,\text{m}}{\text{s}} \cdot \frac{1\,\text{mi}}{1609\,\text{m}} \cdot \frac{60\,\text{s}}{1\,\text{min}} \cdot \frac{60\,\text{min}}{1\,\text{hr}} = 38.0\,\text{mi/hr}$$

1.3 POWER OF 10 CONVERSIONS

SI units are often written with different "prefixes". Each one means a different power of 10. Some of the more common prefixes to be used with motion are milli, centi, and kilo, as shown in Table 1.2.

As an example of prefixes, think about length:

- 1 millimeter (mm) is 0.001 meters

- 1 centimeter (cm) is 0.01 meters

- 1 kilometer (km) is 1000 meters

TABLE 1.2   Unit Prefixes That Are Commonly Used with Motion

Prefix	Symbol	Relative Value
milli	m	$0.001\ (10^{-3})$
centi	c	$0.01\ (10^{-2})$
kilo	k	$1000\ (10^{3})$

Changing from one prefix to another can be thought of in exactly the same way as unit conversions. We'll now do one example of a unit conversion with a prefix in it.

Example 1.3.1: Converting Velocity on a Slower Road

Even though physics uses m/s for units of velocity, velocity is often listed in units of km/hr (kilometers per hour) on road signs in the countries that use SI units. You cross a country border, for example from the US to Canada. If you were going 30.0 mi/hr, what is the equivalent velocity in km/hr?

Solution:
Here, we are converting mi/hr to km/hr. The time units are the same (1/hr), so we're only converting miles to km. We'll do this in two steps, using the conversion factors that we have in Sections 1.2 and 1.3. We'll go from miles to meters using a conversion factor from Section 1.2:

$$\text{Length}: 1609 \text{ m} = 1 \text{ mi} \tag{1.2.2}$$

Then, we will convert from meters to kilometers using 1 km = 1000 m, as shown in Table 1.2 (using the entry for "kilo"). In each case, we will use these conversion factors to cancel out the previous units, as with the examples in Section 1.2:

$$30.0 \frac{\text{mi}}{\text{hr}} \cdot \frac{1609 \text{ m}}{1 \text{ mi}} \cdot \frac{1 \text{ km}}{1000 \text{ m}} = 48.3 \text{ km/hr}$$

Problem to Try Yourself

Convert 1.3 km to feet

Solution:

$$0.300 \text{ km} \cdot \frac{1000 \text{ m}}{1 \text{ km}} \cdot \frac{3.281 \text{ ft}}{1 \text{ m}} = 984 \text{ ft}$$

1.4 SIGNIFICANT FIGURES

Significant figures tell you how well you know a certain value. For example, if you are measuring the amount of time, and your value is 1.25 s, keeping numbers out to the second number after the decimal tells you that the number is known to within about 0.01 seconds. If it was kept out to just one decimal place, for example 1.4 s, you would know that the number is known to within about 0.1 seconds.

When looking at a number, you can count how many significant figures are in it. In general, it is the number of digits shown. Any zero at the start of the number doesn't count, and any number with a zero at the end should have a decimal in it (or at the end, like 100. instead of 100).

When adding or subtracting two numbers, look at the number of significant figures after the decimal point. The smallest amount is what your solution will have.

$$Example : 1.23 + 0.1 = 1.3$$

The first number (1.23) had 2 significant figures after the decimal, while the second number (0.1) had one. Since one is smaller than two, the answer will have one significant figure after the decimal. Note: when you need to cut numbers off, you round. Here, 1.33 is rounded to 1.3.

When multiplying or dividing two numbers, look at the total number of significant figures. The smallest amount is what your solution will have.

$$Example : 1.23 / 2.0 = 0.62$$

The first number (1.23) has three significant figures, and the second number (2.0) has two. Since two is smaller than three, the solution will have two significant figures. Note: the zero in the front of 0.62 does not count as a significant figure, since all zeroes in the front do not count.

If your number is large, but you only have a small number of significant figures, you can use **exponential notation**. For example, 23,000 with two significant figures can be written as $2.3 \cdot 10^4$.

In this textbook, most numbers will have three significant figures. Solutions will occasionally ignore the rules of significant figure addition/subtraction/multiplication/division. In a problem with more than one step, you can choose not to apply the significant figure rules until the very end if you would like.

Problem to Try Yourself

If $n_1 = 1.05$ and $n_2 = 1.889$, give the answer in correct significant figures for:

A. $n_1 + n_2$
B. n_1/n_2

Solution:

A. n_1 has two significant figures after the decimal, and n_2 has three. Take the smaller: two.

$n_1 + n_2 = \underline{2.94}$

B. n_1 has three significant figures, and n_2 has four. Take the smaller: three

$n_1/n_2 = \underline{0.556}$

1.5 CHAPTER 1 SUMMARY

Units help us know how much there is of something.

Unit conversions help to go between different units (for example between seconds and hours), between SI and imperial units (for example between meters and feet), and between different powers of 10 (for example between meters and kilometers). For each conversion, you take one or more **conversion factors** and multiply by them as a fraction (for example 3.281 feet/1 meter).

Significant figures tell you how well you know a certain value. They are the number of digits in a number, except for any zeroes at the start. Every number that ends with zero should have a decimal in it (for example, 100. instead of 100). When adding or subtracting two numbers, look at the number of significant figures after the decimal point, and the smallest amount is what your solution will have. When multiplying or dividing two numbers, look at the total number of significant figures, and the smallest amount is what your solution will have.

Chapter 1 problem types:
Unit conversions (Section 1.2), including using prefixes (Section 1.3)
Problems where you need to answer with the correct number of significant figures (Section 1.4)

Practice Problems

Section 1.2

1. Convert 25.0 ft to m

    Solution:

    $$25.0 \text{ ft} \cdot \frac{1 \text{ m}}{3.281 \text{ ft}} = 7.62 \text{ m}$$

2. Convert 3.00 m to feet

    Solution:

    $$3.00 \text{ m} \cdot \frac{3.281 \text{ ft}}{1 \text{ m}} = 9.84 \text{ ft}$$

3. Convert 55.0 mi/hr to m/s

    Solution:

    $$55.0 \frac{\text{mi}}{\text{hr}} \cdot \frac{1609 \text{ m}}{1 \text{ mi}} \cdot \frac{1 \text{ hr}}{60 \text{ min}} \cdot \frac{1 \text{ min}}{60 \text{ s}} = 24.6 \text{ m/s}$$

4. Convert 20.0 m/s to mi/hr

    Solution:

    $$20.0 \frac{\text{m}}{\text{s}} \cdot \frac{1 \text{ mi}}{1609 \text{ m}} \cdot \frac{60 \text{ s}}{1 \text{ min}} \cdot \frac{60 \text{ min}}{1 \text{ hr}} = 44.7 \text{ mi/hr}$$

5. Convert 20.0 m/s to ft/s

    Solution:

    $$20.0 \frac{\text{m}}{\text{s}} \cdot \frac{3.281 \text{ ft}}{1 \text{ m}} = 65.6 \text{ ft/s}$$

6. Convert 5.00 m to feet

    Solution:

    $$5.00 \text{ m} \cdot \frac{3.281 \text{ ft}}{1 \text{ m}} = 16.4 \text{ ft}$$

7. Convert 25.0 m/s to ft/s

Solution:

$$25.0 \frac{\text{m}}{\text{s}} \cdot \frac{3.281 \text{ ft}}{1 \text{ m}} = 82.0 \text{ ft/s}$$

8. Convert 15.0 ft to m

Solution:

$$15.0 \text{ ft} \cdot \frac{1 \text{ m}}{3.281 \text{ ft}} = 4.57 \text{ m}$$

9. Convert 25.0 mi/hr to m/s

Solution:

$$25.0 \frac{\text{mi}}{\text{hr}} \cdot \frac{1609 \text{ m}}{1 \text{ mi}} \cdot \frac{1 \text{ hr}}{60 \text{ min}} \cdot \frac{1 \text{ min}}{60 \text{ s}} = 11.2 \text{ m/s}$$

10. Convert 12.0 m/s to mi/hr

Solution:

$$12.0 \frac{\text{m}}{\text{s}} \cdot \frac{1 \text{ mi}}{1609 \text{ m}} \cdot \frac{60 \text{ s}}{1 \text{ min}} \cdot \frac{60 \text{ min}}{1 \text{ hr}} = 26.8 \text{ mi/hr}$$

Section 1.3

11. Convert 55.0 mi/hr to km/hr

Solution:

$$55.0 \frac{\text{mi}}{\text{hr}} \cdot \frac{1609 \text{ m}}{1 \text{ mi}} \cdot \frac{1 \text{ km}}{1000 \text{ m}} = 88.5 \text{ km/hr}$$

12. Convert 95.0 km/hr to mi/hr

Solution:

$$95.0 \frac{\text{km}}{\text{hr}} \cdot \frac{1000 \text{ m}}{1 \text{ km}} \cdot \frac{1 \text{ mi}}{1609 \text{ m}} = 59.0 \text{ mi/hr}$$

13. Convert 95.0 km/hr to m/s

Solution:

$$95.0\frac{km}{hr}\cdot\frac{1000\,m}{1\,km}\cdot\frac{1\,hr}{60\,min}\cdot\frac{1\,min}{60\,s}=26.4\,m/s$$

14. Convert 2.50 km to ft

Solution:

$$2.50\,km\cdot\frac{1000\,m}{1\,km}\cdot\frac{3.281\,ft}{1\,m}=8.20\cdot10^3\,ft$$

15. Convert 1.00 ft to cm

Solution:

$$1.00\,ft\cdot\frac{1\,m}{3.281\,ft}\cdot\frac{100\,cm}{1\,m}=30.5\,cm$$

16. Convert 65.0 km/hr to m/s

Solution:

$$65.0\frac{km}{hr}\cdot\frac{1000\,m}{1\,km}\cdot\frac{1\,hr}{60\,min}\cdot\frac{1\,min}{60\,s}=18.1\,m/s$$

17. Convert 25.0 mi/hr to km/hr

Solution:

$$25.0\frac{mi}{hr}\cdot\frac{1609\,m}{1\,mi}\cdot\frac{1\,km}{1000\,m}=40.2\,km/hr$$

18. Convert 4.89 ft to cm

Solution:

$$4.89\,ft\cdot\frac{1\,m}{3.281\,ft}\cdot\frac{100\,cm}{1\,m}=149\,cm$$

19. Convert 1.29 km to ft

Solution:

$$1.29 \text{ km} \cdot \frac{1000 \text{ m}}{1 \text{ km}} \cdot \frac{3.281 \text{ ft}}{1 \text{ m}} = 4.23 \cdot 10^3 \text{ ft}$$

20. Convert 75.0 km/hr to mi/hr

Solution:

$$75.0 \frac{\text{km}}{\text{hr}} \cdot \frac{1000 \text{ m}}{1 \text{ km}} \cdot \frac{1 \text{ mi}}{1609 \text{ m}} = 46.6 \text{ mi}/\text{hr}$$

Section 1.4

21. Using significant figures, what is the answer to 16.8/3.987?

Solution:

For multiplication or division, your answer should have the smaller number of significant figures between the two numbers. One number (16.8) has three significant figures, and one number (3.987) has four. The answer must have three significant figures.

$$16.8 / 3.987 = 4.21$$

22. Using significant figures, what is the answer to 56.16+82.3?

Solution:

For addition or subtraction, your answer should have the smaller number of digits after the decimal. 56.16 has two digits after the decimal, and 82.3 has one digit. The answer should have one digit. In order to get to one digit, we will round.

$$56.16 + 82.3 = 138.5$$

23. Using significant figures, what is the answer to 82.3−56.16?

Solution:

For addition or subtraction, your answer should have the smaller number of digits after the decimal. 56.16 has two digits after the decimal, and 82.3 has one digit. The answer should have one digit. In order to get to one digit, we will round.

$$82.3 - 56.16 = 26.1$$

24. Using significant figures, what is the answer to 16.8·3.987?

Solution:

For multiplication or division, your answer should have the smaller number of significant figures between the two numbers. One number (16.8) has three significant figures, and one number (3.987) has four. The answer must have three significant figures.

$$16.8 \cdot 3.987 = 67.0$$

25. Using significant figures, what is the answer to 82.3/56?

Solution:

For multiplication or division, your answer should have the smaller number of significant figures between the two numbers. One number (82.3) has three significant figures, and one number (56) has two. The answer must have two significant figures.

$$82.3 / 56 = 1.5$$

26. Using significant figures, what is the answer to 2.572−1.37?

Solution:

For addition or subtraction, your answer should have the smaller number of digits after the decimal. 2.572 has three digits after the decimal, and 1.37 has two digits. The answer should have two digits. In order to get to two digits, we will round.

$$2.572 - 1.37 = 1.20$$

27. Using significant figures, what is the answer to 16.78/3.234?

Solution:

For multiplication or division, your answer should have the smaller number of significant figures between the two numbers.

One number (16.78) has four significant figures, and one number (3.234) also has four. The answer must have four significant figures.

$$16.78 / 3.234 = 5.189$$

28. Using significant figures, what is the answer to 10.5+0.725?

Solution:

For addition or subtraction, your answer should have the smaller number of digits after the decimal. 10.5 has one digit after the decimal, and 0.725 has three digits. The answer should have one digit after the decimal. In order to get to one digit, we will round.

$$10.5 + 0.725 = 11.2$$

29. Using significant figures, what is the answer to 16.78·3.1?

Solution:

For multiplication or division, your answer should have the smaller number of significant figures between the two numbers. One number (16.78) has four significant figures, and one number (3.1) has two. The answer must have two significant figures.

$$16.78 \cdot 3.1 = 52$$

30. Using significant figures, what is the answer to 5.22+0.640?

Solution:

For addition or subtraction, your answer should have the smaller number of digits after the decimal. 5.22 has two digits after the decimal, and 0.640 has three digits. The answer should have two digits. In order to get to two digits, we will round.

$$5.22 + 0.640 = 5.86$$

Motion in a Straight Line

2.1 INTRODUCTION: DESCRIBING MOTION IN THE WORLD

One of the most important topics in introductory physics is describing how things move. There are many examples of things moving when you look at the world – for example, cars and trains moving towards a destination, people walking, and acorns falling.

This chapter is about linear motion, which is things that move in a straight line. In Chapter 3 we'll start to cover things that move in a more general way.

2.2 MOTION QUANTITIES

In order to describe something in physics we need **quantities**, which are amounts of something important. When talking about motion in a straight line, there are four important quantities:

- *Displacement* is the change in the object's position. We use the symbol Δx, and the SI units for displacement are meters (abbreviated m). Displacement has two parts to it: a magnitude (amount) and a direction. Note: Δ is the Greek letter delta, and in physics it means that we are looking at the amount of change of something.

- *Velocity* is how quickly something is moving. Velocity has a magnitude (amount) and a direction, just like displacement. In SI units, velocity has units of meters/second, abbreviated m/s. We'll use the symbol v for velocity.

DOI: 10.1201/9781003005049-2

- *Acceleration* is how fast the velocity changes. We'll use the symbol a for acceleration; in SI units acceleration is meters/second^2 (m/s^2). Acceleration has a magnitude and direction, just like displacement and velocity. Note: you'll see in Section 2.5 that acceleration is like velocity/time; this produces the units of (m/s)/s = m/s^2.

- *Time.* When we describe something moving, we'll be looking at the motion between two time points: an initial time point and a final time point. The initial time point is where we start watching the motion, and the final time point is when we stop watching the motion. The amount of time between the initial and final time points is called the time. In this textbook we'll use the symbol Δt. In SI units, time is given in seconds, abbreviated s. Time only has a magnitude.

2.3 DISPLACEMENT, POSITION, AND COORDINATE SYSTEMS

Displacement is one of the most important quantities in physics. It's a change in position, with position being the location of an object. Displacement can be defined as:

$$\Delta x = x_f - x_i \tag{2.3.1}$$

In this equation, Δx is displacement. Δ is the Greek letter delta, and it goes in front of any physics quantity that is a change in some amount. Here, it is the change in position. x_f and x_i are the positions at the final (f) and initial (i) time points of a problem.

Position is always measured on what is called a **coordinate system**. For objects moving in a straight line (the topic of this chapter), the coordinate system has two things:

- An **origin**, which is the center of your coordinate system.

- A positive direction.

A straight-line coordinate system is shown in Figure 2.1, with "to the right" being the positive direction.

When solving a problem, you get to choose the coordinate system, meaning that you get to choose the origin and the positive direction. If you only need the displacement, and not the position, you will only need to choose a positive direction. Almost all (or maybe all) of the problems in

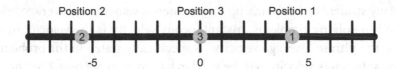

FIGURE 2.1   An example of a straight line coordinate system, with three example positions.

this textbook will only need displacement, not position, and the displacement will often be obvious in the problem – something like "calculate the velocity of a bus that moved 25.0 m". The 25.0 m is the displacement.

Side note: in this chapter, the displacement will have two parts to it:

1. An amount: the distance moved (in units of meters). This is sometimes called the **magnitude**.

2. A sign (+ or –) that indicates the **direction**. As mentioned above, you choose a positive direction when doing a problem. If the displacement is in that direction, it's a + sign; if it's in the opposite direction, it's a – sign.

Having an amount and a direction relates to the displacement being a **vector**, which will be discussed in Chapter 3.

2.4 AVERAGE AND INSTANTANEOUS VELOCITY

When we talk about velocity, we can talk about the velocity at a specific time point, or the average velocity over a range of time points. The **average velocity** is determined using the displacement:

$$v_{avg} = \Delta x / \Delta t \qquad (2.4.1)$$

(v_{avg} is average velocity; Δx is the displacement; Δt is the time between the initial and final time points.)

The velocity at a specific time point is called the **instantaneous velocity**. Many physics problems have two time points – an initial time point and a final time point. The velocity at either time point is an instantaneous velocity.

Similar to displacement, velocity has a magnitude and a direction. The units of velocity are m/s (meters/second).

Many students tend to mix up instantaneous velocity and average velocity. When solving problems, instantaneous velocity is the velocity to use unless the phrase "average velocity" is specifically stated in the problem.

Instantaneous velocity will be used extensively in Section 2.6. The following is a short example of average velocity.

Example 2.4.1: Calculating Average Velocity

A car is sitting at a red light. When the light goes green, the car moves 45.0 meters during a 5.00 second time period. What was the average velocity of the car over those 5.00 seconds?

Solution:

$$v_{avg} = \Delta x / \Delta t = \left(45.0 \text{ m}\right) / \left(5.00 \text{ s}\right) = 9.00 \text{ m/s}$$

Note: the displacement (45.0 m) is given in the problem, and the coordinate system is defined with the direction the car is traveling as positive.

2.5 AVERAGE AND INSTANTANEOUS ACCELERATION

Average acceleration is very similar to average velocity, except that average velocity depends on a change in location (displacement), and average acceleration depends on a change in velocity. The math is very similar:

$$a_{avg} = \Delta v / \Delta t = \left(v_f - v_i\right) / \Delta t \qquad (2.5.1)$$

a_{avg} is average acceleration, Δv is change in velocity between the final and initial time points, v_f is the instantaneous velocity at the final time point, v_i is the instantaneous velocity at the initial time point, and Δt is the amount of time between the initial and final time points. The change in velocity, Δv, is equal to $v_f - v_i$.

Conceptually, an instantaneous acceleration is the acceleration at any specific point in time that you specify. In all (or almost all) of the problems you will do in introductory physics, the acceleration stays constant, meaning that it has the same value at all times during the problem. If that is the case, then the average acceleration and the instantaneous acceleration are equal.

Acceleration has a magnitude and a direction, just like displacement and velocity.

Example 2.5.1: Calculating Average Acceleration

A car is sitting at a red light. When the light goes green, the car begins to move in the direction of east. After 5.00 seconds, the car is traveling to the east at a velocity of 15.0 m/s (roughly 30 miles/hour). What was the average acceleration of the car over those 5.00 seconds?

Solution:

The equation for average acceleration is $a_{avg} = \Delta v/\Delta t = (v_f - v_i)/\Delta t$. The initial time point is when the car begins to move, and the final time point is when the car is traveling at 15.0 m/s. Define east as the positive direction. Final velocity is then 15.0 m/s (positive since it's to the east), and initial velocity is zero (sitting at a red light). Δt is the amount of time between the initial and final time points, which is given as 5.00 seconds.

$$a_{avg} = \left(v_f - v_i\right)/\Delta t = \left(15.0 \, m/s - 0\right)/\left(5.00 \, s\right) = 3.00 \, m/s^2$$

<u>Magnitude: 3.00 m/s², Direction: east (since a_{avg} is positive).</u>

2.6 MOTION ON A LINE WITH CONSTANT ACCELERATION

In this chapter, we have only looked at motion along a straight line. If we make one more assumption, we can use a set of equations that solve many motion problems. That assumption is that, for the length of time described in our problem, the acceleration stays equal to the same value throughout the problem. This is called **constant acceleration**. If we make these assumptions, we can use the following equations:

$$v_f = v_i + a \cdot \Delta t \tag{2.6.1}$$

$$\Delta x = v_i \cdot \Delta t + \frac{1}{2} \cdot a \cdot \left(\Delta t\right)^2 \tag{2.6.2}$$

$$\Delta x = \frac{1}{2} \cdot \left(v_i + v_f \right) \cdot \Delta t \qquad (2.6.3)$$

$$v_f^2 = v_i^2 + 2 \cdot a \cdot \Delta x \qquad (2.6.4)$$

Problems that use these equations are often called **kinematics problems**. A derivation of these equations (that is, using math to show where these equations come from) is outside of the scope of this textbook. However, it's useful to point out a few important features of these equations:

- There are five total variables, and they are all variables we have seen before. The variables are v_f (velocity at the final time point), v_i (velocity at the initial time point), a (acceleration, which stays at the same value throughout the problem since it's constant), Δx (displacement), and Δt (change in time).

- Although it is not obvious, these equations are redundant. Even though there are four equations, you can only use them to solve for two total variables.

- All of these equations can be found starting from two equations that you have already seen – Equations 2.4.1 and 2.5.1.

- Each of the four equations has exactly four variables – one of the five variables is missing in each equation. How can we use this? For most problems, you will know three variables and have one that you want to find. You can usually find one equation with the variable you want, and three variables that you know. The exception: if you know every variable except for v_i and the variable you want to find, you'll have to find v_i first and then find the variable you want.

Side note: these equations (from Section 2.6) don't include position by itself (just displacement), and they don't have average velocity or average acceleration. Instead, they contain quantities that you see much more regularly in your everyday life – velocities at some specific time (instantaneous), how far you moved, etc. Be careful – it's very common for students to get confused and use the equations for average velocity and average acceleration instead of the equations in this section. Usually, you only want to use the equations for average velocity and average acceleration if the word "average" is specifically mentioned in the problem.

Example 2.6.1: The Short Yellow Traffic Light
(Linear Motion with Constant Acceleration)

You're driving with a velocity of 20.0 m/s. Suddenly, a traffic light in front of you turns yellow. You quickly put on the brakes and slide to a stop. Assume that the acceleration is constant. It takes you 3.00 seconds to stop after you start to hit the brakes.

 A. How far do you travel along the road during the 3.00-second braking time?
 B. What is the acceleration (magnitude and direction) of the car during those 3.00 seconds?

Solution:
Part A:
Since the acceleration is constant, we can use Equations 2.6.1–2.6.4 above. We first need to set a coordinate system and determine the initial and final time points. When something is moving in a straight line, it's usually easiest to define the direction it is moving in as positive. Let's make the initial time point to be the time when braking starts, and the final time point to be when the car reaches zero velocity (a stop).

What variables do we know?

 • $v_i = 20.0$ m/s (positive)
 • $\Delta t = 3.00$ s (time between initial and final time points)
 • $v_f = 0$ (the car comes to a stop at the final time point)

Let's solve for Δx first – it's Part A. Looking at Equations 2.6.1–2.6.4, we need an equation that has Δx as well as the variables we know (v_i, Δt, and v_f). Equation 2.6.3 is the correct one to use:

$$\Delta x = \frac{1}{2}\cdot\left(v_i + v_f\right)\cdot\Delta t = \frac{1}{2}\cdot\left(20.0\,\text{m/s} + 0\right)\cdot\left(3.00\,\text{s}\right) = 30.0\,\text{m}$$

Noticing that the sign is positive, we can say that the displacement is 30.0 meters in the same direction as you were traveling.

Part B:
Now, let's solve for acceleration (Part B). Since we now know every variable except acceleration, any equation from 2.6.1–2.6.4 that has

acceleration will work. Let's use Equation 2.6.1 and solve it before plugging in variables:

$$v_f = v_i + a \cdot \Delta t$$

Subtract v_i from both sides: $v_f - v_i = a \cdot \Delta t$

Divide both sides by Δt: $a = \dfrac{v_f - v_i}{\Delta t}$

Put in numbers: $a = \dfrac{v_f - v_i}{\Delta t} = \dfrac{0 - 20.0\,\text{m/s}}{3.00\,\text{s}} = -6.67\,\text{m/s}^2$

Looking at the sign (negative) and the positive direction (the direction the car is traveling in), we can say that the acceleration is 6.67 m/s², in the direction opposite to where you were traveling.

Note: the acceleration being negative means that the car's velocity changes towards the negative direction.

Problem to Try Yourself

A car speeds up from 0 m/s to 20.0 m/s while traveling a distance of 50.0 m, at constant acceleration. What is the acceleration?

Solution:
Define the direction of the car's travel as the positive direction. When the car is starting (from 0 m/s) is the initial time point, and when the car reaches 20.0 m/s is the final time point. We know $v_i = 0$, $v_f = 20.0$ m/s, and $\Delta x = 50.0$ m. We need an equation with the variable we want to find (a) and the variables that we know. Equation 2.6.4 is the correct choice:

$$v_f^2 = v_i^2 + 2 \cdot a \cdot \Delta x$$

To solve for a, first subtract v_i^2 from both sides: $v_f^2 - v_i^2 = 2 \cdot a \cdot \Delta x$

Next, divide both sides by $(2 \cdot \Delta x)$: $a = \dfrac{v_f^2 - v_i^2}{2 \cdot \Delta x}$

Put in numbers: $a = \dfrac{\left(20.0\,\text{m/s}\right)^2 - 0^2}{2 \cdot \left(50.0\,\text{m}\right)} = 4.00\,\text{m/s}^2$

Note: the positive sign in the answer means that the acceleration is in the positive direction, which was defined at the start of the problem solution as the direction that the car is traveling.

2.7 FREE FALL

Free fall is when an object flies through the air in a line that is vertical (straight up and down), and that gravity is the only force acting on it. (We'll talk more about gravity when we talk about forces in Chapter 4.) In a free fall situation, the acceleration is constant and equal to 9.80 m/s^2, with the direction being towards the ground. Since the acceleration is constant, we can use Equations 2.6.1–2.6.4, just as we did in Section 2.6. In physics, 9.80 m/s^2 shows up so often that we give it a symbol and a name: **g, the acceleration due to gravity.**

Students often get stuck on three details related to free fall:

- When an object is traveling upwards and reaches its maximum (highest) height, its velocity is equal to zero. *Physics for Scientists and Engineers* by Tipler and Mosca explains this as something like, the maximum height is right between going up (positive velocity) and going down (negative velocity). Zero is between positive and negative.

- Free fall problems often ask for the velocity of the object just before it hits the ground. It's common for students to answer "zero", because the object stops when it hits the ground, but this is not what the problem is asking. Set the displacement variable equal to the value you would use to mean the ground, since it is just barely before the ground, and then solve for velocity.

- The direction of the acceleration is important. It's downward. In free fall problems, if you make upward the positive direction in your coordinate system, then the acceleration is –9.80 m/s^2. The minus sign means that the acceleration is opposite to your positive direction.

Side note: the value 9.80 m/s^2 comes from combining Newton's law of gravity (Chapter 11) with Newton's second law (Chapter 4). In order to use 9.80 m/s^2 and Equations 2.6.1–2.6.4, we must ignore air resistance so that gravity is the only force.

Example 2.7.1: Falling Tree Branch

A tree branch is initially at rest, a height of 11.0 meters above the ground. Suddenly, it breaks off from the tree and begins to fall. What is the branch's velocity, just before it hits the ground?

Solution:

This is a free fall problem, which is always a constant acceleration problem, so we can use Equations 2.6.1–2.6.4. First, as with all motion problems, let's set the positive direction and time points. Here we will pick "up" as the positive direction. Let's pick the initial time point to be when the tree branch begins to fall, and the final time point to be (just before) when the branch hits the ground.

As in Section 2.6, we have five variables: displacement (Δx), velocity at the initial time point (v_i), velocity at the final time point (v_f), length of time covered in the problem (Δt), and acceleration (a). Just like with Section 2.6, we must know three variables in order to solve the problem. What do we know? Every free fall problem has an acceleration of 9.80 m/s^2, downward. Since we defined up as positive, the acceleration is $a = -9.80$ m/s^2. We know that the branch is at rest initially, so $v_i = 0$. Finally, we know the displacement. We know that the branch falls 11.0 meters downward during the problem, so $\Delta x = -11.0$ m. (Note: even though we are looking at just before the branch hits the ground, we can say that the displacement is as if the branch is already on the ground.) We don't know the final time point velocity or the total amount of time between the initial and final time points.

The problem asks for the velocity just as the branch is about to hit the ground, which is the final time point. So, we are looking for the final velocity (v_f). We'd like to find the equation within Equations 2.6.1–2.6.4 that has final velocity and the three variables that we already know (v_i, a, Δx). Equation 2.6.4 fits this description:

$$v_f^2 = v_i^2 + 2 \cdot a \cdot \Delta x$$

We can solve for v_f by taking the square root of this equation, but don't forget that taking a square root means that the answer could be positive or negative – for example, the square root of 4 is both 2 and –2, since 2 times 2 is 4 and (–2) times (–2) is also 4. See Section

13.4 for more detail on this. We write this using a plus/minus sign in front:

$$v_f = \pm\sqrt{v_i^2 + 2 \cdot a \cdot \Delta x}$$

Now, we can put in the values that we know:

$$v_f = \pm\sqrt{v_i^2 + 2 \cdot a \cdot \Delta x} = \pm\sqrt{0^2 + 2 \cdot \left(-9.80 \, \text{m}/\text{s}^2\right) \cdot \left(-11.0 \, \text{m}\right)} = \pm 14.7 \, \text{m}/\text{s}$$

We now know that the magnitude of the velocity is 14.7 m/s, but how do we know the sign? Remember that the sign is the direction. The branch is falling, so it must be going downward. Since we defined upward as the positive direction, the (downward) velocity must be negative.

$$\underline{v_f = -14.7 \, \text{m/s (downward)}}$$

Note: the square root symbol also took the square root of the units: m²/s² becomes m/s.

We will now do a second example that is more complicated. Parts A and B will have a different set of final time points. The displacement value will also take some thinking about.

Example 2.7.2 Free Fall with Two Different Time Points

Your friend is on a ladder, cleaning out leaves from a gutter. He asks you to throw him a tool. You throw it to him, straight upward and from a height of 1.50 meters off the ground. Unfortunately, your friend misses the catch, and the tool goes right past him. It reaches a maximum height of 4.50 meters off of the ground before falling back down and hitting the ground. You see this coming, and dodge the tool as it comes back to the ground.

A. What was the velocity (magnitude and direction) with which you threw the tool?
B. How much time passes between the tool reaching maximum height, and the tool hitting the ground?

Solution:
Part A:
It's a free fall problem, so we can use Equations 2.6.1–2.6.4. Our variables are Δx, Δt, v_i, v_f, and a. Let's define upward as positive. Here,

the initial time point is the throw (when the tool leaves the hand), and the final time point is maximum height. Note: in Part A we're ignoring all times after maximum height because we have more information if we use maximum height as a time point – we know that maximum height has velocity equal to 0, and we know what the height is.

Variables we know: $v_f = 0$ (it's maximum height) and $a = -9.80$ m/s² (free fall, negative because we defined upward as positive). The displacement is the distance traveled between the throw (initial time) and maximum height (final time). The maximum height is 4.50 meters off the ground, but the initial height was 1.50 meters off of the ground. Thus, the displacement is 3.00 m (the distance covered going from 1.50 to 4.50 m).

Looking at Equations 2.6.1–2.6.4, we need an equation that includes v_i (what we want to find) and the variables that we know (a, Δx, v_f). Equation 2.6.4 fits that description:

$$v_f^2 = v_i^2 + 2 \cdot a \cdot \Delta x$$

First, solve for v_i, keeping in mind that taking the square root introduces the plus/minus sign:

Subtract $2 \cdot a \cdot \Delta x$ from each side: $v_f^2 - 2 \cdot a \cdot \Delta x = v_i^2$

Take the square root of both sides: $v_i = \pm\sqrt{v_f^2 - 2 \cdot a \cdot \Delta x}$

Now, we can put in the values that we know:

$$v_i = \pm\sqrt{v_f^2 - 2 \cdot a \cdot \Delta x} = \pm\sqrt{0^2 - 2 \cdot \left(-9.80\,\text{m}/\text{s}^2\right) \cdot \left(3.00\,\text{m}\right)}$$

$$= \pm 7.67\,\text{m/s}$$

We have to decide the + or – sign. We know the throw is upward, which is the positive direction. The initial velocity is 7.67 m/s, upward.

Part B:
It's still a free fall problem. Let's keep upward as positive. We now have information from three time points and must choose two to use: the throw, maximum height, and (just before) when the tool lands on the ground. Here, we'll choose to use maximum height as the initial time point and tool landing as final point. The reason is that the piece of information requested is the time between maximum height

and tool landing, which is just Δt if we choose those time points as initial and final.

Variables we know: $a = -9.80$ m/s^2, $v_i = 0$ (initial time is maximum height), $\Delta x = -4.50$ m (it falls downward to the ground, from the maximum height of 4.50 meters).

We want to find Δt. From Equations 2.6.1–2.6.4, we want an equation with Δt and the variables that we know (a, v_i, Δx). Equation 2.6.2 fits that description:

$$\Delta x = v_i \cdot \Delta t + \frac{1}{2} \cdot a \cdot \left(\Delta t\right)^2$$

If none of the variables are zero, we'd need the quadratic formula to solve this. (See Section 13.5 for a review on the quadratic formula.) But, since $v_i = 0$, we can simplify this equation:

$$\Delta x = 0 \cdot \Delta t + \frac{1}{2} \cdot a \cdot \left(\Delta t\right)^2$$

$$\Delta x = \frac{1}{2} \cdot a \cdot \left(\Delta t\right)^2$$

Solve this for Δt. Multiply both sides by 2: $2 \cdot \Delta x = a \cdot \left(\Delta t\right)^2$

Divide both sides by a: $\dfrac{2 \cdot \Delta x}{a} = \left(\Delta t\right)^2$

Take the square root of both sides: $\Delta t = \pm\sqrt{\dfrac{2 \cdot \Delta x}{a}}$

The plus/minus sign comes from taking the square root. However, our amount of time will always be positive: $\Delta t = \sqrt{\dfrac{2 \cdot \Delta x}{a}}$

Put in the variables we know:

$$\Delta t = \sqrt{\frac{2 \cdot \Delta x}{a}} = \sqrt{\frac{2 \cdot \left(-4.50\,\text{m}\right)}{\left(-9.80\,\text{m}/\text{s}^2\right)}} = 0.958\,\text{s}$$

Problem to Try Yourself

Your friend kicks a ball straight upward, at a velocity of 8.50 m/s. How high is it (above where it started) 1.50 seconds later?

Solution:

Define upward as positive. The initial time point is when the ball is kicked, and the final time point is 1.50 seconds later. We know

$v_i = 8.50$ m/s (upward and so it's positive), $\Delta t = 1.50$ s, and $a = -9.80$ m/s². We want to find Δx, which is the amount of change from the initial time point. We need an equation with Δx and the variables that we know (v_i, a, Δt). The equation is 2.6.2:

$$\Delta x = v_i \cdot \Delta t + \frac{1}{2} \cdot a \cdot \left(\Delta t\right)^2$$

Since this equation is already solved for Δx, we can put the numbers in immediately:

$$\Delta x = \left(8.50\,\text{m}/\text{s}\right) \cdot \left(1.50\,\text{s}\right) + \frac{1}{2} \cdot \left(-9.80\,\text{m}/\text{s}^2\right) \cdot \left(1.50\,\text{s}\right)^2$$

$$= 1.73\,\text{m}$$

Note: if you solved for v_f, you could use Equation 2.6.1 and find that $v_f = -6.20$ m/s, which means that the ball is on its way down after 1.50 s.

2.8 WHAT IF ACCELERATION IS NOT CONSTANT?

In Sections 2.6 and 2.7, we developed a method for solving motion problems when the acceleration is constant (stays the same). However, the acceleration isn't always constant in real life. In some cases, we will learn other methods for solving these types of problems. In some simple cases, though, it may be possible to break down the motion into pieces such that each piece has just one value for acceleration.

2.9 CHAPTER 2 SUMMARY

Chapter 2 describes motion that is in one dimension, meaning in a straight line.

Motion is described by displacement (change in position, Δx), velocity (v), acceleration (a), and change in time (Δt).

If the acceleration is constant during the problem,

$$v_f = v_i + a \cdot \Delta t \tag{2.6.1}$$

$$\Delta x = v_i \cdot \Delta t + \frac{1}{2} \cdot a \cdot \left(\Delta t\right)^2 \tag{2.6.2}$$

$$\Delta x = \frac{1}{2} \cdot \left(v_i + v_f \right) \cdot \Delta t \qquad (2.6.3)$$

$$v_f^2 = v_i^2 + 2 \cdot a \cdot \Delta x \qquad (2.6.4)$$

If the object is moving in a line vertically, if gravity is approximately the only force, and if we define upward as positive, we can use Equations 2.6.1–2.6.4 and say that acceleration is –9.80 m/s². This is called **acceleration due to gravity**.

Chapter 2 problem types:

Calculations with displacement, average velocity, and average acceleration (Sections 2.3–2.5)

Problems with constant acceleration and motion in one dimension (Section 2.6)

Free fall problems (Section 2.7)

Problems with constant acceleration and more than one time point (Example 2.7.2)

Practice Problems
Section 2.4

1. A car travels 115 m in 5.20 s. What is the average velocity?

    Solution:

    $$v_{avg} = \Delta x / \Delta t \qquad (2.4.1)$$

    Put in numbers: $v_{avg} = \Delta x / \Delta t = (115 \text{ m})/(5.20 \text{ s}) = \underline{22.1 \text{ m/s}}$

    Note: the direction that the car is moving is defined as the positive direction.

2. A motorcycle travels at an average velocity of 15.0 m/s, for 8.75 s. How far does it travel?

    Solution:

    $$v_{avg} = \Delta x / \Delta t \qquad (2.4.1)$$

Solve for Δx by multiplying both sides by Δt: $\Delta x = v_{avg} \cdot \Delta t$

Put in numbers: $\Delta x = (15.0 \text{ m/s}) \cdot (8.75 \text{ s}) = \underline{131 \text{ m}}$

Note: the direction that the motorcycle is moving is defined as the positive direction.

3. A bus travels 185 m with an average velocity of 15.0 m/s. How much time did it take?

Solution:

$$v_{avg} = \Delta x / \Delta t \qquad\qquad (2.4.1)$$

Solve for Δt. Multiply both sides by Δt: $v_{avg} \cdot \Delta t = \Delta x$

Divide both sides by v_{avg}: $\Delta t = \Delta x / v_{avg}$

Put in numbers: $\Delta t = (185 \text{ m})/(15.0 \text{ m/s}) = \underline{12.3 \text{ s}}$

Note: the direction that the bus is moving is defined as the positive direction.

Section 2.5

4. Coming up on a red light, the bus you are on slows from 11.0 m/s to a stop over 4.50 s. What was the average acceleration?

Solution:

$$a_{avg} = (v_f - v_i)/\Delta t \qquad\qquad (2.5.1)$$

Put in numbers: $a_{avg} = (0 - 11.0 \text{ m/s})/(4.50 \text{ s}) = \underline{-2.44 \text{ m/s}^2}$

Note 1: the answer is negative because the final velocity is less than the initial velocity.

Note 2: the direction that the bus is moving is defined as the positive direction.

5. A subway car speeds up. If it speeds up from rest, and has an average acceleration of 1.50 m/s², how fast is it going after 2.50 s?

Solution:

$$a_{avg} = (v_f - v_i)/\Delta t \qquad\qquad (2.5.1)$$

Solve for v_f. Multiply both sides by Δt: $a_{avg} \cdot \Delta t = v_f - v_i$

Add v_i to both sides: $v_f = a_{avg} \cdot \Delta t + v_i$

Put in numbers: $v_f = (1.50 \text{ m/s}^2) \cdot (2.50 \text{ s}) + 0 = \underline{3.75 \text{ m/s}}$

Note: the direction that the subway car is moving is defined as the positive direction.

6. A subway car slows down. If it slows down from 15.0 m/s to rest with an average acceleration of –1.25 m/s, how much time does it take to slow down?

   Solution:

$$a_{avg} = (v_f - v_i)/\Delta t \qquad (2.5.1)$$

Solve for Δt. Multiply both sides by Δt: $a_{avg} \cdot \Delta t = v_f - v_i$

Divide both sides by a_{avg}: $\Delta t = (v_f - v_i)/a_{avg}$

Put in numbers: $\Delta t = (0 - 15.0 \text{ m/s})/(-1.25 \text{ m/s}^2) = \underline{12.0 \text{ s}}$

Note: the direction that the subway car is moving is defined as the positive direction.

7. After getting on a highway, the bus you are on speeds up. If it speeds up to 20.0 m/s over 7.50 s, with an average acceleration of 1.50 m/s², how fast was the bus going before the acceleration started?

   Solution:

$$a_{avg} = (v_f - v_i)/\Delta t \qquad (2.5.1)$$

Solve for v_i. Multiply both sides by Δt: $a_{avg} \cdot \Delta t = v_f - v_i$

Subtract v_f from both sides: $-v_i = a_{avg} \cdot \Delta t - v_f$

Multiply both sides by -1: $v_i = -a_{avg} \cdot \Delta t + v_f$

Put in numbers: $v_i = -(1.50 \text{ m/s}^2) \cdot (7.50 \text{ s}) + 20.0 \text{ m/s} = \underline{8.75 \text{ m/s}}$

Note: the direction that the bus is moving is defined as the positive direction.

Section 2.6

8. After a long day of classes, a student on a bicycle pulls out onto a straight road and begins to accelerate. She starts at rest, and reaches a velocity of 11.0 m/s to the east after 10.0 seconds of constant acceleration.

   A. How far along the road did she travel during those 10.0 seconds?

   B. What was the acceleration? Make "east" positive and make certain to get the sign correct.

Solution:

Part A:

We are given $v_i = 0$ (starts at rest), $v_f = 11.0$ m/s, and $\Delta t = 10.0$ s. We want to find Δx (how far means displacement). The equation with Δx and the variables we know is:

$$\Delta x = \frac{1}{2} \cdot (v_i + v_f) \cdot \Delta t \qquad (2.6.3)$$

Put in numbers: $\Delta x = \frac{1}{2} \cdot (0 + 11.0\,\text{m/s}) \cdot (10.0\,\text{s}) = 55.0\,\text{m}$

Part B:

The situation is the same as in Part A. We want to find acceleration, and we now know every variable except a. We can use any equation with a in it. We'll choose Equation 2.6.1:

$$v_f = v_i + a \cdot \Delta t \qquad (2.6.1)$$

Subtract v_i from both sides: $v_f - v_i = a \cdot \Delta t$

Divide both sides by Δt: $a = \dfrac{v_f - v_i}{\Delta t}$

Put in numbers: $a = \dfrac{11.0\,\text{m/s} - 0}{10.0\,\text{s}} = 1.10\,\text{m/s}^2$

9. On the way to work, you pull onto a straight road in a bus. The bus accelerates at a constant rate, and changes velocity from 5.00 m/s to 10.0 m/s, all while traveling 80.5 meters. What was the value of the

constant acceleration? Hint: you can say that the direction you are traveling on the road is "positive".

Solution:

We are given $v_i = 5.00$ m/s, $v_f = 10.0$ m/s, and $\Delta x = 80.5$ m. We want to find acceleration (a). The equation with a and the variables we know is Equation 2.6.4:

$$v_f^2 = v_i^2 + 2 \cdot a \cdot \Delta x \qquad (2.6.4)$$

Subtract v_i^2 from both sides: $v_f^2 - v_i^2 = 2 \cdot a \cdot \Delta x$

Divide both sides by $(2 \cdot \Delta x)$: $a = \dfrac{v_f^2 - v_i^2}{2 \cdot \Delta x}$

Put in numbers: $a = \dfrac{(10.0\,\text{m/s})^2 - (5.00\,\text{m/s})^2}{2 \cdot (80.5\,\text{m})} = 0.466\,\text{m/s}^2$

10. You're driving in a straight line on the highway, traveling at a velocity of 27.0 m/s. Seeing a big traffic jam, you hit the breaks and slow to a stop while covering a distance of 85.0 meters. Assume that the value of your acceleration is constant.

   A. What is the value of your acceleration? Set the positive direction to be the direction that the car is traveling before you hit the breaks (north).

   B. How much time does it take for your car to stop?

Solution:

Part A:

You are given $v_i = 27.0$ m/s, $v_f = 0$ (it stops), and $\Delta x = 85.0$ m. We want to find acceleration (a). The equation with a and the variables we know is Equation 2.6.4:

$$v_f^2 = v_i^2 + 2 \cdot a \cdot \Delta x \qquad (2.6.4)$$

Subtract v_i^2 from both sides: $v_f^2 - v_i^2 = 2 \cdot a \cdot \Delta x$

Divide both sides by $(2 \cdot \Delta x)$: $a = \dfrac{v_f^2 - v_i^2}{2 \cdot \Delta x}$

Put in numbers: $a = \dfrac{(0)^2 - (27.0\,\text{m}/\text{s})^2}{2 \cdot (85.0\,\text{m})} = -4.29\,\text{m}/\text{s}^2$

Part B:

We will now solve for time (Δt). We can use any of the equations that have Δt in it. We will use Equation 2.6.3:

$$\Delta x = \frac{1}{2} \cdot (v_i + v_f) \cdot \Delta t \qquad (2.6.3)$$

Divide both sides by $\dfrac{1}{2} \cdot (v_i + v_f)$: $\Delta t = \dfrac{\Delta x}{\dfrac{1}{2} \cdot (v_i + v_f)}$

Put in numbers: $\Delta t = \dfrac{85.0\,\text{m}}{\dfrac{1}{2} \cdot (27.0\,\text{m}/\text{s} + 0)} = 6.30\,\text{s}$

11. A subway car starts from rest, and accelerates at 2.25 m/s², in a straight line. How much time does it take to travel 25.0 m?

Solution:

We are given $a = 2.25$ m/s², $v_i = 0$ (starts at rest), and $\Delta x = 25.0$ m. We want to find Δt. The equation with Δt and variables that we know is Equation 2.6.2:

$$\Delta x = v_i \cdot \Delta t + \frac{1}{2} \cdot a \cdot (\Delta t)^2 \qquad (2.6.2)$$

Put $v_i = 0$ into the equation: $\Delta x = \dfrac{1}{2} \cdot a \cdot (\Delta t)^2$

Divide both sides by $\dfrac{1}{2} \cdot a$: $(\Delta t)^2 = \dfrac{\Delta x}{\dfrac{1}{2} \cdot a}$

Take the square root of both sides: $\Delta t = \pm \sqrt{\dfrac{\Delta x}{\dfrac{1}{2} \cdot a}}$

Take the positive root, since the amount of change in time will be positive.

Put in numbers: $\Delta t = +\sqrt{\dfrac{25.0\,\text{m}}{\dfrac{1}{2} \cdot 2.25\,\text{m}/\text{s}^2}} = 4.71\,\text{s}$

12. A subway car starts from 1.50 m/s, and accelerates at 2.25 m/s², in a straight line. How much time does it take to travel 25.0 m?

Solution:

We are given a = 2.25 m/s², v_i = 1.50 m/s, and Δx = 24.0 m. We want to find Δt. The equation with Δt and variables that we know is Equation 2.6.2:

$$\Delta x = v_i \cdot \Delta t + \frac{1}{2} \cdot a \cdot \left(\Delta t\right)^2 \qquad (2.6.2)$$

Since there are no nonzero values here, the quadratic formula will be necessary. The quadratic formula is:

If $a \cdot x^2 + b \cdot x + c = 0$, where a, b, and c are numbers, and x is the variable you want to solve for,

$$x = \frac{-b \pm \sqrt{b^2 - 4 \cdot a \cdot c}}{2 \cdot a}$$

The quadratic formula is reviewed in Section 13.5

Δt is the variable that we want to solve for. First, we will get the equation into the format of "a" $\cdot$ (Δt)² + "b" $\cdot \Delta t$ + "c" = 0. Note: "a" is different from the acceleration a; that is why quotes are being used.

Subtract Δx from both sides, and switch the order of the other two terms:

$$\frac{1}{2} \cdot a \cdot \left(\Delta t\right)^2 + v_i \cdot \Delta t - \Delta x = 0$$

This means that "a" $= \dfrac{1}{2} \cdot a = \dfrac{1}{2} \cdot 2.25$, "b" $= v_i = 1.50$, and "c" $= -\Delta x = -24.0$.

Put this into the quadratic formula:

$$\Delta t = \frac{-b \pm \sqrt{b^2 - 4 \cdot a \cdot c}}{2 \cdot a}$$

$$= \frac{-1.50 \pm \sqrt{(1.50)^2 - 4 \cdot (0.5 \cdot 2.25) \cdot (-24.0)}}{2 \cdot (0.5 \cdot 2.25)}$$

$$= 4.00\,s \text{ and } -5.33\,s$$

We will choose the positive change for amount of time: $\underline{\Delta t = 4.00\ s}$

13. A subway car changes speed from 15.0 m/s to zero, in a straight line, as it stops at a station. If the stop took 8.00 s, what was the acceleration? Pick the direction the car is initially traveling to be the positive direction.

Solution:

We know $v_i = 15.0$ m/s, $v_f = 0$, and $\Delta t = 8.00$ s. We want to find acceleration (a). The equation with a and the variables that we know is Equation 2.6.1:

$$v_f = v_i + a \cdot \Delta t \tag{2.6.1}$$

To solve for a, first subtract v_i from both sides: $v_f - v_i = a \cdot \Delta t$

Divide both sides by Δt: $a = \dfrac{v_f - v_i}{\Delta t}$

Put in numbers: $a = \dfrac{0 - 15.0\,m/s}{8.00\,s} = -1.88\,m/s^2$

Note: the acceleration is negative, and in the opposite direction from the car's motion.

14. A car driving on the highway, in a straight line, and slows from 24.0 m/s to 16.0 m/s at a constant rate, over 10.0 s. How far did it travel during this time?

Solution:

We know $v_i = 24.0$ m/s, $v_f = 16.0$ m/s, and $\Delta t = 10.0$ s. We want to find Δx (displacement). The equation with Δx and the variables that we know is Equation 2.6.3:

$$\Delta x = \frac{1}{2} \cdot \left(v_i + v_f \right) \cdot \Delta t \tag{2.6.3}$$

Since this equation is already solved for Δx, we can put in numbers:

$$\Delta x = \frac{1}{2} \cdot \left(24.0 \, m/s + 16.0 \, m/s \right) \cdot \left(10.0 \, s \right) = 200 \, m$$

15. A bus starts to accelerate on the highway, in a straight line. It was already traveling 15.0 m/s when it began to speed up. If it accelerates at 0.750 m/s², and travels 80.0 m, what velocity is it traveling?

Solution:

We know $v_i = 15.0$ m/s, $a = 0.750$ m/s², and $\Delta x = 80.0$ m. We want to find v_f (velocity at the later time point). The equation with v_f and the variables that we know is Equation 2.6.4:

$$v_f^2 = v_i^2 + 2 \cdot a \cdot \Delta x \tag{2.6.4}$$

Take the square root of both sides: $v_f = \pm \sqrt{v_i^2 + 2 \cdot a \cdot \Delta x}$

Take the positive root, since the velocity will be in the same direction during the entire problem. (We can pick that direction as positive.)

Put in numbers: $v_f = +\sqrt{\left(15.0 \, m/s \right)^2 + 2 \cdot \left(0.750 \, m/s^2 \right) \cdot \left(80.0 \, m \right)}$

$$= 18.6 \, m/s$$

16. A bus slows down to zero at a bus stop, in a straight line. If the acceleration is −1.50 m/s² and the bus travels 10.0 m as it stops, what was the velocity before it began stopping?

Solution:

We know $v_f = 0$ (stops), $a = -1.50$ m/s², and $\Delta x = 10.0$ m. We want to find v_i (velocity at the start). The equation with v_i and the variables that we know is Equation 2.6.4:

$$v_f^2 = v_i^2 + 2 \cdot a \cdot \Delta x \tag{2.6.4}$$

To find v_i, first subtract $2 \cdot a \cdot \Delta x$ from both sides: $v_i^2 = v_f^2 - 2 \cdot a \cdot \Delta x$

Take the square root of both sides: $v_i = \pm\sqrt{v_f^2 - 2 \cdot a \cdot \Delta x}$

Take the positive root – the bus is traveling in the same direction for the entire problem, and we can pick that as the positive direction.

Put in numbers: $v_i = +\sqrt{(0)^2 - 2 \cdot (-1.50 \, \text{m/s}^2) \cdot (10.0 \, \text{m})} = 5.48 \, \text{m/s}$

17. A bus starts to accelerate on the highway, in a straight line. It was already traveling 17.0 m/s when it began to speed up. If it accelerates at 0.750 m/s², and travels for 9.50 s, what velocity is it traveling?

Solution:

We are given $v_i = 17.0$ m/s, $a = 0.750$ m/s², and $\Delta t = 9.50$ s. We want to find v_f. The equation with v_f and the variables that we know is Equation 2.6.1:

$$v_f = v_i + a \cdot \Delta t \qquad (2.6.1)$$

The equation is already solved for v_f, so put in numbers:

$$v_f = (17.0 \, \text{m/s}) + (0.750 \, \text{m/s}^2) \cdot (9.50 \, \text{s}) = 24.1 \, \text{m/s}$$

18. A bus slows down to zero at a bus stop, in a straight line. If the acceleration is –1.50 m/s² and the bus takes 3.50 s to stop, what was the velocity before it began stopping?

Solution:

We know $v_f = 0$ (stops), $a = -1.50$ m/s², and $\Delta t = 3.50$ s. We want to find v_i. The equation with v_i and the variables that we know already is Equation 2.6.1:

$$v_f = v_i + a \cdot \Delta t \qquad (2.6.1)$$

Subtract $a \cdot \Delta t$ from both sides: $v_i = v_f - a \cdot \Delta t$

Put in numbers: $v_i = 0 - (-1.50 \, \text{m/s}^2) \cdot (3.50 \, \text{s}) = 5.25 \, \text{m/s}$

19. A subway car speeds up from rest to 10.0 m/s, with a constant acceleration of 0.850 m/s². How much time did this take to happen?

Solution:

We are given $v_i = 0$ (from rest), $v_f = 10.0$ m/s, and $a = 0.850$ m/s². We want to find Δt. The equation with Δt and variables that we already know is Equation 2.6.1:

$$v_f = v_i + a \cdot \Delta t \qquad\qquad (2.6.1)$$

Subtract v_i from both sides: $v_f - v_i = a \cdot \Delta t$

Divide both sides by a: $\Delta t = \dfrac{v_f - v_i}{a}$

Put in numbers: $\Delta t = \dfrac{\left(10.0\,\text{m}/\text{s}\right) - 0}{0.850\,\text{m}/\text{s}^2} = 11.8\,\text{s}$

20. A car speeds up at a constant rate, while traveling in a straight line. It starts at 10.0 m/s, and the acceleration is 0.900 m/s². It travels 20.0 m. How much time does this take?

Solution:

We are given $a = 0.900$ m/s², $v_i = 10.0$ m/s, and $\Delta x = 20.0$ m. We want to find Δt. The equation with Δt and variables that we know is Equation 2.6.2:

$$\Delta x = v_i \cdot \Delta t + \frac{1}{2} \cdot a \cdot \left(\Delta t\right)^2 \qquad\qquad (2.6.2)$$

Since there are no nonzero values here, the quadratic formula will be necessary. The quadratic formula is:

If $a \cdot x^2 + b \cdot x + c = 0$, where a, b, and c are numbers, and x is the variable you want to solve for,

$$x = \frac{-b \pm \sqrt{b^2 - 4 \cdot a \cdot c}}{2 \cdot a}$$

The quadratic formula is reviewed in Section 13.5

Δt is the variable that we want to solve for. First, we will get the equation into the format of "a" $\cdot (\Delta t)^2 +$ "b" $\cdot \Delta t +$ "c" $= 0$. Note: "a" is different from the acceleration a; that is why quotes are being used.

Subtract Δx from both sides, and switch the order of the other two terms:

$$\frac{1}{2} \cdot a \cdot (\Delta t)^2 + v_i \cdot \Delta t - \Delta x = 0$$

This means that "a" $= \frac{1}{2} \cdot a = \frac{1}{2} \cdot 0.900$, "b" $= v_i = 10.0$, and "c" $= -\Delta x = -20.0$.

Put this into the quadratic formula:

$$\Delta t = \frac{-b \pm \sqrt{b^2 - 4 \cdot a \cdot c}}{2 \cdot a}$$

$$= \frac{-10.0 \pm \sqrt{(10.0)^2 - 4 \cdot (0.5 \cdot 0.900) \cdot (-20.0)}}{2 \cdot (0.5 \cdot 0.900)}$$

$$= 1.85\,s \text{ and } -24.1\,s$$

We take the positive value of <u>1.85 s</u>.

Section 2.7

21. A kickball is kicked straight into the air. If the initial velocity is 10.0 m/s, and gravity is the only force, how high does it get from its initial point? Hint: at its highest height, velocity is zero.

Solution:

The problem tells us that $v_i = 10.0$ m/s (positive; we'll make upward positive), $v_f = 0$ (highest point), and $a = -9.80$ m/s^2 (gravity is the only force, and upward is positive). We want to find Δx. The equation with Δx and the variables that we know is Equation 2.6.4:

$$v_f^2 = v_i^2 + 2 \cdot a \cdot \Delta x \qquad (2.6.4)$$

Subtract v_i^2 from both sides: $v_f^2 - v_i^2 = 2 \cdot a \cdot \Delta x$

Divide both sides by $(2 \cdot a)$: $\Delta x = \dfrac{v_f^2 - v_i^2}{2 \cdot a}$

Put in numbers:
$$\Delta x = \frac{0^2 - (10.0\,\text{m}/\text{s})^2}{2 \cdot (-9.80\,\text{m}/\text{s}^2)}$$

$$= 5.10\,\text{m}$$

22. A fork falls off a table during dinner, straight downward. It falls 1.10 m, starting from rest. How much time does it take to reach the ground? Approximate that gravity is the only force on it.

Solution:

The problem tells us that $v_i = 0$ (starts from rest), $\Delta x = -1.10$ m (negative if we make upward positive), and $a = -9.80$ m/s² (gravity is only force and upward is positive). We want to find Δt. The equation with Δt and the variables that we know is Equation 2.6.2:

$$\Delta x = v_i \cdot \Delta t + \frac{1}{2} \cdot a \cdot (\Delta t)^2 \qquad (2.6.2)$$

Put in that $v_i = 0$: $\Delta x = \frac{1}{2} \cdot a \cdot (\Delta t)^2$ (note: $v_i \cdot \Delta t$ becomes 0)

Divide both sides by $\frac{1}{2} \cdot a$: $(\Delta t)^2 = \dfrac{\Delta x}{\frac{1}{2} \cdot a}$

Take the square root of both sides: $\Delta t = \pm \sqrt{\dfrac{\Delta x}{\frac{1}{2} \cdot a}}$

Take the positive root, since the amount of change in time will be positive.

Put in numbers: $\Delta t = + \sqrt{\dfrac{-1.10\,\text{m}}{\frac{1}{2} \cdot -9.80\,\text{m}/\text{s}^2}} = 0.474\,\text{s}$

23. A basketball is tipped straight into the air during a block, and then caught. If the initial velocity is 4.00 m/s, and gravity is the only force, how much time does it take to be caught? The catch happens when the ball is on the way down, and 1.00 m below where it was tipped.

Solution:

The initial time point is the tip, and the final time point is the catch. Let's make upward the positive direction. We are given that $v_i = 4.00$

m/s, $\Delta x = -1.00$ m, and $a = -9.80$ m/s^2 (gravity is the only force and upward is positive). We want to find Δt. The equation with Δt and the variables that we know is Equation 2.6.2:

$$\Delta x = v_i \cdot \Delta t + \frac{1}{2} \cdot a \cdot (\Delta t)^2 \qquad (2.6.2)$$

Since there are no nonzero values here, the quadratic formula will be necessary. The quadratic formula is:

If $a \cdot x^2 + b \cdot x + c = 0$, where a, b, and c are numbers, and x is the variable you want to solve for,

$$x = \frac{-b \pm \sqrt{b^2 - 4 \cdot a \cdot c}}{2 \cdot a}$$

The quadratic formula is reviewed in Section 13.5

Δt is the variable that we want to solve for. First, we will get the equation into the format of "a" $\cdot (\Delta t)^2 +$ "b" $\cdot \Delta t +$ "c" $= 0$. Note: "a" is different from the acceleration a; that is why quotes are being used.

Subtract Δx from both sides, and switch the order of the other two terms:

$$\frac{1}{2} \cdot a \cdot (\Delta t)^2 + v_i \cdot \Delta t - \Delta x = 0$$

This means that "a" $= \frac{1}{2} \cdot a = \frac{1}{2} \cdot -9.80$, "b" $= v_i = 4.00$, and "c" $= -\Delta x = +1.00$. (Note: for "c", Δx is negative so $-\Delta x$ will be positive.)

Put this into the quadratic formula:

$$\Delta t = \frac{-b \pm \sqrt{b^2 - 4 \cdot a \cdot c}}{2 \cdot a}$$

$$= \frac{-4.00 \pm \sqrt{(4.00)^2 - 4 \cdot (0.5 \cdot -9.80) \cdot (1.00)}}{2 \cdot (0.5 \cdot -9.80)}$$

$$= 1.02\,\text{s and} -0.201\,\text{s}$$

We will choose the positive change for amount of time: $\underline{\Delta t = 1.02 \text{ s}}$

24. Your coworker accidentally drops something. It falls 1.25 m, starting from rest. What is the velocity just before it hits the ground? Approximate that gravity is the only force on it. Hint: be careful of signs during this problem.

Solution:

We are given $v_i = 0$ (starts from rest), $\Delta x = -1.25$ m (we'll make upward positive), and $a = -9.80$ m/s² (gravity is the only force and upward is positive). We want to find v_f. The equation with v_f and the variables that we already know is Equation 2.6.4:

$$v_f^2 = v_i^2 + 2 \cdot a \cdot \Delta x \qquad (2.6.4)$$

Take the square root of both sides: $v_f = \pm\sqrt{v_i^2 + 2 \cdot a \cdot \Delta x}$

In this case, we made upward positive, and the object is going downward. Because of this, we pick the negative sign.

Put in numbers: $v_f = -\sqrt{0^2 + 2 \cdot \left(-9.80\,\text{m/s}^2\right) \cdot \left(-1.25\,\text{m}\right)} = -4.95\,\text{m/s}$

25. A kickball is kicked straight upward. If the initial velocity is 20.0 m/s, and gravity is the only force, what is the velocity 1.50 s later?

Solution:

Make upward positive. We know that $v_i = 20.0$ m/s, $\Delta t = 1.50$ s, and $a = -9.80$ m/s² (gravity is only force and upward is positive). We want to find v_f. The equation with v_f and the variables that we already know is Equation 2.6.1:

$$v_f = v_i + a \cdot \Delta t \qquad (2.6.1)$$

The equation is already solved for v_f, so we can put in numbers:

$$v_f = \left(20.0\,\text{m/s}\right) + \left(-9.80\,\text{m/s}^2\right) \cdot \left(1.50\,\text{s}\right) = 5.3\,\text{m/s}$$

Note: if Δt was larger, v_f could have been negative. This would have meant that the ball was on the way down at that time point.

26. A pinecone falls off a tall tree after some wind moves the branches. It falls straight downward, starting at 1.00 m/s downward. Approximate

that gravity is the only force on it. If the cone takes 1.75 s to reach the ground, how high up was it to start?

Solution:

We are given $v_i = -1.00$ m/s (downward, so negative if we pick upward as the positive direction), $\Delta t = 1.75$ s, and $a = -9.80$ m/s² (gravity is the only force and upward is positive). We want to find Δx (how far it moves when it gets to the ground). The equation with Δx and the variables that we know is Equation 2.6.2:

$$\Delta x = v_i \cdot \Delta t + \frac{1}{2} \cdot a \cdot (\Delta t)^2 \qquad (2.6.2)$$

This equation is already solved for Δx, so put in numbers:

$$\Delta x = \left(-1.00 \,\text{m/s}\right) \cdot \left(1.75 \,\text{s}\right) + \frac{1}{2} \cdot \left(-9.80 \,\text{m/s}^2\right) \cdot \left(1.75 \,\text{s}\right)^2$$

$$= -16.8 \,\text{m}$$

Note: Since the pinecone went down by 16.8 m (with a displacement of –16.8 m), the starting height was 16.8 m.

27. A kickball is kicked straight into the air. Approximate that gravity is the only force on it. 1.00 seconds after the kick, the ball is 4.00 m above where it was kicked from. What was the initial velocity?

Solution:

We are given $\Delta t = 1.00$ s, $\Delta x = 4.00$ m, and $a = -9.80$ m/s² (gravity is the only force, and we'll pick upward as the positive direction). We want to find v_i. The equation with v_i and the variables we already know is Equation 2.6.2:

$$\Delta x = v_i \cdot \Delta t + \frac{1}{2} \cdot a \cdot (\Delta t)^2 \qquad (2.6.2)$$

Subtract $\frac{1}{2} \cdot a \cdot (\Delta t)^2$ from both sides: $\Delta x - \frac{1}{2} \cdot a \cdot (\Delta t)^2 = v_i \cdot \Delta t$

Divide both sides by Δt: $v_i = \dfrac{\Delta x - \dfrac{1}{2} \cdot a \cdot (\Delta t)^2}{\Delta t}$

Put in numbers: $v_i = \dfrac{(4.00\,\text{m}) - \dfrac{1}{2} \cdot (-9.80\,\text{m}/\text{s}^2) \cdot (1.00\,\text{s})^2}{(1.00\,\text{s})} = 8.90\,\text{m}/\text{s}$

28. Note: this is a more difficult problem than the others. An object is sent into the air with an initial velocity of 14.0 m/s, from the ground. Approximate that gravity is the only force on it.

A. How high above the ground does the object get?

B. How much time does it take for the object to come back to the ground? Hint: Part B requires different time points than Part A.

Solution:

Part A:

This problem requires different sets of initial and final time points, much like Example 2.7.2. Here, let's pick the initial time point as when the object goes into the air, and the final time point as when it reaches its highest point. With those initial and final time points, $v_i = 14.0$ m/s, $v_f = 0$ (at the highest height), and a = −9.80 m/s² (gravity is only force, and we'll pick upward as the positive direction). We want to find Δx. The equation with Δx and the variables that we already know is Equation 2.6.4:

$$v_f^2 = v_i^2 + 2 \cdot a \cdot \Delta x \qquad (2.6.4)$$

Subtract v_i^2 from both sides: $v_f^2 - v_i^2 = 2 \cdot a \cdot \Delta x$

Divide both sides by $(2 \cdot a)$: $\Delta x = \dfrac{v_f^2 - v_i^2}{2 \cdot a}$

Put in numbers: $\Delta x = \dfrac{(0)^2 - (14.0\,\text{m}/\text{s})^2}{2 \cdot (-9.80\,\text{m}/\text{s}^2)} = 10.0\,\text{m}$

Part B:

For Part B, we will pick the initial time point as when the object goes into the air, and the final time point as just before it reaches the ground. We could have also made the highest height the initial time point, and it would give us the same answer. With the time points that are picked here, $v_i = 14.0$ m/s, $\Delta x = 0$ (it starts and ends on the ground), and $a = -9.80$ m/s^2 (gravity is the only force, and we'll pick upward as the positive direction). We want to find Δt. The equation with Δt and the variables that we already know is Equation 2.6.2:

$$\Delta x = v_i \cdot \Delta t + \frac{1}{2} \cdot a \cdot (\Delta t)^2 \qquad (2.6.2)$$

If none of the terms were zero, we would have to use the quadratic formula. But, $\Delta x = 0$. Putting this in: $0 = v_i \cdot \Delta t + \frac{1}{2} \cdot a \cdot (\Delta t)^2$

Both terms on the right side have a factor of Δt, so we can rewrite it as:

$$0 = \Delta t \cdot \left(v_i + \frac{1}{2} \cdot a \cdot \Delta t \right)$$

This gives two answers: $\Delta t = 0$ and $v_i + \frac{1}{2} \cdot a \cdot \Delta t = 0$

The first answer ($\Delta t = 0$) basically tells us that the ball is on the ground at the start. The second answer ($v_i + \frac{1}{2} \cdot a \cdot \Delta t = 0$) tells us how much time passes before it is on the ground again – this is the answer we want. This means that our next step is to solve $v_i + \frac{1}{2} \cdot a \cdot \Delta t = 0$ for Δt.

Subtract v_i from both sides: $\frac{1}{2} \cdot a \cdot \Delta t = -v_i$

Divide both sides by $\frac{1}{2} \cdot a$: $\Delta t = \dfrac{-v_i}{\frac{1}{2} \cdot a}$

Put in numbers: $\Delta t = \dfrac{-(14.0\,\text{m/s})}{\frac{1}{2} \cdot (-9.80\,\text{m/s}^2)} = 2.86\,\text{s}$

CHAPTER 3

Motion in Two and Three Dimensions

3.1 INTRODUCTION: THREE DIMENSIONS ARE MORE REALISTIC THAN ONE

Chapter 2 was all about motion in one dimension, which means that all of the motion was along a straight line. Sometimes this works just fine, like a straight road. But in general, we need more than one dimension to accurately describe motion. In this chapter, we'll use similar strategies to Chapter 2 for situations where the motion is not in a straight line.

3.2 DIMENSIONS BEHAVE SEPARATELY

From where you are, right now, you can describe the space around you using a combination of three directions:

- Up/down

- Left/right

- Forward/backward

These directions are what we call dimensions. Left/right is often called "the x direction" in physics. Up/down is often called "the y direction". Forward/backward is often called "the z direction". Which directions are called x, y, and z can be different, and they don't have to be exactly these definitions.

DOI: 10.1201/9781003005049-3

When we look at motion in three dimensions for this course, the motion actually behaves like three separate motion problems, with each one similar to the problems in Chapter 2: a problem in the x direction, a problem in the y direction, and a problem in the z direction. This may not seem obvious at first.

A classic physics class demonstration shows the dimensions behaving separately during motion. The demonstration starts with a contraption holding two identical balls at exactly the same height. When the teacher presses a button, both of the balls start moving at exactly the same time. One of the balls is dropped straight downward (call it the −y direction), while the second ball goes sideways (call it the +x direction). There are two results of this motion:

1. The balls land on the ground at <u>exactly</u> the same time!

2. The ball that went sideways has a displacement in the +x direction, but the ball that was dropped straight down does not.

This result can be explained by modeling the world as if the dimensions behave separately:

- Looking at the up/down direction (y direction), we can model the ball that drops as a free fall problem just like those in Section 2.7. The initial velocity would be zero. Now, think about the second ball (that goes sideways at the start), and focus only on the y direction. The initial velocity was all in the x direction, which means that there was no velocity in the y direction initially. So if you look at the y direction only, the second ball has the same displacement (height that it falls) and the same initial velocity (zero). It turns out that it also has the same acceleration (9.80 m/s^2, downward, due to gravity). You might remember from Sections 2.6 and 2.7 that once you know three of our five motion variables (v_i, v_f, a, t, Δx), you can predict the other two. Since displacement (Δx), initial velocity (v_i), and acceleration (a) are the same in the y direction, the two motion problems are exactly the same in the y direction! That's why they hit the ground at the same exact time.

- Looking at the x direction, the difference between the two balls is that one has zero initial velocity in the x direction, and one ball has an initial velocity that is greater than zero. The one with greater than zero initial velocity will move in the x direction; the one with zero initial velocity will not.

This isn't usually done in the demonstration, but imagine that there was a third ball and that this ball is shot forward, in the positive z direction. This ball would hit the ground at exactly the same time as the other two balls, because the y direction variables would be exactly the same. The ball would have a displacement in the z direction, because it would be the only ball with an initial velocity in the z direction.

The point of this discussion is that, in a two- and three-dimension motion problem, you can look at the x, y, and z dimensions separately and treat each dimension just like the one-dimensional problems that you did in Chapter 2! In the rest of Chapter 3, we cover specific situations where this happens. In order to fully describe the situation, however, we will need to eventually combine the information from the different dimensions. This is done using vectors, which is the topic of the next section.

Side note: the separate behavior of dimensions starts to go away when objects travel at really high speeds, higher than are generally reached on Earth. This behavior is described by Einstein's theory of special relativity.

3.3 VECTOR MATH: COMPONENTS AND MAGNITUDE AND DIRECTION

In this section, we introduce the math concept called a **vector**. We will use vectors to combine information from different dimensions (x, y, and z). In this section, we will do vectors in two dimensions; we'll generalize to three dimensions in Section 3.7.

Vectors can be written in two different ways:

1. Magnitude and Direction. In this form of describing a vector, you give an amount (called the magnitude) and a direction. One example of this would be a velocity; something like 6.83 m/s, at an angle of 37.9 degrees below the +x direction.

    Note: the magnitude is always a positive number.

2. Components. Component means the amount of the vector in each dimension. We'll write this in the following way:

$$\textbf{Vector} = (\text{x component})\ i + (\text{y component})\ j \tag{3.3.1}$$

Notice that the name of the vector is written in **bold** – in this book, any variable written in **bold** is a vector. i shows that a certain part is in the x direction, and j shows that a certain part is in the y direction. i and j are called **unit vectors**.

For a ball that starts off going towards the +x direction with initial velocity 6.00 m/s (similar to the example described in Section 3.2), we could write the initial velocity as:

$$\mathbf{v_i} = 6.00\ \text{m/s}\ i + 0\ \text{m/s}\ j$$

This equation tells us that $\mathbf{v_i}$ is a vector, that the initial velocity in the x direction is 6.00 m/s, and that the initial velocity in the y direction is 0.

You can use either one of the two ways described above to write down a vector. Most of the physics problem solving that you do will use components – essentially solving a one-dimensional motion problem for each dimension. However, a lot of the information exchanged in the world will be in the form of magnitude and direction. It's common to think about using vectors to give directions – have you ever heard someone give you directions using components?

We can convert between the two vector forms of "magnitude and direction" and "components". This is done using equations from trigonometry; trigonometry is reviewed in Section 13.7. The important equations for vectors are:

If you have components, and want magnitude and angle:

$$\left(\text{magnitude}\right) = \sqrt{\left(\text{x component}\right)^2 + \left(\text{y component}\right)^2} \tag{3.3.2}$$

$$\theta = \tan^{-1}(|\text{y component}|/|\text{x component}|) \tag{3.3.3}$$

θ can be from the +x direction or the –x direction, and it can be above or below. See the "important note" below.

If you have magnitude and angle, and want components:

$$(\text{x component magnitude}) = (\text{magnitude}) \cdot \cos(\theta) \tag{3.3.4}$$

$$(\text{y component magnitude}) = (\text{magnitude}) \cdot \sin(\theta) \tag{3.3.5}$$

You have to assign a + or – sign to components yourself. See the "important note" below.

In these equations, "x component" is the number that goes next to i in Equation 3.3.1, "y component" is the number that goes next to j in Equation

3.3.1, the | | signs mean to take the absolute value (eliminate any minus signs), and θ (Greek letter theta) is an angle that is the direction of the vector. The angle is usually from the +x direction (sometimes called the +x axis), or the –x direction (sometimes called the –x axis). In Equations 3.3.4 and 3.3.5, the word "magnitude" next to the x and y components means that you need to put the sign (+ or –) yourself – see the note below.

Important note: these equations do not give minus signs. You need to figure out the minus signs for yourself, based on what the problem says. Two examples of this:

- The problem asks for components of a vector that is 64.0° north of west, where north is the +y direction and east is the +x direction. When you get the components using Equations 3.3.4 and 3.3.5, give the x component a minus sign (since west is the –x direction), and give the y component a plus sign (since north is the +y direction).

- The problem gives you a positive x component and a negative y component. The magnitude will be positive (it always is), and the angle will be below the +x axis. It's from the +x axis because the x component is positive; it's "below" the axis because the y component is negative.

Side note: when talking about the angle, sometimes instead of "above" or "below" an axis, the terms "clockwise from" or "counterclockwise from" are used. Clockwise is the direction that the hands move on a clock – this would be towards "below" the +x axis and "above" the –x axis.

This gives us all of the tools that we need to convert between the components form of vectors, and the magnitude and direction form of vectors. We'll now do one example going each way.

Example 3.3.1: Going from Magnitude and Direction to Components

In a physics problem, the displacement is 94.3 meters, at an angle of 50.0 degrees north of east. If east is the +x direction and north is the +y direction, what are the x and y components?

Solution:
Here, you are given the magnitude (94.3 m) and the direction (50.0°). You want to find the x and y components. This will be Equations 3.3.4 and 3.3.5:

x component: (x component magnitude) = (magnitude) · cos(θ) = (94.3 m) · cos(50.0°) = 60.6 m

y component: (y component magnitude) = (magnitude) · sin(θ) = (94.3 m) · cos(50.0°) = 72.2 m

This gives us the magnitude of each component. You now need to choose the sign (+ or –) for each component. Since "north" is the +y direction, and "east" is the +x direction, both components are positive.

x component is 60.6 m, y component is 72.2 m

You could write the vector as: 60.0 m i + 72.2 m j

Example 3.3.2: Going from Components to Magnitude and Direction

During a physics demonstration, a ball has a velocity of $\mathbf{v}$ = 6.00 m/s i –5.00 m/s j. Define the positive x direction to be "to the right", and "up" to be the positive y direction. What is the magnitude and direction of the velocity?

Solution:

We are given that the x component is 6.00 m/s, and that the y component is –5.00 m/s. We need to find the magnitude and angle. These are Equations 3.3.2 and 3.3.3:

$$(\text{magnitude}) = \sqrt{(\text{x component})^2 + (\text{y component})^2}$$

Put in numbers:

$$(\text{magnitude}) = \sqrt{(6.00 \text{ m/s})^2 + (-5.00 \text{ m/s})^2} = 7.81 \text{ m/s}$$

θ = tan^{-1}(|y component |/|x component|) = tan^{-1}(|–5.00 m/s|/|6.00 m/s|) = 39.8°

We now need to say whether θ is from the +x or –x direction, and above or below. Since the x component is positive, the angle is from the +x axis. Since the y component is negative, it is below.

The vector is 7.81 m/s, at 39.8° below the +x axis.

Problem to Try Yourself

In a physics problem, the acceleration vector is $\mathbf{a}$ = 2.30 m/s² i – 9.80 m/s² j. Convert this to a magnitude and an angle.

Solution:
We are given that the x component is 2.30 m/s², and that the y component is –9.80 m/s². We need to find the magnitude and angle. These are Equations 3.3.2 and 3.3.3:

$$\left(\text{magnitude}\right)=\sqrt{\left(\text{x component}\right)^2+\left(\text{y component}\right)^2}$$

Put in numbers:

$$\left(\text{magnitude}\right)=\sqrt{\left(2.30 \text{ m}/\text{s}^2\right)^2+\left(-9.80 \text{ m}/\text{s}^2\right)^2}=10.1 \text{ m/s}^2$$

$\theta=\tan^{-1}(|\text{y component }|/|\text{x component}|)=\tan^{-1}(|-9.80 \text{ m/s}^2|/|2.30 \text{ m/s}^2|)=76.8°$

We now need to say whether θ is from the +x or –x direction, and above or below. Since the x component is positive, the angle is from the +x axis. Since the y component is negative, it is below.
The vector is 10.1 m/s², at 76.8° below the +x axis.

3.4 DOING BASIC MATH WITH VECTORS: ADDING VECTORS, MULTIPLICATION BY A SCALAR

When adding two vectors together, you just add the x components to each other and add the y components to each other. If subtracting two vectors, subtract the components.

A scalar is a number with no direction. If you multiply (or divide) a vector by a scalar, every component and the magnitude are all multiplied (or divided) by that number.

The specific relationships for addition, subtraction, and multiplication by a scalar are:
If vector $\mathbf{A}=A_x\, i+A_y\, j$, and vector $\mathbf{B}=B_x\, i+B_y\, j$, and n is a scalar,

$$\mathbf{A}+\mathbf{B}=(A_x+B_x)\, i+(A_y+B_y)\, j \tag{3.4.1}$$

$$\mathbf{A}-\mathbf{B}=(A_x-B_x)\, i+(A_y-B_y)\, j \tag{3.4.2}$$

$$n\cdot\mathbf{A}=n\cdot A_x\, i+n\cdot A_y\, j \tag{3.4.3}$$

Starting from the next section, what will happen is that each vector in a problem will be split into components, and math will be done with each component separately. This is also what is happening in the equations above – the x components are added or subtracted or multiplied separately from the y components.

Example 3.4.1: Combining Addition and Subtraction

If $\mathbf{A} = 4.00\,i + 5.00\,j$ and $\mathbf{B} = 3.50\,i - 4.50\,j$, what are the x and y components of $2.00 \cdot \mathbf{A} - 3.00 \cdot \mathbf{B}$?

Solution:

Do the components separately
x component: $2.00 \cdot 4.00 - 3.00 \cdot 3.50 = \underline{-2.50}$
y component: $2.00 \cdot 5.00 - 3.00 \cdot -4.50 = \underline{23.5}$

Note: you could use Equations 3.3.2 and 3.3.3 to convert these components into a magnitude and a direction. They would come to magnitude 23.6, at an angle of 83.9° above the –x axis. (For extra practice, you could try to calculate this yourself.)

Problem to Try Yourself

If $\mathbf{A} = 8.00\,i - 3.00\,j$ and $\mathbf{B} = -7.50\,i + 5.50\,j$, what are the x and y components of $\mathbf{A} + 3.00 \cdot \mathbf{B}$?

Solution:

Do the components separately
x component: $8.00 + 3.00 \cdot -7.50 = \underline{-14.5}$
y component: $-3.00 + 3.00 \cdot 5.50 = \underline{13.5}$

3.5 TWO-DIMENSIONAL CONSTANT ACCELERATION PROBLEMS

In Section 2.6, we defined four equations for a one-dimensional motion that had constant acceleration:

$$v_f = v_i + a \cdot \Delta t \tag{2.6.1}$$

$$\Delta x = v_i \cdot \Delta t + \frac{1}{2} \cdot a \cdot \left(\Delta t\right)^2 \tag{2.6.2}$$

$$\Delta x = \frac{1}{2} \cdot \left(v_i + v_f \right) \cdot \Delta t \qquad (2.6.3)$$

$$v_f^{\,2} = v_i^{\,2} + 2 \cdot a \cdot \Delta x \qquad (2.6.4)$$

As discussed in Chapter 2, there are five quantities contained in these equations: final velocity (v_f), initial velocity (v_i), acceleration (a), displacement (Δx) and time (Δt). **Of these five quantities, all of them except time will become vectors when the problem is two dimensional instead of one dimensional.** The reason for this is that velocity, acceleration, and displacement have amounts (magnitudes) but also have directions. **When the problem is two-dimensional, we compute the x component and the y component separately.** Time is a scalar, and is the same for each component.

In computing the components separately, we may need to go back and forth between x and y components and magnitude and direction. For example, you might compute answers in components, and then combine into magnitude and direction, or you might be given a vector with magnitude and direction in the problem and need to split it into components before you can use it.

Side note: The average velocity and average acceleration equations (2.4.1 and 2.5.1) could be treated in exactly the same way – just compute the x and y components separately.

Example 3.5.1: Two-Dimensional Constant Acceleration Problem

A bus starts from rest. It then speeds up, with acceleration $\mathbf{a} = 1.25$ m/s² $i + 2.00$ m/s² j. When it reaches a velocity of $\mathbf{v_f} = 6.25$ m/s $i + 10.0$ m/s j, what displacement has the car gone through?

Solution:
Displacement will have an x component, and a y component. Let's look at the x component first. We know the initial velocity ($v_i = 0$), the acceleration (1.25 m/s²), and the final velocity ($v_f = 6.25$ m/s). We don't know displacement (Δx) and we don't know time; we want to find displacement. We need an equation from 2.6.1–2.6.4 that has only Δx (what we want to find) and variables that we already know (v_i, v_f, a). Equation 2.6.4 fits:

$$v_f^2 = v_i^2 + 2 \cdot a \cdot \Delta x$$

Solve this equation for Δx:

Subtract v_i^2 from both sides: $v_f^2 - v_i^2 = 2 \cdot a \cdot \Delta x$

Divide both sides by $(2 \cdot a)$: $\Delta x = \dfrac{v_f^2 - v_i^2}{2 \cdot a}$

Put in numbers: $\Delta x = \dfrac{\left(6.25\,\text{m/s}\right)^2 - 0^2}{2 \cdot 1.25\,\text{m/s}^2} = 15.6\,\text{m}$

For the y component, we know the initial velocity ($v_i = 0$), the acceleration (2.00 m/s^2), and the final velocity ($v_f = 10.0$ m/s). We don't know displacement (Δx) and we don't know time; we want to find displacement. We need an equation from 2.6.1–2.6.4 that has only Δx (what we want to find) and variables that we already know (v_i, v_f, a). Here, Equation 2.6.4 still fits:

$$v_f^2 = v_i^2 + 2 \cdot a \cdot \Delta x$$

We just solved this equation for displacement, for the x component: $\Delta x = \dfrac{v_f^2 - v_i^2}{2 \cdot a}$

Putting in numbers for the y component:

$$\Delta x = \frac{\left(10.0\,\text{m/s}\right)^2 - 0^2}{2 \cdot 2.00\,\text{m/s}^2} = 25.0\,\text{m}$$

The final answer is that the displacement has components

$\Delta x = 15.6$ m $i + 25.0$ m j. As with previous answers, we could convert this into magnitude and direction using Equations 3.3.2 and 3.3.3. You would get a magnitude of 29.5 m and a direction of 58.0 degrees above the +x axis. In this problem, the x and y components used the same equation but with different variables; this will not always be the case.

Side note: some textbooks call the two-dimensional displacement vector Δr, and the components Δx and Δy. Some textbooks write the equations with x and y indices, for example $v_{fx} = v_{ix} + a_x \cdot \Delta t$ and $v_{fy} = v_{iy} + a_y \cdot \Delta t$

Problem to Try Yourself

Define east as the +x direction and north as the +y direction. A squirrel starts at rest, then accelerates at $a = 1.55$ m/s^2 $i - 3.00$ m/s^2 j. 2.00 seconds later, what is the displacement of the squirrel?

Solution:

x component: we know $v_i = 0$ (since it starts at rest), $a = 1.55$ m/s^2, and $\Delta t = 2.00$ s. We want to know Δx. The equation with Δx and the variables we know is Equation 2.6.2:

$$\Delta x = v_i \cdot \Delta t + \frac{1}{2} \cdot a \cdot \left(\Delta t\right)^2$$

Put in the numbers: $\Delta x = (0) \cdot (2.00\,s) + \frac{1}{2} \cdot (1.55\,m/s^2) \cdot (2.00\,s)^2 = 3.10$ m

y component: we know $v_i = 0$ (since it starts at rest), $a = -3.00$ m/s^2, and $\Delta t = 2.00$ s. We want to know Δx. The equation with Δx and the variables we know is Equation 2.6.2, just as with the x direction (since the variables known and the variable to solve for are the same).

$$\Delta x = v_i \cdot \Delta t + \frac{1}{2} \cdot a \cdot \left(\Delta t\right)^2 = -6.00 \text{ m}$$

Put in the numbers: $\Delta x = (0) \cdot (2.00\,s) + \frac{1}{2} \cdot (-3.00\,m/s^2) \cdot (2.00\,s)^2$
= 3.10 m*i* – 6.00 m*j*

3.6 PROJECTILE MOTION

Projectile motion is a special case of two-dimensional motion with a constant acceleration. You can think of projectile motion as when an object is flying through the air, fairly close to Earth (not in space), and **gravity** is the only force that is acting significantly on the object.

Gravity is a force that pulls objects towards each other. For an object near Earth, that means that the object is pulled downward, towards Earth. We will talk about gravity in more depth in Chapter 4.

Everything is the same as Section 3.5 immediately before this, with one extra piece of information:

If we define the x direction as "the direction that the object is moving horizontally" and the y direction as "upward", the acceleration in projectile motion is always

$$a = 0\ i - 9.80 \text{ m/s}^2\ j \tag{3.6.1}$$

The number 9.80 m/s² is the same "acceleration due to gravity" mentioned with free fall in Section 2.7. Since we defined upward as the positive y direction, it is negative.

Here are some tips for solving projectile motion problems:

1. In projectile motion problems, it's very common for the initial velocity to be given as a magnitude and angle, instead of as components. You can use Equations 3.3.4 and 3.3.5 to find the components:

$$(\text{x component magnitude}) = (\text{magnitude}) \cdot \cos(\text{angle}) \qquad (3.3.4)$$

$$(\text{y component magnitude}) = (\text{magnitude}) \cdot \sin(\text{angle}) \qquad (3.3.5)$$

Don't forget to assign plus and minus signs yourself when using these equations.

2. It will often be useful to solve for the time it takes for an object to hit the ground, using the y-dimension, and then use that amount of time in the x-direction equations. Remember that time is always the same in both the x and y equations.

3. If you end up in a situation where you are using Equation 2.6.2, and solving for time, if none of the terms in the equation are equal to zero you will need to use the quadratic equation:

If $a \cdot x^2 + b \cdot x + c = 0$, where a, b, and c are numbers and x is the variable you want to solve for,

$$x = \frac{-b \pm \sqrt{b^2 - 4 \cdot a \cdot c}}{2 \cdot a}$$

Using the quadratic equation is reviewed in Section 13.5.

4. If you end up in a situation where you are using Equation 2.6.2, and solving for time, you will get two solutions and need to choose one as the solution. This will depend on context. Imagine an object that starts from the ground, and then hits the ground. If one of the two times is negative or zero, the other time choice is almost certainly the correct one. If both times are positive, you will likely see language in the problem like "how long does it take when the object reaches a height of … meters, on the way up?" The time as the object hits a given height on the way up would be the first time (smaller value of

your two choices); the second time (larger value) would be on the way down.

5. At maximum height, the y component of velocity is zero. The x component is not zero, unless it started as zero. This is because the x component of velocity never changes from its original value, since the acceleration is zero.

6. Problems often ask for the velocity just before the object hits the ground. Students are often tempted to say that the velocity is zero, based on experience. What the problem actually means is, what is the velocity just before the object hits the ground? You can set the change in height as if the object is on the ground, while understanding that it is really just barely above the ground.

7. You may find it very helpful to do a sense check on your answers. If your answer is that a basketball is in the air for 15 minutes, or traveling 600 m/s, you should check your work. One last thing to keep in mind – our equations assume that air resistance is small enough to ignore, but in reality it does have an effect. So you might see some of the velocities and displacements being slightly higher than reality.

Example 3.6.1

A baseball is hit with an initial velocity of 40.0 m/s, at an angle of 25.0 degrees above the ground. It is hit at an initial height of 1.00 m above the ground, and is caught by an outfielder at a height of 1.00 m above the ground.

A. How much time is the ball in the air for?
B. How far horizontally does the ball travel?

Part A Solution:
We have to choose whether to look at the x direction or the y direction first. Define x as horizontal, in the direction the ball is hit, and y as "upward". Define the initial time point as when the ball is hit, and the final time point as when it is caught. We're given the following information:

• x direction: initial velocity (we can find it); final velocity (equal to initial because acceleration is 0); acceleration = 0.

- y direction: initial velocity; acceleration $= -9.80$ m/s²; displacement (0, because it starts and stops at the same height of 1.00 m)

If you look at the x-direction, hoping to find time, it won't work out. Let's try it and see why. You need an equation that has only time and variables that are known. There is one equation that matches this:

$$v_f = v_i + a \cdot \Delta t \qquad (2.6.1)$$

However, since a = 0 for the x direction, the time is "erased" from the equation, leaving us with just $v_i = v_f$, which we already knew because acceleration is zero. The end result is that we can't use the x direction here.

So, we use the y direction. Let's find the initial velocity first. As suggested in the tips above, let's use Equation 3.3.4 or 3.3.5. Equation 3.3.5 turns out to be for the y component:

$$\text{(y component magnitude)} = \text{(magnitude)} \cdot \sin(\text{angle}) \quad (3.3.5)$$

Putting in numbers, we get:

$$v_y = 40.0 \text{ m/s} \cdot \sin(25.0°) = 16.9 \text{ m/s}$$

We choose a positive sign for the velocity, since the vertical part of the velocity is upward.

With this, we now know three of the five variables in the y direction:

- $v_i = 16.9$ m/s
- $a = -9.80$ m/s²
- $\Delta x = 0$ (starts and ends at a height of 1.00 m)

To find time (Δt), we need an equation that has only time and some (or all) of these three variables. That equation is Equation 2.6.2:

$$\Delta x = v_i \cdot \Delta t + \frac{1}{2} \cdot a \cdot \left(\Delta t \right)^2 \qquad (2.6.2)$$

If no terms in this equation are zero, we need to use the quadratic equation – see tip 3 above. However, since $\Delta x = 0$, we can solve for Δt.

$$0 = v_i \cdot \Delta t + \frac{1}{2} \cdot a \cdot \left(\Delta t \right)^2$$

Note: sometimes it helps to put in any values of zero at the start of doing math, because it often makes the math easier.

Factor out one factor of Δt – meaning, pull Δt out of each term and put a factor of Δt in front:

$$0 = \Delta t \cdot \left(v_i + \frac{1}{2} \cdot a \cdot \Delta t \right)$$

You may remember from algebra that an equation like this has two solutions. One of them is $\Delta t = 0$. What the $\Delta t = 0$ means is that the displacement (Δx) is zero when $\Delta t = 0$, meaning that the ball is 1.00 m off of the ground at time 0. That is true, but it doesn't answer the question. The more interesting time point is the second time point, where the term in parentheses is equal to zero:

$$0 = v_i + \frac{1}{2} \cdot a \cdot \Delta t$$

Subtract v_i from both sides: $-v_i = \frac{1}{2} \cdot a \cdot \Delta t$

Divide both sides by $\frac{1}{2} \cdot a$: $\Delta t = -v_i / \left(\frac{1}{2} \cdot a \right)$

Put in numbers: $\Delta t = -\left(16.9 \, m/s \right) / \left(\frac{1}{2} \cdot -9.80 \, m/s^2 \right) = 3.45 s$

This is the answer to Part A. As a sense check, 3.45 seconds seems like a reasonable amount of time for a baseball to be in the air.

Part B solution:

To determine how far the ball travels horizontally, we need to look at the horizontal direction, which is the x direction.

First, let's find the initial velocity in x. We'll use Equation 3.3.4, in the same way that we used Equation 3.3.5 for the y direction:

(x component magnitude) = (magnitude) $\cdot$ cos(angle)   (3.3.4)

Putting in numbers, we get:

$$v_x = 40.0 \, m/s \cdot \cos(25.0°) = 36.3 \, m/s$$

Variables in the x direction that we know:

- $v_i = 36.3$ m/s
- $a = 0$

- $v_f = 36.3$ m/s (because acceleration equals zero)
- $t = 3.45$ s

The displacement, Δx, is the only thing we don't know. We can use any equation out of 2.6.1–2.6.4 that has Δx in it. Let's use 2.6.2:

$$\Delta x = v_i \cdot \Delta t + \frac{1}{2} \cdot a \cdot (\Delta t)^2 \qquad (2.6.2)$$

Put in numbers: $\Delta x = (36.3\,\text{m/s}) \cdot (3.45\,\text{s}) + \frac{1}{2} \cdot 0 \cdot (3.45\,\text{s})^2 = 125\,\text{m}$

125 meters is about 410 feet. As a sense check, it's reasonable for a baseball to go that far. In reality, a ball hit with that initial velocity (40.0 m/s, roughly 90 mph) would travel a little bit less far, because of air resistance.

The example above is a fairly standard example of projectile motion. A problem like this could be harder if:

- The y direction displacement wasn't zero (making the quadratic equation necessary)
- More variables were asked for – for example, the final velocity in the y direction
- The maximum height was asked for – this would require a new final time point (the time when the ball is at maximum height)
- Any of the variables besides time could be requested in magnitude and direction, adding an extra calculation similar to Example 3.3.2, in Section 3.3.
- It's fairly common for a problem like this to skip Part A and go straight to Part B. You would still need to do all of the work in Part A to find the time, but you'd need to figure out that you needed the time yourself, without being prompted to solve for it first.

The problems can also be easier:

- Any situation where you're not using Equation 2.6.2 to solve for time (or not solving for time at all) is usually easier.
- If the object starts out traveling horizontally, the math for finding the time is similar to this example (same equation) but slightly easier because the y component of initial velocity is zero.

Problem to Try Yourself

A basketball is flying through the air, towards the basket. Between when it is released and when it goes through the basket, it increases in height by 1.00 m and travels a horizontal distance of 5.00 m. It is in the air for 1.75 seconds. What are the x and y components of the initial velocity?

Solution:

Let's look at the x components first, to find the x component of initial velocity. We know $\Delta x = 5.00$ m (horizontal displacement), $a = 0$ (for all projectile motion) and $t = 1.75$ s. We want to find v_i. The equation with v_i and the variables that are known is Equation 2.6.2:

$$\Delta x = v_i \cdot \Delta t + \frac{1}{2} \cdot a \cdot (\Delta t)^2$$

To solve for v_i, subtract $\frac{1}{2} \cdot a \cdot (\Delta t)^2$ from both sides:

$$\Delta x - \frac{1}{2} \cdot a \cdot (\Delta t)^2 = v_i \cdot \Delta t$$

Divide both sides by Δt: $v_i = \dfrac{\Delta x - \dfrac{1}{2} \cdot a \cdot (\Delta t)^2}{\Delta t}$

Put in numbers: $v_i = \dfrac{5.00\,\text{m} - \dfrac{1}{2} \cdot 0 \cdot (2.75\,\text{s})^2}{(1.75\,\text{s})} = 2.86\,\text{m/s}$

For the y component, we know $\Delta x = 1.00$ m (vertical displacement), $a = -9.80$ m/s^2 (for all projectile motion) and $t = 1.75$ s. We want to find v_i. The equation with v_i and the variables that are known is Equation 2.6.2:

$$\Delta x = v_i \cdot \Delta t + \frac{1}{2} \cdot a \cdot (\Delta t)^2$$

Solving Equation 2.6.2 for initial velocity (v_i) is exactly the same math as we did for the x component. The solution was $v_i = \dfrac{\Delta x - \dfrac{1}{2} \cdot a \cdot (\Delta t)^2}{\Delta t}$

Putting in the y component numbers:

$$v_i = \frac{1.00\,\text{m} - \frac{1}{2} \cdot \left(-9.80\,\text{m}/\text{s}^2\right) \cdot \left(1.75\,\text{s}\right)^2}{\left(1.75\,\text{s}\right)} = 9.15\,\text{m}/\text{s}$$

3.7 MOTION IN THREE DIMENSIONS

Motion in three dimensions is the same general process as motion in two dimensions, just with slightly more math. You would now have x, y and z components instead of just x and y components.

For the vector math in Section 3.3, there would just be one more component. Equation 3.4, for vector magnitude, would add an extra term for z:

$$\left(\text{magnitude}\right) = \sqrt{\left(x\,\text{component}\right)^2 + \left(y\,\text{component}\right)^2 + \left(z\,\text{component}\right)^2}$$

(3.7.1)

Instead of one angle, between the x and y directions, you would also need to specify an angle between the z direction and another direction.

For basic math (Section 3.4), components would still be treated separately. There would just be one more component to do math with.

For a general problem with constant acceleration (Section 3.5), you would just solve three separate problems instead of two.

For projectile motion (Section 3.6), the motion actually is in two dimensions: up/down (y), and forward/backward (x), where forward/backward is the direction that the object starts out traveling. The object can't take any turns, because projectile motion has no forces except gravity.

3.8 CHAPTER 3 SUMMARY

Two dimensional (and three dimensional) motion is more realistic than one dimensional motion, but it requires more math. In the problems seen in this textbook, the dimensions will have separate behavior (other than time being the same for all of them), so the math can be done separately, like two or three separate one-dimensional problems.

To describe two dimensional motion, we use vectors. Vectors can be written in two different ways:

1. Magnitude and direction. The direction is usually an angle from either the +x direction or the −x direction.

2. x and y components: **Vector** = (x component) i + (y component) j (3.3.1)

   i shows that a certain part is in the x direction, and j shows that a certain part is in the y direction. i and j are called **unit vectors**.

It's often necessary to go between these two ways of writing a vector.
   If you have components, and want magnitude and angle:

$$\left(\text{magnitude}\right) = \sqrt{\left(\text{x component}\right)^2 + \left(\text{y component}\right)^2} \qquad (3.3.2)$$

$$\theta = \tan^{-1}(|\text{y component}|/|\text{x component}|) \qquad (3.3.3)$$

θ can be from the +x direction or the –x direction, and it can be above or below. Use the information in the problem to decide which.
   If you have magnitude and angle, and want components:

$$(\text{x component magnitude}) = (\text{magnitude}) \cdot \cos(\theta) \qquad (3.3.4)$$

$$(\text{y component magnitude}) = (\text{magnitude}) \cdot \sin(\theta) \qquad (3.3.5)$$

You have to assign a + or – sign to components yourself, based on what the problem says.

When doing basic math like addition and subtraction or multiplication by a scalar (non-vector) number, just use components and do the math separately for the x components and for the y components. For example, if adding two vectors, the total x component is adding the x components, and the total y component is adding the y components.

If the acceleration is constant, Equations 2.6.1–2.6.4 can be used. You do two separate problems – once for each dimension. The change in time (Δt) is the only variable shared between the two dimensions.

$$v_f = v_i + a \cdot \Delta t \qquad (2.6.1)$$

$$\Delta x = v_i \cdot \Delta t + \frac{1}{2} \cdot a \cdot \left(\Delta t\right)^2 \qquad (2.6.2)$$

$$\Delta x = \frac{1}{2} \cdot \left(v_i + v_f\right) \cdot \Delta t \qquad (2.6.3)$$

$$v_f^2 = v_i^2 + 2 \cdot a \cdot \Delta x \qquad (2.6.4)$$

A **projectile motion problem** is a type of constant acceleration problem in two dimensions where gravity is assumed to be the only force. In this case, the x component of acceleration is 0 and the y component of acceleration is –9.80 m/s². (Note: the minus sign happens if you make upward the +y direction.)

Chapter 3 problem types:

With vectors, convert between components and magnitude/angle (Section 3.3)

Two-dimensional problems with constant acceleration (Section 3.5)

Projectile motion problems (Section 3.6)

The problems in Sections 3.5 and 3.6 can include multiple variables solved for, and multiple time points (for example, reaching the ground but also highest height).

Practice Problems

Section 3.3

1. A ball is thrown with an initial velocity of magnitude 8.50 m/s, at an angle of 10.0 degrees above the horizontal. What are the x and y components of this velocity?

   Solution:

   $$(\text{x component magnitude}) = (\text{magnitude}) \cdot \cos(\theta) \qquad (3.3.4)$$
   $$(\text{y component magnitude}) = (\text{magnitude}) \cdot \sin(\theta) \qquad (3.3.5)$$

   For signs, we will pick "up" as the +y direction and the direction of horizontal travel as the +x direction. With these direction choices, both the x and y components are positive in this case. With these signs:

   $(\text{x component}) = +(\text{magnitude}) \cdot \cos(\theta) = (8.50 \text{ m/s}) \cdot \cos(10.0°) = \underline{8.37 \text{ m/s}}$

   $(\text{y component}) = +(\text{magnitude}) \cdot \sin(\theta) = (8.50 \text{ m/s}) \cdot \sin(10.0°) = \underline{1.48 \text{ m/s}}$

2. A ball is headed towards the ground for a landing. Its velocity is 5.50 m/s *i* – 2.50 m/s *j*. What is the magnitude and angle (direction) of this vector?

   Solution:

   $$(\text{magnitude}) = \sqrt{(\text{x component})^2 + (\text{y component})^2} \qquad (3.3.2)$$

$$\theta = \tan^{-1}(|\text{y component }|/|\text{x component}|) \qquad (3.3.3)$$

$$(\text{magnitude}) = \sqrt{(5.50\,\text{m}/\text{s})^2 + (-2.50\,\text{m}/\text{s})^2} = 6.04\,\text{m}/\text{s}$$

The angle is below the +x axis. This is because the x component is positive (+x axis) and the y component is negative (below).

$\theta = \tan^{-1}(|-2.50 \text{ m/s}|/|5.50 \text{ m/s}|) = \underline{24.4° \text{ below (clockwise from) the}}$ $\underline{\text{+x axis.}}$

3. The wind blows a leaf, and it moves with an acceleration that is 2.00 m/s^2, at an angle of 50.0° above "to the left". If "to the right" is the +x direction and "up" is the +y direction, what are the x and y components of the acceleration?

Solution:

$$(\text{x component magnitude}) = (\text{magnitude}) \cdot \cos(\theta) \qquad (3.3.4)$$
$$(\text{y component magnitude}) = (\text{magnitude}) \cdot \sin(\theta) \qquad (3.3.5)$$

For signs, we will pick "up" as the +y direction and "to the right" as the +x direction. With these direction choices, the x component is negative and the y component is positive. With these signs:

(x component) $= -(\text{magnitude}) \cdot \cos(\theta) = -(2.00 \text{ m/s}^2) \cdot \cos(50.0°)$ $= \underline{-1.29 \text{ m/s}^2}$

(y component) $= +(\text{magnitude}) \cdot \sin(\theta) = (2.00 \text{ m/s}^2) \cdot \sin(50.0°) = \underline{1.53 \text{ m/s}^2}$

4. A pen is accidentally knocked off a desk. Shortly after, it has a velocity of −1.50 m/s i − 2.50 m/s j. What are the magnitude and angle (direction) of this vector?

Solution:

$$(\text{magnitude}) = \sqrt{(\text{x component})^2 + (\text{y component})^2} \qquad (3.3.2)$$

$$\theta = \tan^{-1}(|\text{y component }|/|\text{x component}|) \qquad (3.3.3)$$

$$(\text{magnitude}) = \sqrt{(-1.50\,\text{m}/\text{s})^2 + (-2.50\,\text{m}/\text{s})^2} = 2.92\,\text{m}/\text{s}$$

The angle is below the –x axis. This is because the x component is negative (–x axis) and the y component is negative (below).
$\theta = \tan^{-1}(|-2.50 \text{ m/s}|/|-1.50 \text{ m/s}|) = \underline{59.0° \text{ below (counter clockwise}}$ from) the –x axis.

5. During a quick stare out the window, you watch a squirrel go a displacement of 30.0 m, at a direction of 40.0° south of west. If "east" is the +x direction and "north" is the +y direction, what are the x and y components of the displacement?

Solution:

$$(\text{x component magnitude}) = (\text{magnitude}) \cdot \cos(\theta) \qquad (3.3.4)$$

$$(\text{y component magnitude}) = (\text{magnitude}) \cdot \sin(\theta) \qquad (3.3.5)$$

With the direction choices in the problem, the x component is negative and the y component is negative. With these signs:

$$(\text{x component}) = -(\text{magnitude}) \cdot \cos(\theta) = -(30.0 \text{ m}) \cdot \cos(40.0°) = \underline{-23.0 \text{ m}}$$

$$(\text{y component}) = -(\text{magnitude}) \cdot \sin(\theta) = -(30.0 \text{ m}) \cdot \sin(40.0°) = \underline{-19.3 \text{ m}}$$

6. If a velocity vector has x component 10.5 m/s, and y component 7.65 m/s, what are the magnitude and the angle (direction) of the vector?

Solution:

$$(\text{magnitude}) = \sqrt{(\text{x component})^2 + (\text{y component})^2} \qquad (3.3.2)$$

$$\theta = \tan^{-1}(|\text{y component}|/|\text{x component}|) \qquad (3.3.3)$$

$$(\text{magnitude}) = \sqrt{(10.5 \text{m/s})^2 + (7.65 \text{m/s})^2} = 13.0 \text{m/s}$$

The angle is above the +x axis. This is because the x component is positive (+x axis) and the y component is positive (above).
$\theta = \tan^{-1}(|7.65|/|10.5|) = \underline{36.1° \text{ above (counterclockwise from) the +x axis.}}$

7. A bird flies through the air. The bird's velocity is 5.50 m/s, at an angle of 17.0° below "to the right". If "to the right" is the +x direction and "up" is the +y direction, what are the x and y components?

Solution:

$$(\text{x component magnitude}) = (\text{magnitude}) \cdot \cos(\theta) \qquad (3.3.4)$$

$$(\text{y component magnitude}) = (\text{magnitude}) \cdot \sin(\theta) \qquad (3.3.5)$$

With the direction choices in the problem, the x component is positive and the y component is negative. With these signs:

$(\text{x component}) = +(\text{magnitude}) \cdot \cos(\theta) = (5.50 \text{ m/s}) \cdot \cos(17.0°)$
$= \underline{5.26 \text{ m/s}}$

$(\text{y component}) = -(\text{magnitude}) \cdot \sin(\theta) = -(5.50 \text{ m/s}) \cdot \sin(17.0°)$
$= \underline{-1.61 \text{ m/s}}$

8. During a projectile motion problem, you find that the components of your answer (a velocity) are –5.00 m/s i + 7.50 m/s j. What are the magnitude and angle (direction) of this vector?

Solution:

$$\left(\text{magnitude}\right) = \sqrt{\left(\text{x component}\right)^2 + \left(\text{y component}\right)^2} \qquad (3.3.2)$$

$$\theta = \tan^{-1}(|\text{y component}|/|\text{x component}|) \qquad (3.3.3)$$

$$\left(\text{magnitude}\right) = \sqrt{\left(-5.00 \text{ m/s}\right)^2 + \left(7.50 \text{ m/s}\right)^2} = 9.01 \text{ m/s}$$

The angle is above the –x axis. This is because the x component is positive (–x axis) and the y component is positive (above).

$\theta = \tan^{-1}(|7.50 \text{ m/s}|/|-5.00 \text{ m/s}|) = \underline{56.3° \text{ above (clockwise from) the}}$
$\underline{-x \text{ axis.}}$

Section 3.4

9. If vector $v_1 = 3.25 \ i - 3.70 \ j$, and vector $v_2 = 4.50 \ i + 4.80 \ j$, what is $v_1 + v_2$? Answer in components.

Solution:

$$\mathbf{A} + \mathbf{B} = (A_x + B_x) \ i + (A_y + B_y) \ j \qquad (3.4.1)$$

$$\mathbf{v_1} + \mathbf{v_2} = (3.25 + 4.50)\ i + (-3.70 + 4.80)\ j = \underline{7.75\ i + 1.10\ j}$$

10. If vector $\mathbf{v_1} = 3.25\ i - 3.70\ j$, and vector $\mathbf{v_2} = 4.50\ i + 4.80\ j$, what is $\mathbf{v_2} - \mathbf{v_1}$? Answer in components.

Solution:

$$\mathbf{A} - \mathbf{B} = (A_x - B_x)\ i + (A_y - B_y)\ j \qquad (3.4.2)$$

$$\mathbf{v_2} - \mathbf{v_1} = (4.50 - 3.25)\ i + (4.80 - -3.70)\ j = \underline{1.25\ i + 8.50\ j}$$

11. If vector $\mathbf{v_1} = -15.2\ i - 15.7\ j$, and vector $\mathbf{v_2} = 18.5\ i + 17.8\ j$, what is $\mathbf{v_1} + \mathbf{v_2}$? Answer in magnitude and angle.

Solution:

First get components, then magnitude and angle.

$$\mathbf{A} + \mathbf{B} = (A_x + B_x)\ i + (A_y + B_y)\ j \qquad (3.4.1)$$

$$\mathbf{v_1} + \mathbf{v_2} = (-15.2 + 18.5)\ i + (-15.7 + 17.8)\ j = 3.30\ i + 2.10\ j$$

To get to magnitude and angle, use Equations 3.3.2 and 3.3.3:

$$(\text{magnitude}) = \sqrt{(x\,\text{component})^2 + (y\,\text{component})^2} \qquad (3.3.2)$$

$$\theta = \tan^{-1}(|y\ \text{component}\ |/|x\ \text{component}|) \qquad (3.3.3)$$

$$(\text{magnitude}) = \sqrt{(3.30)^2 + (2.10)^2} = 3.91$$

The angle will be above the +x axis, since the x component is positive (+x axis) and the y component is positive (above).

$\theta = \tan^{-1}(|2.10|/|3.30|) = \underline{32.5° \text{ above (counterclockwise from) the +x}}$
$\underline{\text{axis}}$

12. If vector $\mathbf{v_1} = 13.2\ i - 15.7\ j$, and vector $\mathbf{v_2} = 18.5\ i + 17.8\ j$, what is $2.35 \cdot \mathbf{v_1} - 1.90 \cdot \mathbf{v_2}$? Answer in components.

Solution:

We are going to use two equations:

$$\mathbf{A} - \mathbf{B} = (A_x - B_x)\, i + (A_y - B_y)\, j \qquad (3.4.2)$$

$$n \cdot \mathbf{A} = n \cdot A_x\, i + n \cdot A_y\, j \qquad (3.4.3)$$

Based on these, $2.35 \cdot \mathbf{v}_1 - 1.90 \cdot \mathbf{v}_2 = (2.35 \cdot v_{1,x} - 1.90 \cdot v_{2,x})\, i + (2.35 \cdot v_{1,y} - 1.90 \cdot v_{2,y})\, j$

Put in numbers: $2.35 \cdot \mathbf{v}_1 - 1.90 \cdot \mathbf{v}_2 = (2.35 \cdot 13.2 - 1.90 \cdot 18.5)\, i + (2.35 \cdot -15.7 - 1.90 \cdot 17.8)\, j$

$2.35 \cdot \mathbf{v}_1 - 1.90 \cdot \mathbf{v}_2 = \underline{-4.13\, i - 70.7\, j.}$

13. If vector $\mathbf{v}_1 = 13.2\ i - 15.7\ j$, and vector $\mathbf{v}_2 = 18.5\ i + 17.8\ j$, what is $2.35 \cdot \mathbf{v}_2 + 1.90 \cdot \mathbf{v}_1$? Answer in components.

Solution:

We are going to use two equations:

$$\mathbf{A} + \mathbf{B} = (A_x + B_x)\, i + (A_y + B_y)\, j \qquad (3.4.1)$$

$$n \cdot \mathbf{A} = n \cdot A_x\, i + n \cdot A_y\, j \qquad (3.4.3)$$

Based on these, $2.35 \cdot \mathbf{v}_2 + 1.90 \cdot \mathbf{v}_1 = (2.35 \cdot v_{2,x} + 1.90 \cdot v_{1,x})\, i + (2.35 \cdot v_{2,y} + 1.90 \cdot v_{1,y})\, j$

Put in numbers: $2.35 \cdot \mathbf{v}_2 + 1.90 \cdot \mathbf{v}_1 = (2.35 \cdot 18.5 + 1.90 \cdot 13.2)\, i + (2.35 \cdot 17.8 + 1.90 \cdot -15.7)\, j$

$2.35 \cdot \mathbf{v}_2 + 1.90 \cdot \mathbf{v}_1 = \underline{68.6\, i + 12.0\, j.}$

14. If vector $\mathbf{v}_1 = 23.4\ i + 25.6\ j$, and vector $\mathbf{v}_2 = 26.5\ i + 28.1\ j$, what is $\mathbf{v}_2 - 1.20 \cdot \mathbf{v}_1$? Answer in magnitude and angle.

Solution:

We will find components first, then magnitude and angle.

For components, we will use two equations:

$$A - B = (A_x - B_x)\, i + (A_y - B_y)\, j \qquad (3.4.2)$$

$$n \cdot A = n \cdot A_x\, i + n \cdot A_y\, j \qquad (3.4.3)$$

Based on these, $v_2 - 1.20 \cdot v_1 = (v_{2,x} - 1.20 \cdot v_{1,x})\, i + (v_{2,y} - 1.20 \cdot v_{1,y})\, j$

Put in numbers: $v_2 - 1.20 \cdot v_1 = (26.5 - 1.20 \cdot 23.4)\, i + (28.1 - 1.20 \cdot 25.6)$
$j = -1.58\, i - 2.62\, j$

To get to magnitude and angle, use Equations 3.3.2 and 3.3.3:

$$(\text{magnitude}) = \sqrt{(x\,\text{component})^2 + (y\,\text{component})^2} \qquad (3.3.2)$$

$$\theta = \tan^{-1}(|y\,\text{component}\,|/|x\,\text{component}|) \qquad (3.3.3)$$

$$(\text{magnitude}) = \sqrt{(-1.58)^2 + (-2.62)^2} = 3.06$$

The angle will be below the –x axis, since the x component is negative (–x axis) and the y component is negative (below).

$\theta = \tan^{-1}(|-2.62|/|-1.58|) = \underline{58.9° \text{ below (counterclockwise from) the}}$
$\underline{\text{–x axis}}$

Section 3.5

15. You are walking through a parking lot, when suddenly you see a takeout restaurant that appeals to you. Just before you saw this, you were walking at 1.00 m/s $i + 1.20$ m/s j. Shortly after you see the restaurant, you've walked toward it with a displacement of –5.00 m $i + 25.0$ m j, and your velocity is now –1.60 m/s $i + 1.80$ m/s j. What are the x and y components of your acceleration, assuming that the acceleration is constant?

Solution:

We'll find the x component by looking at the x direction, and the y component by looking at the y direction.

x direction:

We know $v_i = 1.00$ m/s (x component of initial velocity vector), $\Delta x = -5.00$ m, and $v_f = -1.60$ m/s. We want to find a. The equation with a and the variables that we know is Equation 2.6.4:

$$v_f^2 = v_i^2 + 2 \cdot a \cdot \Delta x \qquad (2.6.4)$$

To solve for a, first subtract v_i^2 from both sides: $v_f^2 - v_i^2 = 2 \cdot a \cdot \Delta x$

Divide both sides by $(2 \cdot \Delta x)$: $a = \dfrac{v_f^2 - v_i^2}{2 \cdot \Delta x}$

Put in numbers: $a = \dfrac{\left(-1.60\,\text{m}/\text{s}\right)^2 - \left(1.00\,\text{m}/\text{s}\right)^2}{2 \cdot \left(-5.00\,\text{m}\right)} = -0.156\,\text{m}/\text{s}^2$ (x component)

y direction:

We know $v_i = 1.20$ m/s (y component of initial velocity vector), $\Delta x = 25.0$ m (y component of displacement vector), and $v_f = 1.80$ m/s. We want to find a. The equation with a and the variables that we know is Equation 2.6.4:

$$v_f^2 = v_i^2 + 2 \cdot a \cdot \Delta x \qquad (2.6.4)$$

We just solved this equation for a in the x direction: $a = \dfrac{v_f^2 - v_i^2}{2 \cdot \Delta x}$

Put in numbers: $a = \dfrac{\left(1.80\,\text{m}/\text{s}\right)^2 - \left(1.20\,\text{m}/\text{s}\right)^2}{2 \cdot \left(25.0\,\text{m}\right)} = 0.0360\,\text{m}/\text{s}^2$ (y component)

Note: you could use Equations 2.6.1–2.6.3 to find time (Δt). For both dimensions you would find that $\Delta t = 16.7$ s. It has to be the same amount, since time is the same in both dimensions.

16. You are riding a bicycle, with velocity 5.00 m/s i + 3.50 m/s j. You see a friend of yours, and accelerate towards them with a constant acceleration. 5.00 s later, you are traveling at 8.00 m/s i + 7.70 m/s j. What is your displacement during these 5.00 s? Answer in components.

Solution:

For each of the x and y directions, you are given v_i (5.00 m/s for x, 3.50 m/s for y), Δt (5.00 s), and v_f (8.00 m/s for x, 7.70 m/s for y). You want to find displacement (Δx). The equation with displacement and the variables that we already know is Equation 2.6.3:

$$\Delta x = \frac{1}{2} \cdot \left(v_i + v_f\right) \cdot \Delta t \qquad (2.6.3)$$

Since the equation is already solved for Δx, we can put in numbers. We will do this once for the x direction and once for the y direction:

x direction: $\Delta x = \dfrac{1}{2}\cdot\left(5.00\,\text{m/s}+8.00\,\text{m/s}\right)\cdot\left(5.00\,\text{s}\right)=32.5\,\text{m}$
(x component)

y direction: $\Delta x = \dfrac{1}{2}\cdot\left(3.50\,\text{m/s}+7.70\,\text{m/s}\right)\cdot\left(5.00\,\text{s}\right)=28.0\,\text{m}$
(y component)

17. You are walking to class, with velocity 1.00 m/s i – 1.25 m/s j. Suddenly, you realize that you are a little later than you thought. 2.50 seconds later, your velocity is 1.50 m/s i – 1.85 m/s j. What was your acceleration, assuming it was constant? Answer in components.

Solution:

For each of the x and y directions, you are given v_i (1.00 m/s for x, −1.25 m/s for y), Δt (2.50 s), and v_f (1.50 m/s for x, −1.85 m/s for y). You want to find acceleration (a). The equation with acceleration and the variables that we already know is Equation 2.6.1:

$$v_f = v_i + a\cdot\Delta t \qquad\qquad (2.6.1)$$

Since we are solving for a in each case, we will solve once and then put in the numbers for each dimension separately.

To solve for a, first subtract v_i from both sides: $v_f - v_i = a\cdot\Delta t$

Divide both sides by Δt: $a = \dfrac{v_f - v_i}{\Delta t}$

Put in numbers for the x direction:

$$a = \frac{1.50\,\text{m/s}-1.00\,\text{m/s}}{2.50\,\text{s}} = 0.200\,\text{m/s}^2 \quad \text{(x component)}$$

Put in numbers for the y direction:

$$a = \frac{-1.85\,\text{m/s}-\left(-1.25\,\text{m/s}\right)}{2.50\,\text{s}} = 0.240\,\text{m/s}^2 \quad \text{(y component)}$$

18. You are walking through a store, when suddenly you see an item that you forgot. Just before you saw this, you were walking at 1.10 m/s i + 1.30 m/s j. Shortly after you see the item, you've walked toward it with a displacement of 5.00 m i + 5.60 m j, and your velocity is now 1.40 m/s i + 1.50 m/s j. What are the x and y components of your acceleration, assuming that the acceleration is constant?

Solution:

For each of the x and y directions, you are given v_i (1.10 m/s for x, 1.30 m/s for y), Δx (5.00 m for x, 5.60 m for y), and v_f (1.40 m/s for x, 1.50 m/s for y). You want to find acceleration (a). The equation with acceleration and the variables that we already know is Equation 2.6.4:

$$v_f^2 = v_i^2 + 2 \cdot a \cdot \Delta x \qquad (2.6.4)$$

Since we are solving for a in each case, we will solve once and then put in the numbers for each dimension separately.

To solve for a, first subtract v_i^2 from both sides: $v_f^2 - v_i^2 = 2 \cdot a \cdot \Delta x$

Divide both sides by $(2 \cdot \Delta x)$: $a = \dfrac{v_f^2 - v_i^2}{2 \cdot \Delta x}$

Put in x direction numbers: $a = \dfrac{\left(1.40\,\text{m/s}\right)^2 - \left(1.10\,\text{m/s}\right)^2}{2 \cdot \left(5.00\,\text{m}\right)} = \underline{0.0750}$
$\underline{\text{m/s}^2 \text{ (x component)}}$

Put in y direction numbers: $a = \dfrac{\left(1.50\,\text{m/s}\right)^2 - \left(1.30\,\text{m/s}\right)^2}{2 \cdot \left(5.60\,\text{m}\right)} = \underline{0.0500}$
$\underline{\text{m/s}^2 \text{ (y component)}}$

19. An object is flying through the air, but unlike in the projectile motion problems we will not assume that gravity is the only force. The object starts on the ground with an initial velocity of 30.0 m/s, at an angle of 27.0° above the ground. It has a constant acceleration of −1.10 m/s² *i* − 9.80 m/s² *j*.

   A. How much time is the object in the air for before it touches the ground? Hint: look at the y direction.

   B. How far horizontally does the object travel before it touches the ground? Hint: look at the x direction.

Solution:

Part A:

In the y direction, we know that v_i = (30.0 m/s) · sin(27.0°) = 13.6 m/s, a = −9.80 m/s², and Δx = 0. The displacement (Δx) is zero because the object starts and ends on the ground. For v_i, we used Equation 3.3.5:

$$(\text{y component magnitude}) = (\text{magnitude}) \cdot \sin(\theta) \qquad (3.3.5)$$

We want to find time (Δt). The equation with Δt and the variables that we already know is Equation 2.6.2:

$$\Delta x = v_i \cdot \Delta t + \frac{1}{2} \cdot a \cdot \left(\Delta t\right)^2 \qquad (2.6.2)$$

Put in that $\Delta x = 0$: $0 = v_i \cdot \Delta t + \frac{1}{2} \cdot a \cdot \left(\Delta t\right)^2$

Since there is at least one factor of Δt in each term on the right side, we can rewrite the right side as: $0 = \Delta t \cdot (v_i + \frac{1}{2} \cdot a \cdot \Delta t)$

This gives two solutions: $\Delta t = 0$, and $(v_i + \frac{1}{2} \cdot a \cdot \Delta t) = 0$. The second answer is how much time the object takes to reach the ground.

$$v_i + \frac{1}{2} \cdot a \cdot \Delta t = 0$$

Subtract v_i from both sides: $\frac{1}{2} \cdot a \cdot \Delta t = -v_i$

Divide both sides by $\frac{1}{2} \cdot a$: $\Delta t = \dfrac{-v_i}{\frac{1}{2} \cdot a}$

Put in numbers: $\Delta t = \dfrac{-\left(13.6 \, \text{m/s}\right)}{\frac{1}{2} \cdot \left(-9.80 \, \text{m/s}^2\right)} = 2.78 \, \text{s}$

Part B:

In the x direction, we know that $v_i = (30.0 \, \text{m/s}) \cdot \cos(27.0°) = 26.7$ m/s, $a = -1.10$ m/s^2, and $\Delta t = 2.78$ s (from Part A). For v_i, we used Equation 3.3.4:

$$(\text{x component magnitude}) = (\text{magnitude}) \cdot \cos(\theta) \qquad (3.3.4)$$

We want to find displacement (Δx). The equation with displacement and the variables that we already know is Equation 2.6.2:

$$\Delta x = v_i \cdot \Delta t + \frac{1}{2} \cdot a \cdot \left(\Delta t\right)^2 \qquad (2.6.2)$$

Since this equation is already solved for Δx, we can put in numbers:

$$\Delta x = \left(26.7\,\text{m/s}\right)\cdot\left(2.78\,\text{s}\right) + \frac{1}{2}\cdot\left(-1.10\,\text{m/s}^2\right)\cdot\left(2.78\,\text{s}\right)^2 = 70.0\,\text{m}$$

Note: the only difference between this problem and a typical projectile motion problem was the value of the x component of acceleration. In a typical projectile motion problem, which is covered in Section 3.6, the x component of acceleration is zero. If the x component of acceleration was zero in this problem, it would have given $\Delta x = 74.2$ m.

Section 3.6

20. Your friend tosses you a set of keys. It's tossed with a velocity of 5.00 m/s, at an angle of 35.0° above the ground. You catch it at the same height that it was thrown at. Assume that gravity is the only force on the keys.

    A. How much time are the keys in the air for?

    B. How far horizontally do the keys travel?

    Solution:

    Part A:

    As shown in the examples with Section 3.6, in general the y direction is what is used to find the time. (Occasionally the x direction is used.) In the y direction for this problem, we know that $v_i = (5.00$ m/s$)\cdot\sin(35.0°) = 2.87$ m/s, a $= -9.80$ m/s², and $\Delta x = 0$. The displacement (Δx) is zero because the keys start and end at the same height. For v_i, we used Equation 3.3.5:

$$(\text{y component magnitude}) = (\text{magnitude})\cdot\sin(\theta) \qquad (3.3.5)$$

We want to find time (Δt). The equation with Δt and the variables that we already know is Equation 2.6.2:

$$\Delta x = v_i\cdot\Delta t + \frac{1}{2}\cdot a\cdot\left(\Delta t\right)^2 \qquad (2.6.2)$$

Put in that $\Delta x = 0$: $0 = v_i\cdot\Delta t + \frac{1}{2}\cdot a\cdot\left(\Delta t\right)^2$

Since there is at least one factor of Δt in each term on the right side, we can rewrite the right side as: $0 = \Delta t \cdot (v_i + \frac{1}{2} \cdot a \cdot \Delta t)$

This gives two solutions: $\Delta t = 0$, and $(v_i + \frac{1}{2} \cdot a \cdot \Delta t) = 0$. The second answer is how much time the object takes to reach the ground.

$$v_i + \frac{1}{2} \cdot a \cdot \Delta t = 0$$

Subtract v_i from both sides: $\frac{1}{2} \cdot a \cdot \Delta t = -v_i$

Divide both sides by $\frac{1}{2} \cdot a$: $\Delta t = \dfrac{-v_i}{\frac{1}{2} \cdot a}$

Put in numbers: $\Delta t = \dfrac{-(2.87 \, \text{m/s})}{\frac{1}{2} \cdot (-9.80 \, \text{m/s}^2)} = 0585 \text{s}$

Part B:
In the x direction, we know that $v_i = (5.00 \text{ m/s}) \cdot \cos(35.0°) = 4.10 \text{ m/s}$, $a = 0 \text{ m/s}^2$, and $\Delta t = 0.585 \text{ s}$ (from Part A). a (acceleration) is zero in projectile motion problems, where gravity is the only force. For v_i, we used Equation 3.3.4:

$$(\text{x component magnitude}) = (\text{magnitude}) \cdot \cos(\theta) \qquad (3.3.4)$$

We want to find displacement (Δx). The equation with displacement and the variables that we already know is Equation 2.6.2:

$$\Delta x = v_i \cdot \Delta t + \frac{1}{2} \cdot a \cdot (\Delta t)^2 \qquad (2.6.2)$$

Since this equation is already solved for Δx, we can put in numbers:

$$\Delta x = (4.10 \, \text{m/s}) \cdot (0.585 \text{s}) + \frac{1}{2} \cdot (0 \, \text{m/s}^2) \cdot (0.585 \text{s})^2 = 2.40 \text{m}$$

Note: 2.40 m is roughly 7 to 8 feet; this seems OK for a distance for tossed keys.

21. You toss an object to your friend. It's tossed with a velocity of 4.50 m/s, at an angle of 50.0° above the ground. Your friend catches the object 0.500 m below the height that it was thrown at. Assume that gravity is the only force on the object.

    A. What are the x and y components of the final velocity, meaning just before you catch the object?

    B. What is the magnitude and angle (direction) of the final velocity?

Solution:

Part A – x component:

In the x direction, we know that $v_i = (4.50$ m/s$) \cdot \cos(50.0°) = 2.89$ m/s, $a = 0$, and $v_i = v_f$. We know that the initial and final velocities are equal because $a = 0$, which means that the change in velocity (per time) is zero. Remember that the x component of acceleration is zero in all projectile motion problems (object flying through the air and gravity is the only force) For v_i, we used Equation 3.3.4:

$$\text{(x component magnitude)} = \text{(magnitude)} \cdot \cos(\theta) \qquad (3.3.4)$$

We want to find final velocity (v_f). As stated above, $v_i = v_f$, so the answer will be <u>2.89 m/s</u>, but let's look for a minute at where this comes from. Look at Equation 2.6.1:

$$v_f = v_i + a \cdot \Delta t \qquad (2.6.1)$$

If $a = 0$, this equation becomes: $v_f = v_i$

Part A – y component:

In the y direction, we know that $v_i = (4.50$ m/s$) \cdot \sin(50.0°) = 3.45$ m/s, $a = -9.80$ m/s², and $\Delta x = -0.500$ m (caught 0.500 m below where it started). For v_i, we used Equation 3.3.5:

$$\text{(y component magnitude)} = \text{(magnitude)} \cdot \sin(\theta) \qquad (3.3.5)$$

We want to find final velocity (v_f). The equation with v_f and the variables that we already know is Equation 2.6.4:

$$v_f^2 = v_i^2 + 2 \cdot a \cdot \Delta x \qquad (2.6.4)$$

Take the square root: $v_f = \pm\sqrt{v_i^2 + 2 \cdot a \cdot \Delta x}$

Here, the y component is negative, because the object will be traveling downward in the y direction if it goes below the height where it was thrown from. This gives:

$$v_f = -\sqrt{v_i^2 + 2 \cdot a \cdot \Delta x}$$

Put in numbers: $v_f = -\sqrt{\left(3.45\,\text{m/s}\right)^2 + 2 \cdot \left(-9.80\,\text{m/s}^2\right) \cdot \left(-0.500\,\text{m}\right)}$
$= -\,4.66\text{m/s}$

Part B:

We have the two components of final velocity, from Part A: 2.89 m/s for x, and –4.66 m/s for y. We now use Equations 3.3.2 and 3.3.3 to go from components to magnitude and angle:

$$\left(\text{magnitude}\right) = \sqrt{\left(x\,\text{component}\right)^2 + \left(y\,\text{component}\right)^2} \qquad (3.3.2)$$

$$\theta = \tan^{-1}(|y\,\text{component}\,|/|x\,\text{component}|) \qquad (3.3.3)$$

(θ can be from the +x direction or the –x direction, and it can be above or below. Use what the problem gives you to decide which.)

Put numbers into these equations:

$$\left(\text{magnitude}\right) = \sqrt{\left(2.89\,\text{m/s}\right)^2 + \left(-4.66\,\text{m/s}\right)^2} = 5.48\,\text{m/s}$$

$\theta = \tan^{-1}(|{-4.66}\text{ m/s}|/|2.89\text{ m/s}|) = 58.2°$

The angle will be below (clockwise from) the +x axis, since the x component is positive (+x axis) and the y component is negative (below).
58.2° below (clockwise from) the +x axis

22. A basketball is passed between two teammates. The pass is sent with a velocity of 8.00 m/s, at an angle of 10.0° above the ground. The pass is caught at the same height that it is thrown at. Assume that gravity is the only force on the ball.

A. How much time is the ball in the air for?

B. How far horizontally does the ball travel?

Solution:

Part A:

As shown in the examples with Section 3.6, in general the y direction is what is used to find the time. (Occasionally the x direction is used.) In the y direction for this problem, we know that $v_i = (8.00$ m/s) $\cdot \sin(10.0°) = 1.39$ m/s, a $= -9.80$ m/s^2, and $\Delta x = 0$. The displacement (Δx) is zero because the ball starts and ends at the same height. For v_i, we used Equation 3.3.5:

$$(y \text{ component magnitude}) = (\text{magnitude}) \cdot \sin(\theta) \qquad (3.3.5)$$

We want to find time (Δt). The equation with Δt and the variables that we already know is Equation 2.6.2:

$$\Delta x = v_i \cdot \Delta t + \frac{1}{2} \cdot a \cdot (\Delta t)^2 \qquad (2.6.2)$$

Put in that $\Delta x = 0$: $0 = v_i \cdot \Delta t + \frac{1}{2} \cdot a \cdot (\Delta t)^2$

Since there is at least one factor of Δt in each term on the right side, we can rewrite the right side as: $0 = \Delta t \cdot (v_i + \frac{1}{2} \cdot a \cdot \Delta t)$

This gives two solutions: $\Delta t = 0$, and $(v_i + \frac{1}{2} \cdot a \cdot \Delta t) = 0$. The second answer is how much time the object takes to reach the ground.

$$v_i + \frac{1}{2} \cdot a \cdot \Delta t = 0$$

Subtract v_i from both sides: $\frac{1}{2} \cdot a \cdot \Delta t = -v_i$

Divide both sides by $\frac{1}{2} \cdot a$: $\Delta t = \dfrac{-v_i}{\frac{1}{2} \cdot a}$

Put in numbers: $\Delta t = \dfrac{-(1.39 \, \text{m/s})}{\frac{1}{2} \cdot (-9.80 \, \text{m/s}^2)} = 0.284 \, \text{s}$

Part B:

In the x direction, we know that $v_i = (8.00 \text{ m/s}) \cdot \cos(10.0°) = 7.88$ m/s, $a = 0$ m/s², and $\Delta t = 0.284$ s (from Part A). a (acceleration) is zero in projectile motion problems, where gravity is the only force. For v_i, we used Equation 3.3.4:

$$(x \text{ component magnitude}) = (\text{magnitude}) \cdot \cos(\theta) \qquad (3.3.4)$$

We want to find displacement (Δx). The equation with displacement and the variables that we already know is Equation 2.6.2:

$$\Delta x = v_i \cdot \Delta t + \frac{1}{2} \cdot a \cdot (\Delta t)^2 \qquad (2.6.2)$$

Since this equation is already solved for Δx, we can put in numbers:

$$\Delta x = (7.88 \text{ m/s}) \cdot (0.284 \text{ s}) + \frac{1}{2} \cdot (0 \text{ m/s}^2) \cdot (0.284 \text{ s})^2 = 2.23 \text{ m}$$

Note: 2.23 m is roughly 7 feet; this seems OK for a distance for a basketball pass.

23. A basketball is passed between two players, but the second player misses it and it lands on the ground. The pass is sent with a velocity of 8.00 m/s, at an angle of 10.0° above the ground, at a height of 1.75 m above the ground. Assume that gravity is the only force on the ball.

    A. How much time passes until the ball touches the ground?

    B. How far horizontally does the ball travel before hitting the ground?

Solution:

The major difference between this problem and problems 20 and 22 is that the displacement in the y direction is no longer zero. This means that we will need to use the quadratic equation.

Part A:

As shown in the examples with Section 3.6, in general the y direction is what is used to find the time. (Occasionally the x direction is used.) In the y direction for this problem, we know that $v_i = (8.00$ m/s)$\cdot \sin(10.0°) = 1.39$ m/s, $a = -9.80$ m/s^2, and $\Delta x = -1.75$ m (going from 1.75 m above ground to the ground). For v_i, we used Equation 3.3.5:

$$(\text{y component magnitude}) = (\text{magnitude}) \cdot \sin(\theta) \qquad (3.3.5)$$

We want to find time (Δt). The equation with Δt and the variables that we already know is Equation 2.6.2:

$$\Delta x = v_i \cdot \Delta t + \frac{1}{2} \cdot a \cdot (\Delta t)^2 \qquad (2.6.2)$$

Since there are no nonzero values here, the quadratic formula will be necessary. The quadratic formula is:

If $a \cdot x^2 + b \cdot x + c = 0$, where a, b, and c are numbers and x is the variable you want to solve for,

$$x = \frac{-b \pm \sqrt{b^2 - 4 \cdot a \cdot c}}{2 \cdot a}$$

The quadratic formula is reviewed in Section 13.5

Δt is the variable that we want to solve for. First, we will get the equation into the format of "a" $\cdot (\Delta t)^2 +$ "b" $\cdot \Delta t +$ "c" $= 0$. Note: "a" is different from the acceleration a; that is why quotes are being used.

Subtract Δx from both sides, and switch the order of the other two terms:

$$\frac{1}{2} \cdot a \cdot (\Delta t)^2 + v_i \cdot \Delta t - \Delta x = 0$$

This means that "a" $= \frac{1}{2} \cdot a = \frac{1}{2} \cdot -9.80$, "b" $= v_i = 1.39$, and "c" $= -\Delta x = -(-1.75)$.

Put this into the quadratic formula:

$$\Delta t = \frac{-b \pm \sqrt{b^2 - 4 \cdot a \cdot c}}{2 \cdot a} = \frac{-1.39 \pm \sqrt{(1.39)^2 - 4 \cdot (0.5 \cdot -9.80) \cdot (1.75)}}{2 \cdot (0.5 \cdot -9.80)} =$$

0.756 s and −0.472 s

We will choose the positive change for amount of time: $\underline{\Delta t = 0.756 \text{ s}}$

Part B:

In the x direction, we know that $v_i = (8.00 \text{ m/s}) \cdot \cos(10.0°) = 7.88 \text{ m/s}$, $a = 0 \text{ m/s}^2$, and $\Delta t = 0.756 \text{ s}$ (from Part A). a (acceleration) is zero in projectile motion problems, where gravity is the only force. For v_i, we used Equation 3.3.4:

$$(\text{x component magnitude}) = (\text{magnitude}) \cdot \cos(\theta) \qquad (3.3.4)$$

We want to find displacement (Δx). The equation with displacement and the variables that we already know is Equation 2.6.2:

$$\Delta x = v_i \cdot \Delta t + \frac{1}{2} \cdot a \cdot (\Delta t)^2 \qquad (2.6.2)$$

Since this equation is already solved for Δx, we can put in numbers:

$$\Delta x = (7.88 \text{ m/s}) \cdot (0.756 \text{s}) + \frac{1}{2} \cdot (0 \text{ m/s}^2) \cdot (0.756 \text{s})^2 = 5.96 \text{ m}$$

Note: 5.96 m is roughly 20 feet; this seems OK for a distance for a basketball pass.

24. During a timeout at a game, foam balls are launched into the crowd. 3.00 seconds after launch, one ball lands in the upper levels of the stadium – a vertical distance of 8.00 meters and a horizontal distance of 35.0 meters from where the ball is launched. Ignore air resistance. Assume that gravity is the only force on the ball.

    A. What is the x-component of the initial velocity of the ball?

    B. What is the y-component of the initial velocity of the ball?

Solution:

For each of the x and y directions, we are given time (3.00 s for both), displacement (35.0 m in the x direction, 8.00 m in the y direction), and acceleration (0 m/s^2 in the x direction, -9.80 m/s^2 in the y direction). The acceleration is known because this is a projectile motion problem (object flying in the air, gravity the only force). In both cases, we know Δt, Δx, and a, and want to know v_i. The equation with v_i and the variables that we already know is Equation 2.6.2:

$$\Delta x = v_i \cdot \Delta t + \frac{1}{2} \cdot a \cdot (\Delta t)^2 \qquad (2.6.2)$$

Subtract $\frac{1}{2} \cdot a \cdot (\Delta t)^2$ from both sides: $\Delta x - \frac{1}{2} \cdot a \cdot (\Delta t)^2 = v_i \cdot \Delta t$

Divide both sides by Δt: $v_i = \dfrac{\Delta x - \frac{1}{2} \cdot a \cdot (\Delta t)^2}{\Delta t}$

A: put in numbers for the x direction:

$$v_i = \frac{(35.0\,\text{m}) - \frac{1}{2} \cdot (0\,\text{m/s}^2) \cdot (3.00\,\text{s})^2}{3.00\,\text{s}} = 11.7\,\text{m/s}$$

B: put in numbers for the y direction:

$$v_i = \frac{(8.00\,\text{m}) - \frac{1}{2} \cdot (-9.80\,\text{m/s}^2) \cdot (3.00\,\text{s})^2}{3.00\,\text{s}} = 17.4\,\text{m/s}$$

25. A basketball player takes a shot, towards the basket. 1.20 s after the shot is released, the ball's position has changed by 6.00 m i + 1.00 m j. Assume that gravity is the only force on the ball.

 A. What are the x and y components of the initial velocity?

 B. What are the magnitude and angle (direction) of the initial velocity?

Solution:

Part A:

For each of the x and y directions, we are given time (1.20 s for both), displacement (6.00 m in the x direction, 1.00 m in the y direction), and acceleration (0 m/s² in the x direction, −9.80 m/s² in the y direction). The acceleration is known because this is a projectile motion problem (object flying in the air, gravity the only force). In both cases, we know Δt, Δx, and a, and want to know v_i. The equation with v_i and the variables that we already know is Equation 2.6.2:

$$\Delta x = v_i \cdot \Delta t + \frac{1}{2} \cdot a \cdot (\Delta t)^2 \qquad (2.6.2)$$

Subtract $\frac{1}{2} \cdot a \cdot (\Delta t)^2$ from both sides: $\Delta x - \frac{1}{2} \cdot a \cdot (\Delta t)^2 = v_i \cdot \Delta t$

Divide both sides by Δt: $v_i = \dfrac{\Delta x - \dfrac{1}{2} \cdot a \cdot (\Delta t)^2}{\Delta t}$

x component: Put in numbers for the x direction:

$$v_i = \frac{(6.00\,\text{m}) - \dfrac{1}{2} \cdot (0\,\text{m}/\text{s}^2) \cdot (1.20\,\text{s})^2}{1.20\,\text{s}} = 5.00\,\text{m/s}$$

y component: Put in numbers for the y direction:

$$v_i = \frac{(1.00\,\text{m}) - \dfrac{1}{2} \cdot (-9.80\,\text{m}/\text{s}^2) \cdot (1.20\,\text{s})^2}{1.20\,\text{s}} = 6.71\,\text{m/s}$$

Part B:

To get from components to magnitude and angle, use Equations 3.3.2 and 3.3.3:

$$(\text{magnitude}) = \sqrt{(\text{x component})^2 + (\text{y component})^2} \qquad (3.3.2)$$

$$\theta = \tan^{-1}(|\text{y component}|/|\text{x component}|) \qquad (3.3.3)$$

$$(\text{magnitude}) = \sqrt{(5.00\,\text{m/s})^2 + (6.71\,\text{m/s})^2} = 8.37\,\text{m/s}$$

The angle will be above the +x axis, since the x component is positive (+x axis) and the y component is positive (above).

$\theta = \tan^{-1}(|6.71 \text{ m/s}|/|5.00 \text{ m/s}|) = \underline{53.3° \text{ above (counterclockwise from)}}$ $\underline{\text{the +x axis}}$

26. A ball is kicked into the air, from the ground. The initial velocity is 6.50 m/s, at an angle of 40.0° above the ground. Assume that gravity is the only force on the ball. What is the largest height above the ground that the ball reaches?

Solution:

To find height, we need to look at the y direction. We make the kick the initial time point, and the largest height the final time point. In the y direction, we know that $v_i = (6.50 \text{ m/s}) \cdot \sin(40.0°) = 4.18$

m/s, a = −9.80 m/s², and $v_f = 0$. v_f is zero because the vertical velocity is zero at the highest height reached, a is −9.80 m/s² for projectile motion problems (object flying through the air with gravity as the only force), and v_i comes from Equation 3.3.5:

$$(\text{y component magnitude}) = (\text{magnitude}) \cdot \sin(\theta) \qquad (3.3.5)$$

We want to find displacement, since the ball is kicked from the ground and we want to know how far above the ground it gets. The equation with displacement (Δx) and the variables that we already know is Equation 2.6.4:

$$v_f^2 = v_i^2 + 2 \cdot a \cdot \Delta x \qquad (2.6.4)$$

Subtract v_i^2 from both sides: $v_f^2 - v_i^2 = 2 \cdot a \cdot \Delta x$

Divide both sides by (2 · a): $\Delta x = \dfrac{v_f^2 - v_i^2}{2 \cdot a}$

Put in numbers: $\Delta x = \dfrac{0^2 - \left(4.18 \, \text{m/s}\right)^2}{2 \cdot \left(-9.80 \, \text{m/s}^2\right)} = 0.891 \, \text{m}$

27. A volleyball is hit into the air, from the ground. The initial velocity is 10.5 m/s, at an angle of 70.0° above the ground. Assume that gravity is the only force on the ball. What is the largest height above the ground that the ball reaches?

Solution:

To find height, we need to look at the y direction. We make the time of the volleyball being hit the initial time point, and the time of the largest height the final time point. In the y direction, we know that $v_i = (10.5 \text{ m/s}) \cdot \sin(70.0°) = 9.87$ m/s, a = −9.80 m/s², and $v_f = 0$. v_f is zero because the vertical velocity is zero at the highest height reached, a is −9.80 m/s² for projectile motion problems (object flying through the air with gravity as the only force), and v_i comes from Equation 3.3.5:

$$(\text{y component magnitude}) = (\text{magnitude}) \cdot \sin(\theta) \qquad (3.3.5)$$

We want to find displacement, since the ball is kicked from the ground and we want to know how far above the ground it gets. The equation with displacement (Δx) and the variables that we already know is Equation 2.6.4:

$$v_f^2 = v_i^2 + 2 \cdot a \cdot \Delta x \qquad (2.6.4)$$

Subtract v_i^2 from both sides: $v_f^2 - v_i^2 = 2 \cdot a \cdot \Delta x$

Divide both sides by $(2 \cdot a)$: $\Delta x = \dfrac{v_f^2 - v_i^2}{2 \cdot a}$

Put in numbers: $\Delta x = \dfrac{0^2 - (9.87\,\text{m/s})^2}{2 \cdot (-9.80\,\text{m/s}^2)} = 4.97\,\text{m}$

28. Note: this problem is more challenging. A baseball is hit into the air, with an initial velocity of 35.0 m/s, at an angle of 35.0° above the ground. Assume that gravity is the only force on the ball. The ball is caught at the same height that it was hit at. We will now look at two later time points – when the ball is as high as it gets, and when the ball is caught. (This is similar to Example 2.7.2, which was done in one dimension.)

A. How much time is the ball in the air for?

B. How far horizontally does the ball travel before it is caught?

C. What is the largest height reached by the ball, above where it was hit from? Hint: this part will require a different final time point than Parts A and B.

Solution:

This problem is basically combining the problem type of problems 20, 22, and 23 (Parts A and B) with the problem type of problems 26 and 27 (Part C).

Part A:

In Part A, the baseball being hit will be the initial time point, and the ball being caught will be the final time point. We'll look at the y direction to determine time. In the y direction for this problem, we know that $v_i = (35.0 \text{ m/s}) \cdot \sin(35.0°) = 20.1$ m/s, $a = -9.80$ m/s^2, and $\Delta x = 0$. The displacement (Δx) is zero because the ball starts and ends at the same height, and a is -9.80 m/s^2 because this is a projectile motion problem (object flying through the air with gravity the only force on it). For v_i, we used Equation 3.3.5:

$$(\text{y component magnitude}) = (\text{magnitude}) \cdot \sin(\theta) \qquad (3.3.5)$$

We want to find time (Δt). The equation with Δt and the variables that we already know is Equation 2.6.2:

$$\Delta x = v_i \cdot \Delta t + \frac{1}{2} \cdot a \cdot (\Delta t)^2 \qquad (2.6.2)$$

Put in that $\Delta x = 0$: $0 = v_i \cdot \Delta t + \frac{1}{2} \cdot a \cdot (\Delta t)^2$

Since there is at least one factor of Δt in each term on the right side, we can rewrite the right side as: $0 = \Delta t \cdot (v_i + \frac{1}{2} \cdot a \cdot \Delta t)$

This gives two solutions: $\Delta t = 0$, and $(v_i + \frac{1}{2} \cdot a \cdot \Delta t) = 0$. The second answer is how much time the object takes to reach the ground.

$$v_i + \frac{1}{2} \cdot a \cdot \Delta t = 0$$

Subtract v_i from both sides: $\frac{1}{2} \cdot a \cdot \Delta t = -v_i$

Divide both sides by $\frac{1}{2} \cdot a$: $\Delta t = \dfrac{-v_i}{\frac{1}{2} \cdot a}$

Put in numbers: $\Delta t = \dfrac{-(20.1 \, \text{m/s})}{\frac{1}{2} \cdot (-9.80 \, \text{m/s}^2)} = 4.10 \, \text{s}$

Part B:

In the x direction, we know that $v_i = (35.0 \, \text{m/s}) \cdot \cos(35.0°) = 28.7 \, \text{m/s}$, $a = 0 \, \text{m/s}^2$, and $\Delta t = 4.10 \, \text{s}$ (from Part A). a (acceleration) is zero in projectile motion problems, where gravity is the only force. For v_i, we used Equation 3.3.4:

$$(\text{x component magnitude}) = (\text{magnitude}) \cdot \cos(\theta) \qquad (3.3.4)$$

We want to find displacement (Δx). The equation with displacement and the variables that we already know is Equation 2.6.2:

$$\Delta x = v_i \cdot \Delta t + \frac{1}{2} \cdot a \cdot (\Delta t)^2 \tag{2.6.2}$$

Since this equation is already solved for Δx, we can put in numbers:

$$\Delta x = (28.7\,m/s) \cdot (4.10s) + \frac{1}{2} \cdot (0m/s^2) \cdot (4.10s)^2 = 118m$$

Note: 118 m is about 390 ft, which makes sense for a baseball.

Part C:

We now change the final time point to be the time where the ball is at its largest height. This means that v_f is now known to be zero, but the time and displacements are different. The initial velocity stays the same, since we keep the same initial time point. The acceleration stays the same because it is still a projectile motion problem. Let's look at the y direction. We know that $v_i = 20.1$ m/s, $a = -9.80$ m/s^2, and $v_f = 0$. We want to find displacement (Δx). The equation with displacement and the variables that we already know is Equation 2.6.4:

$$v_f^2 = v_i^2 + 2 \cdot a \cdot \Delta x \tag{2.6.4}$$

Subtract v_i^2 from both sides: $v_f^2 - v_i^2 = 2 \cdot a \cdot \Delta x$

Divide both sides by $(2 \cdot a)$: $\Delta x = \dfrac{v_f^2 - v_i^2}{2 \cdot a}$

Put in numbers: $\Delta x = \dfrac{0^2 - (20.1m/s)^2}{2 \cdot (-9.80m/s^2)} = 20.6m$

29. Note: this problem is more challenging. A ball is tossed into the air with initial velocity 5.50 m/s, at an angle of 47.0° above the ground. It lands in a tree, at the same height as it was thrown from. Assume that gravity is the only force on the ball. We will now look at two later time points – when the ball is as high as it gets, and when the ball lands in the tree. (This is similar to Example 2.7.2, which was done in one dimension.)

A. How much time passes before the ball lands in the tree?

B. How far horizontally does the ball travel before it lands in the tree?

C. What is the largest height (above the initial throw point) that the ball gets to?

Solution:

This problem is basically combining the problem type of problems 20, 22, and 23 (Parts A and B) with the problem type of problems 26 and 27 (Part C).

Part A:

In Part A, the ball being tossed will be the initial time point, and the ball landing in the tree will be the final time point. We'll look at the y direction to determine time to reach the same height that it started it. In the y direction for this problem, we know that $v_i = (5.50$ m/s$) \cdot \sin(47.0°) = 4.02$ m/s, $a = -9.80$ m/s^2, and $\Delta x = 0$. The displacement (Δx) is zero because the ball starts and end at the same height, and a is -9.80 m/s^2 because this is a projectile motion problem (object flying through the air with gravity the only force on it). For v_i, we used Equation 3.3.5:

$$(y \text{ component magnitude}) = (\text{magnitude}) \cdot \sin(\theta) \qquad (3.3.5)$$

We want to find time (Δt). The equation with Δt and the variables that we already know is Equation 2.6.2:

$$\Delta x = v_i \cdot \Delta t + \frac{1}{2} \cdot a \cdot (\Delta t)^2 \qquad (2.6.2)$$

Put in that $\Delta x = 0$: $0 = v_i \cdot \Delta t + \frac{1}{2} \cdot a \cdot (\Delta t)^2$

Since there is at least one factor of Δt in each term on the right side, we can rewrite the right side as: $0 = \Delta t \cdot (v_i + \frac{1}{2} \cdot a \cdot \Delta t)$

This gives two solutions: $\Delta t = 0$, and $(v_i + \frac{1}{2} \cdot a \cdot \Delta t) = 0$. The second answer is how much time the object takes to reach the ground.

$$v_i + \frac{1}{2} \cdot a \cdot \Delta t = 0$$

Subtract v_i from both sides: $\frac{1}{2} \cdot a \cdot \Delta t = -v_i$

Divide both sides by $\frac{1}{2} \cdot a$: $\Delta t = \dfrac{-v_i}{\frac{1}{2} \cdot a}$

Put in numbers: $\Delta t = \dfrac{-(4.02\,\text{m}/\text{s})}{\frac{1}{2} \cdot (-9.80\,\text{m}/\text{s}^2)} = 0.821\,\text{s}$

Part B:

In the x direction, we know that $v_i = (5.50 \text{ m/s}) \cdot \cos(47.0°) = 3.75$ m/s, $a = 0$ m/s^2, and $\Delta t = 0.821$ s (from Part A). a (acceleration) is zero in projectile motion problems, where gravity is the only force. For v_i, we used Equation 3.3.4:

$$\text{(x component magnitude)} = \text{(magnitude)} \cdot \cos(\theta) \qquad (3.3.4)$$

We want to find displacement (Δx). The equation with displacement and the variables that we already know is Equation 2.6.2:

$$\Delta x = v_i \cdot \Delta t + \frac{1}{2} \cdot a \cdot (\Delta t)^2 \qquad (2.6.2)$$

Since this equation is already solved for Δx, we can put in numbers:

$$\Delta x = (3.75\,\text{m}/\text{s}) \cdot (0.821\,\text{s}) + \frac{1}{2} \cdot (0\,\text{m}/\text{s}^2) \cdot (0.821\,\text{s})^2 = 3.08\,\text{m}$$

Note: 3.08 m is about 10 ft, which roughly makes sense for a ball tossed at 5.50 m/s (about 12 miles/hour).

Part C:

We now change the final time point to be the time where the ball is at its largest height. This means that v_f is now known to be zero, but the time and displacements are different. The initial velocity stays the same, since we keep the same initial time point. The acceleration stays the same because it is still a projectile motion problem. Let's look at the y direction. We know that $v_i = 4.02$ m/s, $a = -9.80$ m/s^2, and $v_f = 0$. We want to find displacement (Δx), which is the position change from the initial time point. The equation with displacement and the variables that we already know is Equation 2.6.4:

$$v_f^2 = v_i^2 + 2 \cdot a \cdot \Delta x \qquad\qquad (2.6.4)$$

Subtract v_i^2 from both sides: $v_f^2 - v_i^2 = 2 \cdot a \cdot \Delta x$

Divide both sides by $(2 \cdot a)$: $\Delta x = \dfrac{v_f^2 - v_i^2}{2 \cdot a}$

Put in numbers: $\Delta x = \dfrac{0^2 - \left(4.02 \, \text{m}/\text{s}\right)^2}{2 \cdot \left(-9.80 \, \text{m}/\text{s}^2\right)} = 0.826 \, \text{m}$

Introduction to Forces

4.1 INTRODUCTION: FORCES CAUSE MOTION

A force is something you experience every day, for example:

1. Gravity, pulling things downward.

2. Using friction against the ground to push yourself forward.

3. A normal force from a chair, keeping you from falling through the chair.

In Chapters 2 and 3, we studied motion – position, displacement, velocity, and acceleration. We related those concepts to each other, particularly for the case where acceleration was constant (not changing) during the problem. Chapters 4 and 5 talk about forces, because **forces cause motion**. It turns out that we can know a lot more about a given situation if we understand what is causing the motion.

There are many different types of forces in the world. In Chapter 4, we will talk about how forces affect motion. In Chapter 5, we will learn about several different types of forces.

4.2 NEWTON'S SECOND LAW

Isaac Newton published three laws of motion. The first two relate forces and acceleration, and the third talks about how a force involves two objects. We will start with Newton's second law:

$$\mathbf{F}_{net} = m \cdot \mathbf{a} \tag{4.2.1}$$

DOI: 10.1201/9781003005049-4

In words, this equation says that the net force (F_{net}) is equal to the object's mass (m) times the object's acceleration (a). This might be the most used equation in this entire textbook.

You probably already have a sense of what mass is from everyday life – the heavier something is, the more mass it has. There are physics definitions that are more specific and/or have a wider meaning, but this is the simplest way to think of it.

The units of force are contained in this equation. They are:

- (units of mass)·(units of acceleration) = kg·m/s^2.

- The combination kg·m/s^2 is usually called a "Newton", after Isaac Newton, and abbreviated "N".

The net force is found by adding every force on the object. Forces are vectors. You'll notice that they are **bolded** in Equation 4.2.1. When you add them to get the sum, you just add components, as seen in Section 3.4. If there are three forces $\mathbf{F_1} = F_{x1}\, i + F_{y1}\, j$, $\mathbf{F_2} = F_{x2}\, i + F_{y2}\, j$, and $\mathbf{F_3} = F_{x3}\, i + F_{y3}\, j$, the sum ($F_{net}$) would be equal to $(F_{x1} + F_{x2} + F_{x3})\, i + (F_{y1} + F_{y2} + F_{y3})\, j$. In words, add the x components together to get the total x component, and add the y components together to get the total y component. Some components will be negative – for example, if "up" is the positive y direction and the force is downward.

If your problem is in two dimensions, just apply this equation separately for each of the x and y components. In words:

$$\text{(sum of all x components of forces)} = \text{(mass)·(x component of acceleration)} \quad (4.2.2)$$

$$\text{(sum of all y components of forces)} = \text{(mass)·(y component of acceleration)} \quad (4.2.3)$$

Basically all of Chapter 4 and Chapter 5 will be problems that use this one equation. The difference between each time will be which specific forces are included in each problem. In Chapter 5, we will learn more about several different types of forces.

For now, we'll do an example where the forces are known and given. As we go through Chapter 5, more forces will be introduced.

Example 4.2.1: Airplane

An airplane is in the process of taking off. It has a mass of 1.20 10⁵ kg. It has the following forces on it:

1. Force from the engines – unknown; call it $F_E = (F_{E,x}, F_{E,y})$.
2. Gravity: 0 i + –1.18 10⁶ N j (Remember, N is Newtons, the unit of forces).
3. Air drag: –3.00 10⁴ N i + –3.00 10⁴ N j.

As it takes off, the airplane has an acceleration of a = 1.50 m/s² i + 2.50 m/s² j. What are the x and y components of the engine force?

Solution:
We need to apply Equation 4.2.1 in the x and y directions, separately.
x direction: (sum of x component of forces) = (mass)·(x component of acceleration)

The engine force's x component is $F_{E,x}$; the x component of gravity is 0, and the x component of air drag is –3.00 10⁴ N. The mass is given as 1.20 10⁵ kg, and the x component of acceleration is 1.50 m/s². Thus, the engine force is the only unknown in the equation:

$$F_{E,x} + 0 - 3.00\ 10^4\ N = (1.20\ 10^5\ kg)\ (1.50\ m/s^2)$$

Multiply out the right hand side, and remove 0 from the sum on the left side:

$$F_{E,x} - 3.00\ 10^4\ N = 1.80\ 10^5\ N$$

In the last step, remember that the unit 1 N = 1 kg m/s². Add 3.00 10⁴ N to each side.

$$\underline{F_{E,x} = 2.10\ 10^5\ N}$$

Now, we do the same thing for the y direction:

y direction: (sum of y component of forces) = (mass) (y component of acceleration)

Putting in numbers from the problem statement:

$F_{E,y} - 1.18 \ 10^6 \text{ N} - 3.00 \ 10^4 \text{ N} = (1.20 \ 10^5 \text{ kg})(2.50 \text{ m/s}^2)$

Add $1.18 \ 10^6$ N and $3.00 \ 10^4$ N to both sides:

$F_{E,y} = 1.18 \ 10^6 \text{ N} + 3.00 \ 10^4 \text{ N} + (1.20 \ 10^5 \text{ kg})(2.50 \text{ m/s}^2) = \underline{1.51 \ 10^5 \text{ N}}$

The only major ways in which this problem differs from the other problems in Chapters 4 and 5 is that the amount of forces was given to you. Once you learn more about various types of forces, you will be required to (1) recognize that they are there, and (2) calculate them before putting them into the sum of forces in Equation 4.2.1.

There are other ways that this problem could be extended by relating it to topics covered in Chapters 2 and 3. This problem could have just as easily given you the engine force, and asked you to solve for acceleration. If the acceleration was constant during the problem, and if you were given some information about velocity, displacement, or time, you could then use the values for acceleration to do a two-dimensional constant acceleration problem just like the example problem in Section 3.5. Finally, you could have been asked to find the magnitude and direction (angle) of any of these vectors, just like Section 3.3.

Problem to Try Yourself

You drag an object with mass 4.39 kg across the ground – maybe a tarp with leaves on it after some raking in the fall. The object has four forces on it: force from the ground (upward, 25.0 N), gravity (downward), friction (12.0 N, to the left), and a pull force from you of 18.0 N i + 22.0 N j. The object slides along the flat ground, which means that the y component of acceleration is equal to zero.

A. What is the magnitude of gravity?
B. What is the magnitude of the x component of acceleration?

Hint: look at the y direction for Part A, and the x direction for Part B.

Solution:

Define upward as the +y direction, and "right" as the +x direction.

Part A:

Gravity is in the y direction (it's downward, so actually the −y direction). Let's look at the forces in the y direction first. There is the force upward from the ground (+25.0 N), gravity (negative since downward), and the y component of the pull from you (+18.0 N). We are also given that the acceleration in the y direction is zero, because the object keeps the same y position (ground) at all times.

Newton's second law is $F_{net} = m \cdot a$

Applying this to the y direction: 25.0 N − (gravity) + 18.0 N = (4.39 kg)·0

4.39·0 is equal to zero, and 25 + 18 = 43: 43.0 N − (gravity) = 0

Add (gravity) to both sides: (gravity) = <u>43.0 N</u>

Part B:

Look at the forces in the x direction. There is friction (−12.0 N; negative because to the left) and the x component of the pull from you (+22.0 N).

Newton's second law is $F_{net} = m \cdot a$

Applying this to the x direction: −12.0 N + 22.0 N = (4.39 kg)·a

Add the terms on the left side: −12 + 22 = 10: 10.0 N = (4.39 kg)·a

Divide both sides by 4.39 kg: a = <u>2.28 m/s²</u>

Notes: this problem could be made slightly harder if the force from you was given as a magnitude and direction – you'd use Equations 3.3.4 and 3.3.5 to get the x and y components. With units, remember that 1 N = 1 kg·m/s².

4.3 NEWTON'S FIRST AND THIRD LAWS

Newton's first law is just a special case of Newton's second law.

Newton's second law: $F_{net} = m \cdot a$

Newton's first law: If $F_{net} = 0$, then $a = 0$

Mathematically, this might seem like an unnecessary statement to make. Thinking about things without numbers reveals the importance. If the sum of forces (also called net force) on an object is zero, it means that the acceleration is zero. When the acceleration is zero, the velocity stays exactly the same – same amount, same direction. If the object is stopped, it won't move. If the object is moving, it will continue to move in exactly the way it has been.

Newton's third law is related to the fact that forces in everyday life all require two objects to happen. For example, the force of gravity that you feel comes from a force interaction between you and the earth. When a basketball is sent towards a basket, the force on the ball probably came from a hand. Newton's third law is the following:

> When a force happens, it happens between two objects. Each of the two objects gets the same amount of force on it, but in the opposite direction.

Think about moving an object, like a chair next to a table. You would feel a force in the opposite direction that you are moving the chair. When a car's wheel turns to move the car forwards, the ground is moved backwards.

How do these laws come into our physics problems? Newton's first law applies in any problem where the velocity stays constant. Newton's third law is helpful for later, when you will be figuring out which forces are part of a given problem. Occasionally, you may even solve a problem with two objects (examined separately) where Newton's third law comes into the problem directly.

4.4 CHAPTER 4 SUMMARY

Newton's second law relates forces to motion, specifically acceleration

$$F_{net} = m \cdot a \qquad (4.2.1)$$

F_{net} (net force) involves adding all of the forces in a given direction, after defining a positive direction. All forces in the positive direction are added, and all forces in the negative direction are subtracted. For a two-dimensional problem, get all forces into components, and use Equation 4.2.1 twice – one for each dimension, with acceleration then being the component for that dimension:

(sum of all x components of forces) = (mass)·(x component of acceleration)   (4.2.2)

(sum of all y components of forces) = (mass)·(y component of acceleration)   (4.2.3)

Newton's first law is that, if total force is zero, then acceleration is zero. Newton's third law is that forces happen between two objects, and when a force happens both objects get the same force, just in opposite directions.

Chapter 4 Problem Types:
Force problems that use Newton's second law, either in one dimension or in two dimensions

Practice Problems
Section 4.2

1. A toothbrush brushes against a set of teeth. Define the x direction as the direction along the teeth that the brush is going, and the y direction as towards/away from the teeth. The brush has three forces on it: force from a hand (not given), friction (2.00 N, towards the –x axis), and a force from touching the teeth (2.50 N, towards the +y axis). The acceleration in the x direction is 2.00 m/s², all in the +x direction; the acceleration in the y direction is zero. The mass of the toothbrush is 0.100 kg.

   A. What is the x component of the force from the hand?

   B. What is the y component of the force from the hand?

   C. Convert the results from Parts A and B into a magnitude and angle (direction).

   Solution:

   Part A:

   The forces in the x direction are the x component of the hand force and friction (2.00 N in the negative direction). The x component of the acceleration is 2.00 m/s². Putting this into Newton's second law:

$$F_{net} = m·a \qquad (4.2.1)$$

(x component of hand force) − 2.00 N = (0.100 kg)·(2.00 m/s²)

Add 2.00 N to both sides:

(x component of hand force) = 2.00 N + (0.100 kg)·(2.00 m/s²) = 2.20 N

Part B:

The forces in the x direction are the y component of the hand force and the force from the teeth (2.50 N in the positive direction). The y component of acceleration is 0 m/s², since the acceleration is only in the x direction. Putting this into Newton's second law:

$$F_{net} = m \cdot a \qquad (4.2.1)$$

(y component of hand force) + 2.50 N = (0.100 kg)·(0 m/s²)

Subtract 2.50 N from both sides:

(y component of hand force) = −2.50 N + (0.100 kg)·(0 m/s²) = −2.50 N

Part C:

To get from components to magnitude and angle, use Equations 3.3.2 and 3.3.3:

$$(\text{magnitude}) = \sqrt{(\text{x component})^2 + (\text{y component})^2} \qquad (3.3.2)$$

$$\theta = \tan^{-1}(|\text{y component}|/|\text{x component}|) \qquad (3.3.3)$$

$$(\text{magnitude}) = \sqrt{(2.20\,\text{N})^2 + (-2.50\,\text{N})^2} = 3.33\,\text{N}$$

The angle will be below the +x axis, since the x component is positive (+x axis) and the y component is negative (below).

$\theta = \tan^{-1}(|-2.50\,\text{N}|/|2.20\,\text{N}|) = 48.7°$ below (clockwise from) the +x axis

Note: in the third dimension (which would be up and down), there would be two forces: gravity (downward) and more force from the hand (upward, equal to gravity so the toothbrush doesn't go up and down). We will ignore that here.

2. A kite flies in the air. It has a mass of 0.100 kg. It has three forces on it: force from the wind (0.500 N i + 1.600 N j), force from a string on the kite (−0.400 N i − 0.400 N j), and gravity (0 N i − 0.980 N j). What are the x and y components of the acceleration?

Solution:

Apply Newton's second law (F_{net} = m·a, Equation 4.2.1) to each of the x and y directions separately.

x direction: 0.500 N − 0.400 N + 0 N = (0.100 kg)·a

Combine the numbers on the left side: 0.100 N = (0.100 kg)·a

Divide both sides by 0.100 kg: <u>a = 1.00 m/s² (x component)</u>

y direction: 1.600 N − 0.400 N − 0.980 N = (0.100 kg)·a

Combine the numbers on the left side: 0.220 N = (0.100 kg)·a

Divide both sides by 0.100 kg: <u>a = 2.20 m/s² (y component)</u>

3. Your friend moves an object out of the way, using their foot. The object has a mass of 2.00 kg. The object has three forces on it: force from the floor (0 N i + 19.8 N j), force from the foot moving it (8.00 N i + 0 j), and gravity (magnitude not given, but the direction is the −y direction). The acceleration is 4.00 m/s² i + 0 j. (The 0 for y component of acceleration is because there is no motion up or down.)

   A. What is the magnitude of the gravity force? Hint: use the y direction; no knowledge of the equations for gravity is needed here.

   B. What is the mass of the object? Hint: use the x direction.

Solution:

Part A:

Apply Newton's second law (F_{net} = m·a, Equation 4.2.1) to the y direction:

$$19.8 \text{ N} + 0 - (\text{magnitude of gravity}) = m \cdot 0$$

The right side is 0, since the y component of acceleration is 0. The magnitude of gravity is negative because it is in the –y direction.

Add (magnitude of gravity) to both sides: <u>19.8 N = magnitude of gravity</u>.

Part B:
Apply Newton's second law (F_{net} = m·a, Equation 4.2.1) to the x direction:

$$0 + 8.00 \text{ N} + 0 = m \cdot (4.00 \text{ m/s}^2)$$

The third force is zero because gravity is only in the y direction.

Divide both sides by 4.00 m/s²: m = (8.00 N)/(4.00 m/s²) = <u>2.00 kg</u>

4. Two friends move a heavy object. They lift it a bit, but the bottom still drags a bit against the floor as it moves, making for some friction. The object has a mass of 40.0 kg. The first person puts a force of 196 N, at an angle of 75.0° above the +x axis. The second person puts a force of 153 N, at an angle of 79.0° above the +x axis. There is a force from the floor, of 65.0 N in the +y direction. The gravity force is 392 N, in the –y direction. There is a friction force of 30.0 N, in the –x direction. What are the x and y components of acceleration?

Solution:

Apply Newton's second law (F_{net} = m·a, Equation 4.2.1) to each of the x and y directions separately. To do this, first split the forces from the two friends into components using Equations 3.3.4 and 3.3.5:

$$(x \text{ component magnitude}) = (\text{magnitude}) \cdot \cos(\theta) \qquad (3.3.4)$$

$$(y \text{ component magnitude}) = (\text{magnitude}) \cdot \sin(\theta) \qquad (3.3.5)$$

(don't forget to choose + or – direction for each component)

Friend 1 x component: (196 N)·cos(75.0°) = 50.7 N
(positive since +x axis named)

Friend 1 y component: (196 N)·sin(75.0°) = 189 N
(positive since above +x axis)

Friend 2 x component: $(153 \text{ N}) \cdot \cos(79.0°) = 29.2 \text{ N}$
(positive since +x axis named)

Friend 2 y component: $(153 \text{ N}) \cdot \sin(79.0°) = 150 \text{ N}$
(positive since above an axis)

The x direction has the two forces from the friends, plus a 30.0 N force in the negative direction.

Applying Newton's second law ($F_{net} = m \cdot a$, Equation 4.2.1):

$$50.7 \text{ N} + 29.2 \text{ N} - 30.0 \text{ N} = (40.0 \text{ kg}) \cdot a$$

Add the numbers on the left side: $49.9 \text{ N} = (40.0 \text{ kg}) \cdot a$

Divide both sides by 40.0 kg: $\underline{a = 1.25 \text{ m/s}^2 \text{ (x component)}}$

In the y direction, there are the forces from the friends plus two additional forces: gravity (392 N, negative direction) and force from the floor (65.0 N, positive direction).
Applying Newton's second law ($F_{net} = m \cdot a$, Equation 4.2.1):

$$189 \text{ N} + 150 \text{ N} + 65.0 \text{ N} - 392 \text{ N} = (40.0 \text{ kg}) \cdot a$$

Add the numbers on the left side: $12 \text{ N} = (40.0 \text{ kg}) \cdot a$

Divide both sides by 40.0 kg: $\underline{a = 0.30 \text{ m/s}^2 \text{ (y component)}}$

Note: the addition of forces would have left no significant figures after the decimal. This is why the last two steps have numbers with only two significant figures.

5. A ball with mass of 0.200 kg flies though the air. The ball has two forces on it: gravity (0 N i – 0.123 N j) and air resistance (–0.100 N i + 0 j). What are the x and y components of acceleration?

Solution:
Apply Newton's second law ($F_{net} = m \cdot a$, Equation 4.2.1) to each of the x and y directions separately.

x direction: $0 \text{ N} - 0.100 \text{ N} = (0.200 \text{ kg}) \cdot a$

Combine the numbers on the left side: $-0.100 \text{ N} = (0.200 \text{ kg}) \cdot a$

Divide both sides by 0.200 kg: <u>a = −0.500 m/s² (x component)</u>

y direction: −0.123 N + 0 N = (0.200 kg)·a

Divide both sides by 0.200 kg: <u>a = −0.615 m/s² (y component)</u>

6. You push a baby stroller across the ground, without realizing that the brakes are still on, and so the wheels aren't rolling. The stroller has four forces on it: force from you (107 N, in the +x direction), force from the ground (155 N, in the +y direction), force from friction (80.0 N, in the −x direction), and gravity (155 N, in the −y direction). The x component of acceleration is 1.35 m/s².

A. What is the mass of the stroller?

B. What is the y component of acceleration?

Solution:
Part A:
Apply Newton's second law ($F_{net} = m·a$, Equation 4.2.1) to the x direction:

$$107 \text{ N} - 80.0 \text{ N} = m·(1.35 \text{ m/s}^2)$$

Note: the forces that are only in the y direction are not included in the x direction equation.
Combine the forces on the left side: 27 N = m·(1.35 m/s²)
Divide both sides by 1.35 m/s²: m = (27 N)/(1.35 m/s²) = <u>20.0 kg</u>

Part B:

Apply Newton's second law ($F_{net} = m·a$, Equation 4.2.1) to the y direction:

$$155 \text{ N} - 155 \text{ N} = (20.0 \text{ kg})·a$$

Note: the forces that are only in the x direction are not included in the y direction equation.

Combine the forces on the left side: 0 N = (20.0 kg)·a

Divide both sides by (20.0 kg) a = (0 N)/(20.0 kg) = <u>0 m/s²</u>

Specific Types of Forces, and Force Problems

5.1 INTRODUCTION: THERE ARE DIFFERENT TYPES OF FORCES

There are many different types of forces in the world. In this chapter, we'll look at the types of forces that are most commonly discussed in introductory physics courses, as well as looking at three specific types of force problems. But before that, we need to talk about the very useful topic of free body diagrams.

5.2 FREE BODY DIAGRAMS

A free body diagram is a very useful tool to help you do physics problems. For each object in a given problem, making a free body diagram is as simple as this:

1. Draw something to represent each object in the problem. Usually there is only one object; occasionally there are two or three in a complicated problem. Tipler's text (cited in the acknowledgements) suggests using just a dot instead of drawing a full picture.

2. Set a coordinate system, similar to what we talked about earlier in the textbook, for example in Chapters 2 and 3. For each dimension (likely two dimensions, but possibly one for specific problems), have

DOI: 10.1201/9781003005049-5

a positive direction. Most often, "to the right" or "east" can be positive x, and "up" or "north" can be positive y.

3. Spend some time thinking about what forces are on the object. For every force that you think is on the object, draw that force. Label it with the amount of force (magnitude), and draw it in the direction it is going.

   • Sometimes you may not know the magnitude or direction of a force; in that case you can give that force two components (one in the + x direction and one in the +y direction) and figure out later what those forces are.

Once a correct free body diagram is made, it's easy to add up the forces in the x and y directions. Remember from Chapter 4 that the sum of all x components of force equals mass times the x component of acceleration, and the sum of all y components of force equals mass times the y component of acceleration. You will need to break some forces into components (using Equations 3.3.4 and 3.3.5) before adding the components.

$$\text{(x component magnitude)} = \text{(magnitude)}\cdot\cos(\theta) \qquad (3.3.4)$$
$$\text{(y component magnitude)} = \text{(magnitude)}\cdot\sin(\theta) \qquad (3.3.5)$$

Figure 5.1 shows an example of a free body diagram. In this situation, up is the +y (positive y) direction, right is the +x direction, and there are three forces. One force (F_1) is in the +x direction, one force (F_2) is in the −y direction, and one force (F_3) is in the both the −x and +y directions. F_3 would need to be split into two components in order to compute the net force in the x and y directions. The x component would be negative, and the y component would be positive.

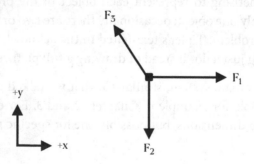

FIGURE 5.1 An example of a free body diagram, with a coordinate system included in the bottom left.

5.3 WHAT FORCES ARE ON MY OBJECT?

It can be very intimidating to try and think of every single force that is on an object. Here is a guide to the types of forces that will typically be seen in a problem for an introductory physics course:

1. Gravity (Section 5.4). In all of the problems with an object located near Earth, gravity is downward.

2. The "Normal Force" (many sections). In this case, the word "normal" means "at a 90° angle", not "usual". If your object is on another object (for example, the ground) then the other object will exert a force at a 90° angle to itself – for example, upward from horizontal ground.

3. Friction (Sections 5.5 and 5.6). There will be friction if the object is moving across the ground (kinetic friction), or if another force is trying to move the object across the ground and it is staying still anyway (static friction). This is sometimes ignored.

   a. Note: air resistance is similar, except that it comes from the air instead of the ground. It is almost always ignored in basic problems.

4. Spring force (Section 5.7) There will be a spring force if the object is touching a spring.

5. Look at what is touching your object. Is something pulling on or pushing on it?

6. If you eventually study electricity and magnetism, you'll see that magnetic and electric fields can cause a force on objects. That is a more advanced topic, and we will not cover it here.

5.4 GRAVITY (NEAR EARTH)

Gravity is a force that pulls two objects together. Any two objects that have mass are pulled together by gravity. However, the force of gravity is only big enough to take into account when at least one of the two objects is huge – most commonly, if one of the objects is Earth. For this section, we are going to make two assumptions:

1. The force is between Earth and your object. That way one of the two objects (Earth) is large enough.

2. The object is fairly close to Earth's surface. Even an airplane's height above Earth would be fine.

When we make those two assumptions, the force of gravity is always equal to the following:

- Magnitude of gravity force: m·g   (5.4.1)

  - m = mass of your object. g = 9.80 m/s², which is called the "acceleration due to gravity".

- Direction of gravity force: straight downward

Remember that all forces, including gravity, are vectors. That means that they have a magnitude and a direction.

Side note: we will do a more general case of gravity in Chapter 11, including showing where 9.80 m/s² comes from, and where "straight down" really means towards the center of the earth.

Example 5.4.1: Calculating force of gravity, near Earth.

If a person has a mass of 75.0 kg, and is on Earth's surface, what is the force of gravity on that person?

Solution:
Magnitude of force = m·g = (75.0 kg)·(9.80 m/s²) = 735 Newtons
Direction: straight downward
Note: remember from Chapter 4 that Newtons are the units of force and that 1 Newton equals 1 kg·m/s².

Example 5.4.2: Gravity and Normal Force

The same person from Example 5.4.1 (75.0 kg mass) is on flat ground. The person stays still. What is the net force, and what is the normal force, if these are the only two forces and the normal force is upward?

Solution:
A free body diagram of the person would have two forces: gravity downward, and normal force upward.

The net force is zero, because the person isn't moving. Remember that Newton's second law is

$$F_{net} = m \cdot a \qquad (4.2.1)$$

If the person isn't moving, their acceleration is zero, and so $F_{net} = m \cdot 0 = 0$.

Let's now write out an equation for the net force that uses the forces in the problem:

$$F_{net} = (\text{normal force}) - (\text{gravity})$$

Add gravity to both sides: $(\text{normal force}) = F_{net} + (\text{gravity})$
As in Example 5.4.1, gravity is 735 Newtons, downward.
Put in numbers: $(\text{normal force}) = 0 + 735 \text{ N} = \underline{735 \text{ N}}$

Note: if the person had been moving up and down, the acceleration may not have been zero, and the normal force could have been a different value than being equal to gravity! If there were forces in the x direction, they would be handled separately, like Example 4.2.1

Problem to Try Yourself

You are standing at a bus stop, waiting with a bag full of groceries. The bag has a mass of 11.2 kg. The bag has three forces on it: gravity (downward), normal force from the ground (upward), and a small amount of force from you holding the bag (15.0 N, upward). The bag is staying still. What is the magnitude of the normal force?

Solution:

$$F_{net} = m \cdot a$$

We know that $a = 0$, and gravity $= m \cdot g$ (downward). Define upward as the positive direction.

Put the forces into F_{net}, and 0 in for a: $(\text{normal force}) - m \cdot g + 15.0 \text{ N} = m \cdot 0$

Add m·g to both sides, and subtract 15.0 N from both sides: $(\text{normal force}) = m \cdot g - 15.0 \text{ N}$

Put in numbers for m and g: $(\text{normal force}) = (11.2 \text{ kg}) \cdot (9.80 \text{ m/s}^2) - 15.0 \text{ N} = \underline{94.8 \text{ N}}$

5.5 KINETIC FRICTION

Kinetic friction happens when two objects rub together during motion. It goes against the motion, making things harder to move. Kinetic friction is also the reason that something sliding across a surface eventually comes to a stop.

Like any force, kinetic friction is a vector, and so it has a magnitude and a direction. The magnitude of the kinetic friction force is:

$$F_{kf} = \mu_k \cdot F_n \tag{5.5.1}$$

In this equation, F_{kf} is the magnitude of the kinetic friction force, μ_k is the coefficient of kinetic friction, and F_n is the magnitude of the normal force. The coefficient of kinetic friction (μ_k) is a measure of how rough the two surfaces rubbing are. If you rub sandpaper against wood, μ_k will be a high value, perhaps around 1; if you slide something on ice μ_k will be smaller.

The direction of the kinetic friction force is usually in the direction opposite of the object's motion.

Side note: more technically, from *College Physics* by Tammaro, the direction of kinetic friction is the opposite direction of the component of velocity that is parallel to the surfaces that are rubbing.

Example 5.5.1

You are dragging a trash can back, after trash day. The empty trash can has mass 2.50 kg. The forces on the trash can are gravity (downward), friction (to the left), normal force (upward), and the pull force from you on the can (25.0 N, 30.0° above "to the right"). The trash can stays on the (flat) ground, so that there is only motion in the x direction. The coefficient of kinetic friction between the trash can and the ground is 0.500

    A. What is the magnitude of the normal force?
    B. What is the magnitude of the kinetic friction force?
    C. What is the x component of acceleration?

Solution:
Part A:
Since the normal force is in the y direction, let's look at the y direction. Define upward as positive. The forces are gravity (m·g, downward),

normal force (F_n, upward), and the y component of the pull force. The y component of the pull force comes from Equation 3.3.5:

$$\text{(y component magnitude)} = \text{(magnitude)} \cdot \sin(\theta) \qquad (3.3.5)$$

$$25.0 \text{ N} \cdot \sin(30.0°) = 12.5 \text{ N}$$

Acceleration is zero, since there is no motion in the y direction. Now use Newton's second law:

$$F_{net} = m \cdot a$$

Put the forces into F_{net}, and 0 in for a: $-m \cdot g + F_n + 12.5 \text{ N} = m \cdot 0$
Add m·g to both sides, and subtract 12.5 N from both sides:

$$F_n = m \cdot g - 12.5 \text{ N}$$

Put in numbers for m and g: $F_n = (2.50 \text{ kg}) \cdot (9.80 \text{ m/s}^2) - 12.5 \text{ N} = \underline{12.0 \text{ N}}$

Part B:

$$F_{kf} = \mu_k F_n \qquad (5.5.1)$$

Put in numbers: $F_{kf} = (0.500) \cdot (12.0 \text{ N}) = \underline{6.00 \text{ N}}$

Part C:
Use $F_{net} = m \cdot a$ for the x direction. Make "to the right" positive. The forces are kinetic friction (6.00 N, to the left) and the x component of the pull. For the x component of the pull, use Equation 3.3.4:

$$\text{(x component magnitude)} = \text{(magnitude)} \cdot \cos(\theta) \qquad (3.3.4)$$

$$25.0 \text{ N} \cdot \cos(30.0°) = 21.7 \text{ N}$$

Put the forces in: $-6.00 \text{ N} + 21.7 \text{ N} = m \cdot a$
Divide both sides by m, and add the two numbers for the forces:
$a = (15.7 \text{ N})/m$
Put in number for mass: $a = (15.7 \text{ N})/(2.50 \text{ kg}) = \underline{6.28 \text{ m/s}^2}$

Notes: this is a fairly difficult problem with friction in it. The acceleration is pretty high – about two-thirds the acceleration of something falling. You would probably start pulling less hard after seeing this. The y component of acceleration is zero, so that x component is the magnitude of acceleration.

Problem to Try Yourself

You are still pulling a 2.50 kg trash can, but this time you pull it completely sideways. (This makes the math easier.) The forces on the trash can are now: gravity (downward), normal force (upward), kinetic friction, and a 15.0 N pull force to the right. The trash can moves on flat ground, with a coefficient of kinetic friction of 0.500.

A. What is the magnitude of the normal force?
B. What is the magnitude of the kinetic friction force?
C. What is the acceleration?

Solution:

Part A:
Look at forces in the y direction: normal force (F_n) (upward) and gravity ($m \cdot g$) (downward). Since the trash can moves on flat ground, the y component of acceleration is zero.

$$F_{net} = m \cdot a$$

Put in the forces in F_{net}, and that $a = 0$: $F_n - m \cdot g = m \cdot 0$
Add $m \cdot g$ to both sides: $F_n = m \cdot g$
Put in numbers: $F_n = (2.50 \text{ kg}) \cdot (9.80 \text{ m/s}^2) = \underline{24.5 \text{ N}}$

Part B:

$$F_{kf} = \mu_k \cdot F_n \qquad (5.5.1)$$

Put in numbers: $F_{kf} = (0.500) \cdot (24.5 \text{ N}) = \underline{12.3 \text{ N}}$

Part C:
Forces in x: kinetic friction (12.3 N to the left) and pull (15.0 N to the right)

$$F_{net} = m \cdot a$$

Put in the forces in the net force: $-12.3 \text{ N} + 15.0 \text{ N} = m \cdot a$

Divide both sides by m, and add the numbers on the left side: $a = 2.7 \text{ N/m}$

Put in the mass: $a = 2.7 \text{ N}/(2.50 \text{ kg}) = \underline{1.1 \text{ m/s}^2}$

5.6 STATIC FRICTION

Static friction is similar to kinetic friction, except that it involves a situation where friction is keeping something from moving. (You can think of kinetic friction as slowing down something that is moving.)

The trick to static friction is that it has an equation that looks like the kinetic friction equation, but instead of giving the magnitude of the friction force, it gives the maximum possible amount of static friction. _**The static friction force can be any value between zero and the maximum static friction force. If the object isn't moving, the amount of static friction is exactly enough to cancel out the other forces.**_

The equation for maximum static friction force is:

$$F_{\text{maximum static friction}} = \mu_s \cdot F_n \qquad (5.6.1)$$

(static friction force) = (just enough to not move, if less than

maximum is needed)   (5.6.2)

In these equations, $F_{\text{maximum static friction}}$ is the largest possible magnitude of static friction. μ_s is the coefficient of static friction (which is similar to the coefficient of kinetic friction), and F_n is the magnitude of the normal force.

The direction of the static friction force can be thought of as being opposite to the direction in which you are trying to move an object. So, if you try to move it to the right, static friction is to the left. (Note: the full definition is a little more technical; see the side note in Section 5.5.)

Example 5.6.1

You are dragging a full trash can, with mass 15.0 kg. The coefficient of static friction is 0.500. The forces on the trash can are: normal force (upward), gravity (downward), static friction (to the left) and a pull force (to the right). The trash can is not moving.

A. What is the highest possible amount of static friction? Put another way, how hard can you pull without moving the can? Hint: find normal force first.

B. What is the static friction force if you pull with half of your answer to Part A?

Solution:
Part A:

$$F_{\text{maximum static friction}} = \mu_s \cdot F_n \qquad (5.6.1)$$

We need to find the normal force (F_n) in order to calculate this. Look at the forces in the y direction: F_n (normal force, up) and m·g (gravity, down). Since the trash can is not moving, a = 0. Make upward the positive direction.

$$F_{\text{net}} = m \cdot a$$

Put in the forces, and a = 0: $F_n - m \cdot g = m \cdot 0 = 0$
Add m·g to both sides: $F_n = m \cdot g$
Put in numbers: $F_n = (2.50 \text{ kg}) \cdot (9.80 \text{ m/s}^2) = 24.5 \text{ N}$
Now calculate the highest static friction: $F_{\text{maximum static friction}} = \mu_s \cdot F_n = (0.500) \cdot (24.5 \text{ N}) = \underline{12.3 \text{ N}}$

Part B:

(static friction force) = (just enough to not move,
if less than maximum is needed) (5.6.2)

If the maximum is 12.3 N, and you pull with half of that (call it 6.15 N), the static friction force will also be 6.15 N. The reason for this is Equation 5.6.2: the static friction force is just enough to keep the object from moving, as long as the amount required for that is less than the highest possible.

Notes on this problem: many students get stuck assuming that $F_n = m \cdot g$ in every problem. If there are any other vertical forces, like in Example 5.5.1, this is not true. This problem could be made harder by having the pull force be in two dimensions, like Example 5.5.1.

Problem to Try Yourself

You are trying to move a heavy rock into a better-looking location, but it is not moving. You put a force of 125 N on it, to the right, and it almost moves. We'll say that 125 N is the highest possible amount of static friction. The coefficient of static friction is 0.500.

A. What is the mass of the rock?

B. What is the static friction force if you use a force of 115 N instead of 125 N?

Solution:

Part A:

$$F_{maximum\ static\ friction} = \mu_s \cdot F_n \qquad (5.6.1)$$

We again need to find F_n (normal force). Look at the forces in y: F_n (upward), m·g (gravity, downward). Make upward the positive direction. The acceleration is zero, because the rock isn't moving. Newton's second law gives

$$F_n - m \cdot g = m \cdot 0 = 0.$$

Add m·g to both sides: $F_n = m \cdot g$

We can't put in numbers yet because mass is the unknown variable. Instead, we'll put m·g in for F_n in Equation 5.6.1:

$$F_{maximum\ static\ friction} = \mu_s \cdot m \cdot g$$

To solve for m, divide both sides by ($\mu_s \cdot g$): m = ($F_{maximum\ static\ friction}$)/ ($\mu_s \cdot g$)

Put in numbers: m = (125 N)/(0.500·9.80 m/s²) = 25.5 kg

Part B:

$$(\text{static friction force}) = (\text{just enough to not move,}$$
$$\text{if less than maximum is needed}) \qquad (5.6.2)$$

If the highest possible is 125 N, and you use 115 N, the static friction force will also be 115 N. The reason for this is Equation 5.6.2: the static friction force is just enough to keep the object from moving, as long as the amount required for that is less than the highest possible.

5.7 SPRING FORCES

Springs are often used in introductory physics problems. Springs can be compressed (made shorter in length) or stretched (made longer in length). When a spring is not being compressed or stretched, it has a certain length, called its **equilibrium length**.

If a spring is compressed or stretched, it tries to get back to its original length. As this happens, any objects touching the spring may feel a force. The force is related to Hooke's law, and is given here as the following:

$$F_{spring} = k \cdot \Delta x \qquad (5.7.1)$$

F_{spring} is the magnitude of the spring force on an object that is touching the spring, k is a number called the spring constant, and Δx is the amount by which the spring is compressed or stretched away from its equilibrium length.

The direction of the spring force, on an object touching the spring, is always back towards the spring's equilibrium length. So, for example, if a spring is stretched or compressed in the direction of "to the right", the spring will try to move anything attached to it back to the left as it tries to get back to its equilibrium length.

The spring constant is different for different springs. It is smaller for springs that are easier to stretch and compress. The units of k are Newtons/meter (N/m) – if you look at Equation 5.7.1, you can see that the force has units of Newtons and Δx has units of meters, so k must have units of N/m.

Example 5.7.1

While trying to replace the batteries in an electronic device, you compress the little spring that the "negative" end of the battery is supposed to touch. You compress the spring by 0.00500 meters. If the spring has a spring constant of 40.0 N/m, what is the magnitude of the force on the battery from the spring?

Solution:

Here, you can just apply Equation 5.7.1:

$$F_{spring} = k \cdot \Delta x = (40.0 \text{ N/m}) \cdot (0.00500 \text{ m}) = \underline{0.200 \text{ Newtons}}$$

(Note: 0.200 Newtons is roughly equivalent to 0.5 pounds of force.)

The direction of the force would be whatever direction the spring needed to go in order to reach its equilibrium length. For example, if the spring was compressed to the left, the force on the battery would be to the right.

Problem to Try Yourself

On a playground, a child sits on top of a toy unicorn, and the toy unicorn is on top of a large spring. The child and unicorn have a combined mass of 25.0 kg, and the spring is compressed by a length of 0.100 meters. The child, unicorn, and spring are not moving at this point. What is the spring constant of the spring? Hint: look at the child and unicorn as one combined object, and look at the forces on that object.

Solution:

Look at the forces on the child/unicorn object:

- Upward: spring force. The spring is compressed downward, so it will try to get back upward to its original length.
- Downward: gravity (m·g)

Applying Newton's second law to the vertical (y) direction, and making "upward" the positive direction, we get:

(magnitude of spring force) – (magnitude of gravity) = m·a

We can put in Equation 5.7.1 for the spring force's magnitude, and m·g for gravity.

$$k \cdot \Delta x - m \cdot g = m \cdot a$$

a = 0, because the object is not moving (no motion means that the velocity isn't changing, so the acceleration is zero).
This gives: $k \cdot \Delta x - m \cdot g = 0$.
Adding m·g to both sides gives: $k \cdot \Delta x = m \cdot g$
To solve for k, divide both sides by Δx: $k = m \cdot g / \Delta x$
Put in numbers: $k = (25.0 \text{ kg}) \cdot (9.81 \text{ m/s}^2)/(0.100 \text{ m}) = \underline{2.45 \cdot 10^3 \text{ N/m}}$

5.8 COMMON FORCE PROBLEMS, TYPE 1: RAMP PROBLEMS

In Sections 5.3–5.7, we have discussed some of the types of forces that are often seen in physics problems. Section 5.3 provides an outline of these forces, and Sections 5.4 through 5.7 provide some detail on gravity, friction, and springs. In the next four sections, we will talk about four common force problems. Each one can be difficult at first, but for each type of problem you will see that the solution is always very close to being the same, so after a few tries these problems will be easier to do.

The first problem type is ramp problems. This usually involves something moving up or down a ramp. A typical setup situation is shown in Figure 5.2. There are usually three to four forces on the ramp, including:

- Gravity, vertically downward

- Normal force, outward from the ramp (90° from the direction of the ramp)

- Friction force. The friction force will be directed along the ramp, going in the direction of "against whatever motion would otherwise happen". The friction force is static friction if the object isn't moving, and kinetic friction if the object is moving.

- Sometimes there is an extra force, usually in the direction of the ramp. For example, a person might be moving something up a hill, in which case the force from the person would be in the direction of "up the ramp" and there would be a kinetic friction force in the

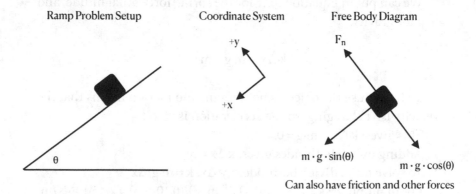

Ramp Problem Setup   Coordinate System   Free Body Diagram

$+y$

$+x$

F_n

$m \cdot g \cdot \sin(\theta)$

$m \cdot g \cdot \cos(\theta)$

Can also have friction and other forces

θ

FIGURE 5.2   A typical ramp problem setup.

direction of "down the ramp" to go against the box as it travels up the ramp. (Note: "up the ramp" means along the ramp, towards the top; "down the ramp" means along the ramp, towards the bottom.)

The thing that makes ramp problems tricky is that the movement is horizontal and vertical at the same time. One way to handle this is to change how we define the x and y directions. The x and y directions need to be 90° apart from each other, but there is no requirement that they have to be horizontal (x) and vertical (y).

The easiest change to the x and y directions is the do the following:

The +x direction becomes "down the ramp".

The +y direction becomes "outward from the ramp" – specifically, the direction of the normal force.

When this change in x and y directions are made, the motion is always in just the x direction, the normal force is always in just the +y direction, and the friction force is always in the x direction. However, gravity is now in both the x and y directions. The way to deal with gravity is to split it into x and y components, using the changed x and y directions. This can be done using geometry, and the answer is:

$$\text{x component of gravity: } m{\cdot}g{\cdot}\sin(\theta) \text{, in the +x direction} \qquad (5.8.1)$$

$$\text{y component of gravity: } m{\cdot}g{\cdot}\cos(\theta) \text{, in the -y direction} \qquad (5.8.2)$$

In these equations, m is mass, g is acceleration due to gravity (9.80 m/s²), and θ (Greek letter "theta") is the angle that the ramp makes from the horizontal direction. You might notice that this is the first (and only) time in this textbook where cosine doesn't go with the x direction and sine doesn't go with the y direction.

Making these changes, the problem should now work like our previous two-dimensional force problems.

Example 5.8.1

In a factory, a box slides down a ramp on its way to a delivery truck. The box has a mass of 7.60 kg, and the ramp makes an angle of 25.2°

from the horizontal. There is friction between the ramp's surface and the box, with the coefficient of kinetic friction being equal to 0.410. Define "down the ramp" as the +x direction, and "outward from the ramp" as the +y direction, and assume that the box is always touching the ramp.

A. What is the y component of the acceleration of the box? Hint: no math is needed.
B. What is the magnitude of the normal force?
C. What is the x component of the acceleration of the box?

Solution to Part A:

With the different definition of x and y directions for the ramp problems, the y direction is 90° from the ramp. Since the box always stays on the ramp, there is no motion in the y direction. If there is no motion, there is zero velocity; if there is no velocity, then there is zero acceleration.

Solution to Part B:

As mentioned in this section, the normal force is in the +y direction in ramp problems. So, to find the magnitude of the normal force, it makes sense to look in the y direction. The forces in the y direction are:

- Normal force, in the +y direction
- y component of gravity, $m \cdot g \cdot \cos(\theta)$, in the −y direction

Using Newton's second law for the y direction gives:

(magnitude of normal force) − $m \cdot g \cdot \cos(\theta) = m \cdot a$

a is the y component of acceleration

Our answer to Part A was that a = 0, so:

(magnitude of normal force) − $m \cdot g \cdot \cos(\theta) = m \cdot 0 = 0$

Adding $m \cdot g \cdot \cos(\theta)$ to both sides of the equation gives:

(magnitude of normal force) = $m \cdot g \cdot \cos(\theta)$

Plugging in numbers gives:

(magnitude of normal force) = (7.60 kg)·(9.80 m/s²)·cos(25.2) = <u>67.4 Newtons</u>

Solution to Part C:

To find the x component of acceleration, we'll use Newton's second law on the x direction. The forces in the x direction are:

x component of gravity, m·g·sin(θ), in the +x direction

It is the –x direction since the motion is in the +x direction. Since the box is moving, the friction is kinetic. The equation for kinetic friction is Equation 5.5.1, and is equal to $\mu_k \cdot F_n$. We found F_n (magnitude of the normal force) as the solution to Part B.

Putting the forces above into Newton's second law, we'd get:

- (x component of gravity) – (friction force) = m·a_x
- Friction force, in the –x direction

The x component of gravity is positive because it's in the +x direction; the friction force is negative because friction in the –x direction. Putting in the equations given for these forces in the paragraphs above gives:

- m·g·sin(θ) – $\mu_k \cdot F_n$ = m·a

To solve for a, just divide both sides by m:

- a = (m·g·sin(θ) – $\mu_k \cdot F_n$)/m

Now, put in values for each variable:

- a = ((7.60 kg)·(9.80 m/s²)·sin(25.2) – 0.410·(67.4 Newtons))/(7.60 kg)

- a = 0.537 m/s²

A few notes on this solution. Remember that 1 Newton = 1 kg·m/s², for units. An acceleration this small would mean that the box speeds up a little bit as it goes down the ramp, but not a lot.

Problem to Try Yourself

A 910 kg car (approximately one ton) is parked on a hill, and is not moving. The hill is shaped like a ramp with angle 11.5° above horizontal. It turns out that the static friction is at its highest possible amount.

A. What is the magnitude of the normal force?
B. What is the coefficient of static friction?

Solution:

Part A:
Look at the y direction forces: F_n (normal force, +y direction) and $m \cdot g \cdot \cos(\theta)$ (y component of gravity, −y direction). The acceleration in the y direction is zero, because the y direction is above and below the ramp, and the car stays on the ramp.

$F_{net} = m \cdot a$

Put in the forces for F_{net}, and a = 0: $F_n - m \cdot g \cdot \cos(\theta) = m \cdot 0$

Add $m \cdot g \cdot \cos(\theta)$ to both sides: $F_n = m \cdot g \cdot \cos(\theta)$

Put in numbers: $F_n = (910 \text{ kg}) \cdot (9.80 \text{ m/s}^2) \cdot \cos(11.5°) = \underline{8.74 \cdot 10^3}$ <u>Newtons</u>

Part B:
Look at the x direction forces: $m \cdot g \cdot \sin(\theta)$ (x component of gravity, +x direction), highest possible static friction force ($\mu_s \cdot F_n$, −x direction). a = 0 because there is no motion.

$F_{net} = m \cdot a$

Put in the forces for F_{net}, and a = 0: $-\mu_s \cdot F_n + m \cdot g \cdot \sin(\theta) = m \cdot 0$

Add $\mu_s \cdot F_n$ to both sides: $\mu_s \cdot F_n = m \cdot g \cdot \sin(\theta)$

Divide both sides by F_n: $\mu_s = m \cdot g \cdot \sin(\theta)/F_n$

Put in numbers: $\mu_s = (910 \text{ kg}) \cdot (9.80 \text{ m/s}^2) \cdot \sin(11.5°)/(8.74 \cdot 10^3$ N) = $\underline{0.203}$

5.9 COMMON FORCE PROBLEMS, TYPE 2: PROBLEMS WITH ELEVATORS

Another common type of force problem is problems with elevators. These problems typically have a person in an elevator that is accelerating (either upward or downward), and usually the question asks for the value of the normal force on the person, from the elevator's floor.

To solve these elevator problems, you use Newton's second law on the person in the elevator. The forces on the person are:

– Gravity, downward (call it the –y direction)

– Normal force, upwards (call it the +y direction)

Using Newton's second law gives:

$$F_n - m{\cdot}g = m{\cdot}a \qquad (5.9.1)$$

In the equation above, F_n is the magnitude of the normal force. Adding m·g to both sides gives a solution for the normal force:

$$F_n = m{\cdot}g + m{\cdot}a \qquad (5.9.2)$$

If the elevator is accelerating upward, m·a is a positive number and the normal force ends up being greater than m·g. If the elevator is accelerating downward, m·a is a negative number and the normal force ends up being less than m·g.

Side note: some textbooks talk about having a scale on the floor, and asking what the reading on the scale would be. The number on the scale is the same as the amount (magnitude) of normal force.

Example 5.9.1

A person with mass 70.0 kg is in an elevator. If the elevator begins to accelerate upward with acceleration 1.50 m/s², what is the magnitude of the normal force on the person from the elevator's floor?

Solution:

In this case, we can apply Newton's second law to the person. The forces on the person are gravity (downward; call it –y) and normal force (upward, call it +y). Newton's second law gives:

$F_n - m \cdot g = m \cdot a$

Next, add m·g to both sides of the equation:

$F_n = m \cdot g + m \cdot a$

Put in the appropriate values:

$F_n = (70.0 \text{ kg}) \cdot (9.80 \text{ m/s}^2) + (70.0 \text{ kg}) \cdot (1.50 \text{ m/s}^2) = \underline{791 \text{ Newtons}}$

Note 1: remember that 1 Newton = 1 kg·m/s². Note 2: note that the acceleration was positive because it was upward. Note 3: we could have just started with Equation 5.9.2 here, but the problem starts from Newton's second law instead (Equation 5.9.1) to emphasize the point that you can solve any problem in Chapter 5 using Newton's second law as the starting point.

Problem to Try Yourself

An elevator accelerates downward at 2.50 m/s². A person standing in the elevator feels a normal force of 1.10·10³ Newtons. What is the person's mass?

Solution:
One major point in this problem is that the acceleration is downward. If we pick upward as positive, then the acceleration is negative.

$F_n - m \cdot g = m \cdot a$

Add m·g to both sides: $F_n = m \cdot g + m \cdot a$

You can rewrite the right side, since m is part of both terms: $F_n = m \cdot (g + a)$

Divide both sides by (g + a): $m = F_n / (g + a)$

Put in numbers, remembering that a is negative: $m = (1.10 \cdot 10^3 \text{ N}) / (9.80 \text{ m/s}^2 - 2.50 \text{ m/s}^2) = \underline{151 \text{ kg}}$

5.10 COMMON FORCE PROBLEMS, TYPE 3: PROBLEMS WITH A ROPE AND PULLEY

A typical force problem with a rope has a setup similar to the following: two masses are connected by a string, which is pulled tight. One mass (call it m_A) is on a horizontal surface; the other mass (call it m_B) is hanging vertically. The rope changes direction from horizontal to vertical as it goes over a pulley. In a typical physics 1 problem, we usually ignore the mass of the pulley and of the string, as well as any friction related to them, and focus on the motion of the two masses. What we expect to have happen, at least at first, is for the mass that is hanging vertically to drop, and pull the mass on the horizontal surface to the right. The presence of the rope and pulley add four changes to the problem, compared to usual:

1. There are two moving objects, and we will apply Newton's second law to each of them.

2. The two blocks move with the same acceleration

3. The rope has what is called a **tension** force in it. This force is the same at both ends, and pulls towards the middle of the rope. In this problem setup, m_A is pulled to the right by the tension force, and m_B is pulled upward.

4. The motion of the two blocks are tied together, and their motion should be labeled as being in the same direction. The mass that is hanging (m_B) will move downward, and the mass that is on a horizontal surface (m_A) will move to the right. What this means is that "to the right" will be positive for m_A, and "down" will be positive for m_B. Having "down" be positive is unusual, but here it is necessary.

Side note: there can be variations on this problem – for example, a third mass hanging to the left.

Example 5.10.1

Two masses are connected by a rope and pulley, with one mass hanging vertically and one on a horizontal surface. Let $m_A = 10.0$ kg and $m_B = 15.0$ kg. Ignore any friction, including between m_A and the horizontal surface. The two masses move with the same acceleration. What is the value of that acceleration, and what is the amount of tension in the rope?

Solution:

Because these objects are connected by a rope that is pulled tight, which goes over a pulley, we can say that the acceleration (call it a) and the rope force (call it F_T for magnitude of tension) are the same. Let's look at the forces on each mass. Let's start with m_B, which is simpler.

Looking at the forces on m_B, there are two forces:

1. gravity, $m_B \cdot g$, downward. Remember from the introduction to rope/pulley problems above that downward is the positive direction for m_B.
2. Tension force (F_T), upward. This will be the negative direction here, since downward is positive.

Applying Newton's second law to m_B gives:

$$m_B \cdot g - F_T = m_B \cdot a$$

Looking at the forces on m_A, there are three forces:

1. gravity, $m_A \cdot g$, downward
2. normal force, upward
3. rope force, F_T, to the right

(Note: if we had included friction, it would have been a fourth force and to the left)

Mass m_A will only move to the right, not up and down. Because there is no motion up and down, we can safely say that the one downward force ($m_A \cdot g$) and the one upward force (normal force) are equal to each other. (We could use Newton's second law applied to the vertical direction here, and we'd get the same result; knowing the normal force would help if we needed to calculate friction.)

For the purpose of this problem, we need to look at the direction that the block is moving, which is the horizontal direction for m_A. Applying Newton's second law, with "to the right" as positive, we get:

$$F_T = m_A \cdot a$$

We now have two equations (the one for m_A and the one for m_B) and we have two unknowns (a and F_T). We can now solve. This problem is similar to the type of math reviewed in Section 13.6. What we want to do first is modify one of the two equations to have either just a and not F_T, or just F_T and not a. One way to do this is to use the equation with m_A ($F_T = m_A \cdot a$), and put in $m_A \cdot a$ where the equation for m_B contains F_T.

The equation with m_B is:

$$m_B \cdot g - F_T = m_B \cdot a$$

If we plug in $m_A \cdot a$ for F_T (using the equation ($F_T = m_A \cdot a$), we get:

$$m_B \cdot g - m_A \cdot a = m_B \cdot a$$

We now want to solve for a. So we get all of the terms with a in it onto one side. Add $m_A \cdot a$ to both sides, to get:

$$m_B \cdot g = m_A \cdot a + m_B \cdot a$$

Next, we can factor out a, because it is in both terms on the right side of the equation. This looks like:

$$m_B \cdot g = (m_A + m_B) \cdot a$$

Next, we can divide by $(m_A + m_B)$ to get a by itself:

$$m_B \cdot g/(m_A + m_B) = a$$

Finally, we can put in numbers for a:

$$a = m_B \cdot g/(m_A + m_B) = (15.0 \text{ kg}) \cdot (9.80 \text{ m/s}^2)/(10.0 \text{ kg} + 15.0 \text{ kg}) = \underline{5.88}$$
$\underline{\text{m/s}^2}$

We now have an answer for a, and just need an answer for F_T. One of our two equations was already in the form $|F_T| = $ something:

$$F_T = m_A \cdot a$$

We can put in numbers to solve it now:

$$F_T = m_A \cdot a = (10.0 \text{ kg}) \cdot (5.88 \text{ m/s}^2) = \underline{58.8 \text{ Newtons}}$$

Problem To Try Yourself

Note: this problem is designed to have a similar physics setup but easier math than the very long example above. Two masses are connected by a rope and a pulley, with one mass (m_A) on a horizontal surface and one mass (m_B) hanging vertically. The tension in the rope is 30.0 N, and the acceleration is 1.00 m/s². What are the values of m_A and m_B? Assume no friction.

Solution:

Look at m_A first, with "to the right" the positive direction. Newton's second law gives: $F_T = m_A \cdot a$

Divide both sides by a and put in numbers: $m_A = F_T/a = 30.0 \text{ N}/1.00$ $\text{m/s}^2 = \underline{30.0 \text{ kg}}$.

Now look at m_B, with "down" the positive direction. Newton's second law gives:

$$m_B \cdot g - F_T = m_B \cdot a$$

Add F_T to both sides, and subtract $m_B \cdot a$ from both sides:

$$m_B \cdot g - m_B \cdot a = F_T$$

Since both terms on the right side have m_B in them, we can rewrite the right side:

$$m_B \cdot (g - a) = F_T$$

Divide both sides by $(g - a)$: $m_B = F_T/(g - a)$

Put in numbers: $m_B = (30.0 \text{ N})/(9.80 \text{ m/s}^2 - 1.00 \text{ m/s}^2) = \underline{3.41 \text{ kg}}$

5.11 COMMON FORCE PROBLEMS, TYPE 4: COMBINING FORCE AND KINEMATICS PROBLEMS

The acceleration (a) in Newton's second law is the same acceleration in the kinematics equations used in Chapters 2 and 3:

$$v_f = v_i + a \cdot \Delta t \tag{2.6.1}$$

$$\Delta x = v_i \cdot \Delta t + \frac{1}{2} \cdot a \cdot (\Delta t)^2 \tag{2.6.2}$$

$$\Delta x = \frac{1}{2} \cdot (v_i + v_f) \cdot \Delta t \tag{2.6.3}$$

$$v_f^2 = v_i^2 + 2 \cdot a \cdot \Delta x \tag{2.6.4}$$

If the forces are constant during the problem, so is the acceleration, and if the acceleration is constant we can use Equations 2.6.1–2.6.4.

These problems are like doing two problems in one. They can go in either order:

1. A kinematics problem to get acceleration, and then using acceleration in a force problem.

2. A force problem to get acceleration, and then using acceleration in a kinematics problem.

Example 5.11.1

A 0.550 kg bird begins to fly horizontally, starting from rest (on a tree branch). The bird accelerates at a constant rate, and is traveling 5.00 m/s after traveling a distance of 7.00 m.

    A. What is the acceleration of the bird?
    B. What is the net force on the bird?

Solution:

Part A is a kinematics problem. We're given $v_i = 0$ (starts at rest), $v_f = 5.00$ m/s, and $\Delta x = 7.00$ m. We want to find a. The equation with a and the variables we know is Equation 2.6.4:

$$v_f^2 = v_i^2 + 2 \cdot a \cdot \Delta x$$

Subtract v_i^2 from both sides: $v_f^2 - v_i^2 = 2 \cdot a \cdot \Delta x$

Divide both sides by $(2 \cdot \Delta x)$: $a = \dfrac{v_f^2 - v_i^2}{2 \cdot \Delta x}$

Put in numbers: $a = \dfrac{\left(5.00 \text{m/s}\right)^2 - 0^2}{2 \cdot \left(7.00 \text{m}\right)} = 1.79 \text{m/s}^2$

Part B is using Newton's second law: $F_{net} = m \cdot a$

Put in numbers: $F_{net} = (0.550 \text{ kg}) \cdot (1.79 \text{ m/s}^2) = \underline{0.982 \text{ Newtons}}$

Note: this problem could have been more difficult by adding an additional step: finding a specific force, given that we now know what F_{net} is.

Problem to Try Yourself

You pull a 4.50 kg dinner platter across a table, to get it closer to you. The platter has four forces on it: normal force (upward), gravity (downward), kinetic friction (to the left, $\mu_k = 0.500$), and the pull (23.0 N, to the right). The platter starts at rest, and only moves horizontally.

A. What is the normal force on the platter?
B. What is the magnitude of the kinetic friction force?
C. What is the x component of acceleration?
D. How far has the platter moved, 2.00 seconds later?

Solution:

Part A:

We eventually want the acceleration in x, but because friction is one of the forces, we need to look at the y direction in order to find the normal force.

Look at the forces in the y direction: $F_n - m \cdot g = m \cdot a$. The acceleration in the y direction is zero, so $F_n - m \cdot g = m \cdot 0$, which is 0.

Add $m \cdot g$ to both sides: $F_n = m \cdot g = (4.50 \text{ kg}) \cdot (9.80 \text{ m/s}^2) = \underline{44.1 \text{ N}}$

Part B:

Now find the friction force in x using Equation 5.5.1:

$$F_{kf} = \mu_k \cdot F_n \qquad\qquad (5.5.1)$$

Put in numbers: $F_{kf} = (0.500) \cdot (44.1 \text{ N}) = \underline{22.1 \text{ N}}$

Part C:

Now look at forces in the x direction. Pick "right" as positive. Pull is to the right, friction is to the left.

$23.0 \text{ N} - 22.1 \text{ N} = m \cdot a$

$0.9 \text{ N} = m \cdot a$

Divide both sides by m, and put in the numbers: $a = 0.9 \text{ N/m} = 0.9 \text{ N}/4.50 \text{ kg} = \underline{0.2 \text{ m/s}^2}$

Part D:
Now do the kinematics problem. We know $v_i = 0$, $a = 0.2$ m/s², and $t = 2.00$ s. We want to find Δx. The equation with Δx and the variables we know is Equation 2.6.2:

$$\Delta x = v_i \cdot \Delta t + \frac{1}{2} \cdot a \cdot (\Delta t)^2$$

Put in numbers: $\Delta x = (0) \cdot (0.500\,\text{s}) + \frac{1}{2} \cdot (0.2\,\text{m}/\text{s}^2)$

$\cdot (2.00\,\text{s})^2 = 0.4\,\text{m}$

5.12 CHAPTER 5 SUMMARY

Chapter 5 covers several types of forces that are often seen, and four types of physics problems that are often seen.

For problems happening in everyday life, **gravity** is straight downward and m·g, where g = 9.80 m/s².

Normal force is perpendicular (90° angle) to the surface where two objects touch (for example, an object and the ground).

Friction comes from surfaces rubbing. It goes in the opposite direction of motion that is happening (kinetic friction), or in the opposite direction of motion that would happen if the object was moving (static friction). Kinetic friction is equal to:

$$F_{kf} = \mu_k \cdot F_n \tag{5.5.1}$$

In this equation, F_{kf} is the magnitude of the kinetic friction force, μ_k is the coefficient of kinetic friction, and F_n is the magnitude of the normal force.

The largest possible (maximum) amount of static friction is a similar equation. However, the static friction can also be a smaller amount, if only a smaller amount is needed in order to keep an object from moving.

$$F_{\text{maximum static friction}} = \mu_s \cdot F_n \tag{5.6.1}$$

$$\text{(static friction force)} = \text{(just enough to not move,}$$
$$\text{if less than maximum is needed)} \tag{5.6.2}$$

In these equations, $F_{\text{maximum static friction}}$ is the largest possible magnitude of static friction. μ_s is the coefficient of static friction (which is similar to the coefficient of kinetic friction), and F_n is the magnitude of the normal force.

Springs have a certain length that they are without any other forces on them, called the **equilibrium length**. If the spring is compressed or stretched, it will put a force on to try and move back to the equilibrium length:

$$F_{\text{spring}} = k \cdot \Delta x \qquad (5.7.1)$$

F_{spring} is the magnitude of the spring force on an object that is touching the spring, k is the spring constant, and Δx is the amount by which the spring is compressed or stretched away from its equilibrium length. The direction of the force is always back towards the equilibrium length.

The first type of force problem covered was a ramp problem, where an object is on a ramp. In a ramp problem, different x and y directions can be used, for example with x "down the ramp" and y perpendicularly out from the ramp. Forces involved usually include normal force (+y direction), friction (x direction), and gravity (split into x and y components):

$$\text{x component of gravity: } m \cdot g \cdot \sin(\theta), \text{ in the +x direction} \qquad (5.8.1)$$

$$\text{y component of gravity: } m \cdot g \cdot \cos(\theta), \text{ in the –y direction} \qquad (5.8.2)$$

Additional forces could also be part of the problem.

The second type of force problem covered was problems using an elevator that is accelerating. For this situation, gravity and normal force are the two forces, and Newton's second law gives:

$$F_n - m \cdot g = m \cdot a \qquad (5.9.1)$$

The third type of force problem covered was problems using a rope and two masses. Mass m_A moves to the right, and mass m_B moves downward. The rope had a **tension** that pulls m_A to the right, and m_B upward. Downward is positive for m_B, so that when the two masses move they are both in the positive direction. The two masses are each part of the problem, and each will need Newton's second law used separately on them. The acceleration is the same for both masses.

The fourth type of force problem covered was when the forces are constant, and the acceleration from a force problem can be used in combination with the kinematics problems from Chapters 2 and 3.

$$v_f = v_i + a \cdot \Delta t \qquad (2.6.1)$$

$$\Delta x = v_i \cdot \Delta t + \frac{1}{2} \cdot a \cdot (\Delta t)^2 \qquad (2.6.2)$$

$$\Delta x = \frac{1}{2} \cdot (v_i + v_f) \cdot \Delta t \qquad (2.6.3)$$

$$v_f^2 = v_i^2 + 2 \cdot a \cdot \Delta x \qquad (2.6.4)$$

Chapter 5 problem types:
Force problems that use Newton's second law with gravity (Section 5.4), friction (Sections 5.5–5.6), springs (Section 5.7), the normal force, and any other forces in the problem.

Ramp problems (Section 5.8)

Problems with an elevator (Section 5.9)

Rope and pulley problems (Section 5.10)

Problems with forces and kinematics (Section 5.11)

Practice Problems:
Section 5.4 (Gravity near Earth)

1. If the force of gravity on a lifting weight is 107 Newtons (near Earth), what is the mass of the object?

   Solution:

   Magnitude of gravity force: m·g, where g = 9.80 m/s² $\qquad$ (5.4.1)

   107 N = m·(9.80 m/s²)

   Divide both sides by 9.80 m/s²: m = (107 N)/(9.80 m/s²) = <u>10.9 kg</u>

2. If the mass of your lunch is 0.500 kg, what is the force of gravity on it?

Solution:

Magnitude of gravity force: m·g, where g = 9.80 m/s²   (5.4.1)

(0.500 kg)·(9.80 m/s²) = 4.90 N

3. A shopping cart (with its wheels stuck and with some groceries inside) has four forces on it: gravity (downward), force from the floor (upward), force from the person shopping (86.0 N, to the right) and friction (75.0 N, to the left). The acceleration is 0.750 m/s² to the right (and zero in the up/down direction).

A.   What is the mass of the cart? Hint: look at the x direction.

B.   What is the force of gravity on the cart?

C.   What is the force from the floor on the cart? Note: this is also called the normal force.

Solution:

Part A:

Apply Newton's second law (F_{net} = m·a, Equation 4.2.1) to the x direction:

0 N + 0 N + 86.0 N − 75.0 N = m·(0.750 m/s²)

The first two forces are both zero because gravity and the force from the floor are only in the y direction.

Add the forces on the left side: 11.0 N = m·(0.750 m/s²)

Divide both sides by 0.750 m/s²: m = (11.0 N)/(0.750 m/s²) = 14.7 kg

Part B:

Magnitude of gravity force: m·g, where g = 9.80 m/s²   (5.4.1)

(14.7 kg)·(9.80 m/s²) = 144 N

Part C:

Apply Newton's second law (F_{net} = m·a, Equation 4.2.1) to the y direction:

−144 N + (force from the floor) + 0 N + 0 N = m·(0 m/s²)

Notes: the third and fourth forces are only in the x direction, so the y component is zero; the y component of acceleration is also zero; gravity is negative because it is downward, which means that this solution is using "upward" as the +y direction.

Add 144 N to both sides: (force from the floor) = 144 N + m·(0 m/ s²) = 144 N

Note: the positive sign on the answer means that it is going upward.

4. A pile of leaves and a tarp are dragged across the yard. Define "to the right" as the +x direction and "up" as the +y direction. It has four forces on it: pull from a person (50.0 N i + 40.0 N j), gravity (0 N i − 80.0 N j), normal force from the ground (0 i + 40.0 N j), and friction (−30.0 N i + 0 N j).

A. What is the mass of the tarp and leaves (as one object)?

B. What is the x component of acceleration of the tarp and leaves?

C. What is the y component of acceleration of the tarp and leaves?

Solution:

Part A:

Magnitude of gravity force: m·g, where g = 9.80 m/s²   (5.4.1)

From the information we are given, the magnitude of the gravity force is 80.0 N.

80.0 N = m·(9.80 m/s²)

Divide both sides by 9.80 m/s²: m = (80.0 N)/(9.80 m/s²) = 8.16 kg

Part B:

Apply Newton's second law (F_{net} = m·a, Equation 4.2.1) to the x direction.

x direction: 50.0 N + 0 N + 0 N − 30.0 N = (8.16 kg)·a

Note: the two 0 N forces are gravity and the normal force, which are only in the y direction.

Combine the numbers on the left side: 20.0 N = (8.16 kg)·a

Divide both sides by 8.16 kg: $\underline{a = (20.0\ N)/(8.16\ kg) = 2.45\ m/s^2\ (x\ component)}$

Part C:

Apply Newton's second law ($F_{net} = m \cdot a$, Equation 4.2.1) to the y direction.

y direction: 40.0 N – 80.0 N + 40.0 N + 0 N = (8.16 kg)·a

Note: the 0 N force is friction, which is only in the x direction.

Combine the numbers on the left side: 0 N = (8.16 kg)·a

Divide both sides by 8.16 kg: $\underline{a = (0\ N)/(8.16\ kg) = 0\ m/s^2\ (y\ component)}$

5. A bag of books from the library is in your hand, as your wait for the bus. The bag is not moving. The bag and books have a mass of 7.50 kg. There are three forces on the bag: gravity (downward), normal force from the ground (50.0 N, upward), and a force from your hand (upward).

   A. What is the magnitude of the force of gravity?

   B. What is the magnitude of the force from your hand? Hint: if the bag is not moving, acceleration is zero.

Solution:

Part A:

   Magnitude of gravity force: m·g, where g = 9.80 m/s²   (5.4.1)

(7.50 kg)·(9.80 m/s²) = $\underline{73.5\ N}$

Part B:

Apply Newton's second law ($F_{net} = m \cdot a$, Equation 4.2.1) to the y direction.

y direction: –73.5 N + 50.0 N + (force from hand) = (7.50 kg)·(0 m/s²)

Notes: The acceleration is zero because the bag is not moving – if it is staying still, velocity is staying at zero and so the acceleration is zero as well. Gravity is negative because it is downward.

Combine the two numbers on the left side, and multiply the numbers on the right side:

−23.5 N + (force from hand) = 0

Add 23.5 N to both sides: <u>(force from hand) = 23.5 N</u>

Section 5.5 (Kinetic friction)

6. The normal force on an object is 207 Newtons. It slides across the ground, with a coefficient of kinetic friction of 0.600. What is the magnitude of the kinetic friction force?

Solution:

$$F_{kf} = \mu_k \cdot F_n \qquad\qquad (5.5.1)$$

Put in numbers: $F_{kf} = (0.600) \cdot (207 \text{ N}) = \underline{124 \text{ N}}$

7. An object slides across the ground, with a coefficient of kinetic friction of 0.800, and the result is a kinetic friction force with magnitude 15.0 Newtons. What was the magnitude of the normal force on the object?

Solution:

$$F_{kf} = \mu_k \cdot F_n \qquad\qquad (5.5.1)$$

Divide both sides by μ_k: $F_n = F_{kf}/\mu_k = (15.0 \text{ N})/(0.800) = \underline{18.8 \text{ N}}$

8. A 900. kg car slides to a stop, as it sees a red light. The forces on the car are gravity (−y direction), normal force (+y direction), and kinetic friction (−x direction). If the coefficient of kinetic friction is 0.600, and the car stays in the same position in the y direction (stays on flat ground),

A. What is the component of acceleration in the y direction?

B. What is the magnitude of the gravity force?

C. What is the magnitude of the kinetic friction force? Hint: find the normal force.

D. What is the magnitude of the acceleration in the x direction?

Solution:

Part A:

There is no motion in the y direction, so the <u>y component of acceleration is zero.</u>

Part B:

Magnitude of gravity force: $m \cdot g$, where $g = 9.80$ m/s²   (5.4.1)

$(900.$ kg$) \cdot (9.80$ m/s²$) = \underline{8.82 \cdot 10^3 \text{ N}}$

Part C:

First find the normal force, then find the kinetic friction force using $F_{kf} = \mu_k \cdot F_n$ (Equation 5.5.1)

To find the normal force, apply Newton's second law ($F_{net} = m \cdot a$, Equation 4.2.1) to the y direction.

$-8.82 \cdot 10^3$ N $+ F_n + 0$ N $+ 0$ N $= (900.$ kg$) \cdot (0$ m/s²$)$

Note: the two zero forces are only in the x direction. F_n is the normal force magnitude.

Add $8.82 \cdot 10^3$ N to both sides, and multiply $(900.$ kg$) \cdot (0$ m/s²$)$:

$F_n = 8.82 \cdot 10^3$ N

The kinetic friction force is $F_{kf} = \mu_k \cdot F_n = (0.600) \cdot (8.82 \cdot 10^3$ N$) = \underline{5.29 \cdot 10^3}$
<u>N</u>

Part D:

Apply Newton's second law ($F_{net} = m \cdot a$, Equation 4.2.1) to the x direction. There is only one force (friction).

$-5.29 \cdot 10^3$ N $= (900$ kg$) \cdot a$

Divide both sides by 900. kg: $a = (-5.29 \cdot 10^3$ N$)/(900.$ kg$) = \underline{-5.88 \text{ m/s}^2}$
(x component)

9. You move a 2.00 kg bag of rice across the kitchen counter. The forces on the bag are gravity (–y direction), normal force (+y direction), pull force from you (12.5 N, +x direction), and friction. If the coefficient of static friction is 0.500, and the counter is flat,

   A. What is the y direction component of the acceleration?

   B. What is the magnitude of the gravity force?

C. What is the magnitude of the kinetic friction force?

D. What is the x direction component of the acceleration?

Solution:

Part A:

There is no motion in the y direction, so the <u>y component of accelera-tion is zero.</u>

Part B:

$$\text{Magnitude of gravity force: } m{\cdot}g, \text{ where } g = 9.80 \text{ m/s}^2 \qquad (5.4.1)$$

$(2.00 \text{ kg}){\cdot}(9.80 \text{ m/s}^2) = \underline{19.6 \text{ N}}$

Part C:

First find the normal force, then find the kinetic friction force using $F_{kf} = \mu_k{\cdot}F_n$ (Equation 5.5.1)

To find the normal force, apply Newton's second law ($F_{net} = m{\cdot}a$, Equation 4.2.1) to the y direction.

$-19.6 \text{ N} + F_n + 0 \text{ N} + 0 \text{ N} = (2.00 \text{ kg}){\cdot}(0 \text{ m/s}^2)$

Note: the two zero forces are only in the x direction. F_n is the normal force magnitude.

Add 19.6 N to both sides, and multiply $(2.00 \text{ kg}){\cdot}(0 \text{ m/s}^2)$:

$F_n = 19.6 \text{ N}$

The kinetic friction force is $F_{kf} = \mu_k{\cdot}F_n = (0.500){\cdot}(19.6 \text{ N}) = \underline{9.80 \text{ N}}$

Part D:

Apply Newton's second law ($F_{net} = m{\cdot}a$, Equation 4.2.1) to the x direc-tion. There is a pull force and friction.

$12.5 \text{ N} - 9.80 \text{ N} = (2.00 \text{ kg}){\cdot}a$

Divide both sides by 2.00 kg: $a = (12.5 \text{ N} - 9.80 \text{ N})/(2.00 \text{ kg}) = \underline{1.4 \text{ m/}}$ <u>s²</u> (x component)

Note: 12.5–9.80 = 2.7 using significant figures; this is why the answer to Part D has two significant figures.

10. You pull a 2.00 kg chair across the floor, with force 9.00 N $i + 8.00$ N j. The other three forces on the chair are gravity, friction, and the normal force. The acceleration of the chair is 1.50 m/s^2 $i + 0$ m/s^2 j.

A. What is the magnitude of the gravity force?

B. What is the magnitude of the normal force?

C. What is the magnitude of the kinetic friction force?

D. What is the coefficient of kinetic friction?

Solution:

Part A:

Magnitude of gravity force: m·g, where g = 9.80 m/s^2 (5.4.1)

(2.00 kg)·(9.80 m/s^2) = <u>19.6 N</u>

Part B:

Apply Newton's second law (F_{net} = m·a, Equation 4.2.1) to the y direction. There is no motion in the y direction, so the y component of acceleration is zero.

8.00 N − 19.6 N + F$_n$ = (2.00 kg)·(0 m/s^2)

Add the two numbers on the left side, and multiply the two numbers on the right side:

−11.6 N + F$_n$ = 0

Add 11.6 N to both sides: <u>F$_n$ = 11.6 N</u>

Part C:

Apply Newton's second law (F_{net} = m·a, Equation 4.2.1) to the x direction. Gravity and the normal force are only in the y direction, and have an x component of zero. Friction is in the −x direction.

9.00 N + 0 N − F$_{kf}$ + 0 N = (2.00 kg)·(1.50 m/s^2)

Add F$_{kf}$ to both sides, and subtract (2.00 kg)·(1.50 m/s^2) from both sides:

F$_{kf}$ = 9.00 N − (2.00 kg)·(1.50 m/s^2) = <u>6.00 N</u>

Part D:

$$F_{kf} = \mu_k F_n \qquad (5.5.1)$$

Divide both sides by F_n: $\mu_k = F_{kf}/F_n$

Put in numbers: $\mu_k = (6.00 \text{ N})/(11.6 \text{ N}) = \underline{0.517}$

11. A 905 kg car comes to a stop, with an acceleration of -3.50 m/s^2. The three forces on the car are kinetic friction ($-x$ direction), normal force from the ground ($+y$ direction) and gravity ($-y$ direction). The car stays on a flat road the entire time, so that there is no motion in the y direction. What is the coefficient of kinetic friction?

Solution:

Since the kinetic friction force is in the x direction, let's try applying Newton's second law ($F_{net} = m \cdot a$, Equation 4.2.1) to the x direction. Only the kinetic friction force is in the x direction.

$-(\text{kinetic friction}) = (905 \text{ kg}) \cdot (-3.50 \text{ m/s}^2)$

The friction force is given a minus sign because it is in the $-x$ direction. Let's now use the equation for kinetic friction force ($F_{kf} = \mu_k \cdot F_n$, Equation 5.5.1) and put it in for kinetic friction:

Divide both sides by $(-F_n)$: $\mu_k = (905 \text{ kg}) \cdot (-3.50 \text{ m/s}^2)/(-F_n)$

Based on this, we can find the coefficient of kinetic friction (μ_k) only if we find normal force.

Let's find the normal force by applying Newton's second law ($F_{net} = m \cdot a$, Equation 4.2.1) to the y direction. The kinetic friction force is only in the x direction, so the y component of friction is zero. There is no motion in the y direction (meaning no up and down motion), so the y component of acceleration is zero.

$0 \text{ N} + F_n - (\text{gravity}) = (905 \text{ kg}) \cdot (0 \text{ m/s}^2)$

Since $905 \cdot 0 = 0$, this gives $F_n - (\text{gravity}) = 0$

Add gravity to both sides: $F_n = (\text{gravity})$.

Gravity is $m \cdot g$, where $g = 9.80 \text{ m/s}^2$ (Equation 5.4.1)

$F_n = m \cdot g = (905 \text{ kg}) \cdot (9.80 \text{ m/s}^2) = 8.87 \cdot 10^3 \text{ N}$

Now that we know the normal force, we can go back to the equation for the coefficient of kinetic friction that we found using Newton's second law and the x direction:

$\mu_k = (905 \text{ kg}) \cdot (-3.50 \text{ m/s}^2)/(-F_n)$

Put in the value of F_n that we found: $\mu_k = (905 \text{ kg}) \cdot (-3.50 \text{ m/s}^2)/(-8.87 \cdot 10^3 \text{ N}) = \underline{0.357}$

Section 5.6 (Static Friction)

12. A potted plant has a mass of 15.0 kg. It has four forces on it: gravity (–y direction), normal force from the ground (+y direction), a force attempting to move the object (+x direction) and a static friction force (–x direction; coefficient of static friction 0.500).

   A. What is the gravity force on the plant?

   B. If the acceleration in the y direction is zero, what is the magnitude of the normal force?

   C. How large can the static friction force get?

Solution:

Part A:

   Magnitude of gravity force: m·g, where g = 9.80 m/s²   (5.4.1)

$(15.0 \text{ kg}) \cdot (9.80 \text{ m/s}^2) = \underline{147 \text{ N}}$

Part B:

Apply Newton's second law ($F_{net} = m \cdot a$, Equation 4.2.1) to the y direction. Two forces are only in the x direction, and so they have y components of zero.

$-147 \text{ N} + F_n + 0 \text{ N} + 0 \text{ N} = (15.0 \text{ kg}) \cdot (0 \text{ m/s}^2)$

Since 15.0·0 = 0, this gives $F_n - 147 \text{ N} = 0$

Add 147 N to both sides: $F_n = \underline{147 \text{ N}}$

Part C:

$$F_{\text{maximum static friction}} = \mu_s \cdot F_n \qquad \qquad (5.6.1)$$

Put in numbers: $F_{\text{maximum static friction}} = (0.500) \cdot (147 \text{ N}) = \underline{73.5 \text{ N}}$

13. A 45.0 kg object has four forces on it: gravity (down), normal force (up), force trying to move the object (to the right) and static friction (to the left). The object is not moving, so the acceleration is zero in both the x and y direction. The coefficient of static friction is not needed. What is the magnitude of the static friction force, if the force trying to move the object is 175 Newtons?

Solution:

Apply Newton's second law ($F_{\text{net}} = m \cdot a$, Equation 4.2.1) to the x direction:

$0 \text{ N} + 0 \text{ N} + 175 \text{ N} - (\text{static friction force}) = (45.0 \text{ kg}) \cdot (0 \text{ m/s}^2)$

This simplifies to: $175 \text{ N} - (\text{static friction force}) = 0$

Add (static friction force) to both sides: <u>(static friction force) = 175 N</u>

Note: this is one of the cases where the maximum static friction force is not needed. Instead, it's the situation in Equation 5.6.2:

(static friction force) = (just enough to not move, if less than maximum is needed)

If the coefficient of friction was 0.500, you could apply Newton's second law to the y direction, you'll find that $F_n = (45.0 \text{ kg}) \cdot (9.80 \text{ m/s}^2) = 441 \text{ N}$, and using this with Equation 5.6.1 gives a maximum static friction force of $(0.500) \cdot (441 \text{ N}) = 221 \text{ N}$. The actual amount needed is smaller than the maximum possible amount.

14. At someone's house, you are filling a glass of water using the front of a refrigerator. To fill the glass with water, you leave the glass so that it is pushing in against a lever. If the glass is too light, the lever will move the glass and the water will shut off. The glass of water has four forces on it: gravity, normal force, a force from the lever (15.0 N)

and a static friction force. If the coefficient of static friction is 0.500, what is the smallest mass of the glass (including some water) before it won't move? Hint: at the smallest mass that won't move, the static friction is at its largest value.

Solution:

When the static friction is at large as it gets, the amount is given by

$$F_{\text{maximum static friction}} = \mu_s \cdot F_n \qquad (5.6.1)$$

This will be the amount of the static friction force. Apply Newton's second law ($F_{net} = m \cdot a$, Equation 4.2.1) to the x direction:

$0\,N + 0\,N + 15.0\,N - \mu_s \cdot F_n = m \cdot (0\ m/s^2)$

Note: the 0 N forces are forces that are only in the y direction (gravity and normal). Add $\mu_s \cdot F_n$ to both sides, and take out the terms that are zero:

$\mu_s \cdot F_n = 15.0\,N$

Divide both sides by μ_s: $F_n = (15.0\ N)/\mu_s = (15.0\ N)/(0.500) = 30.0\,N$

Now, to go from F_n to the mass, apply Newton's second law ($F_{net} = m \cdot a$) to the y direction:

$-(\text{gravity}) + F_n + 0\,N + 0\,N = m \cdot (0\ m/s^2)$

The 0 N forces are only in the x direction. For gravity, let's put in the amount as m·g, from Equation 5.4.1 (Magnitude of gravity force: m·g, with g = 9.80 m/s²):

$-m \cdot g + F_n = 0$

Subtract F_n from both sides: $-m \cdot g = -F_n$

Divide both sides by –g: $m = -F_n/(-g)$

Put in numbers: $m = -30.0\ N/(-9.80\ m/s^2) = \underline{3.06\ kg}$

15. A heavy object has a normal force (from the ground) of 470 Newtons. What is the highest possible amount of static friction force in this situation, if the coefficient of static friction is 0.600?

Solution:

$$F_{\text{maximum static friction}} = \mu_s \cdot F_n \qquad (5.6.1)$$

Put in numbers: $(0.600) \cdot (470 \text{ N}) = \underline{282 \text{ N}}$

16. While shopping, you try to move a heavy bag of concrete mix along the flat floor. The bag has a mass of 30.0 kg. You put a force of 50.0 N of force on the bag, in the horizontal direction. The bag doesn't move. The coefficient of kinetic friction is not needed. What is the magnitude of the static friction force?

Solution:

Apply Newton's second law ($F_{\text{net}} = m \cdot a$, Equation 4.2.1) to the x direction:

$0 \text{ N} + 0 \text{ N} + 50.0 \text{ N} - (\text{static friction force}) = (30.0 \text{ kg}) \cdot (0 \text{ m/s}^2)$

This simplifies to: $50.0 \text{ N} - (\text{static friction force}) = 0$

Add (static friction force) to both sides: (static friction force) = 50.0 N

Note: this is one of the cases where the maximum static friction force is not needed. Instead, it's the situation in Equation 5.6.2:

(static friction force) = (just enough to not move, if less than maximum is needed)

If the coefficient of friction was 0.500, you could apply second law to the y direction, you'll find that $F_n = (30.0 \text{ kg}) \cdot (9.80 \text{ m/s}^2) = 294 \text{ N}$, and using this with Equation 5.6.1 gives a maximum static friction force of $(0.500) \cdot (294 \text{ N}) = 147 \text{ N}$. The actual amount needed is smaller than the maximum possible amount.

17. At a construction site, a bulldozer attempts to move a very heavy rock horizontally along the ground. The rock has four forces on it: gravity, normal force, bulldozer force (+x direction) and friction (−x direction). It turns out that static friction can block up to $1.80 \cdot 10^3$ Newtons of force from the bulldozer. The rock is not moving, as the bulldozer puts $1.80 \cdot 10^3$ Newtons of force on the rock. If the coefficient of static friction is 0.500, what is the mass of the rock?

Solution:

When the static friction is at large as it gets, the amount is given by

$$F_{\text{maximum static friction}} = \mu_s \cdot F_n \qquad (5.6.1)$$

This will be the amount of the static friction force. Apply Newton's second law ($F_{net} = m \cdot a$, Equation 4.2.1) to the x direction:

$$0 \text{ N} + 0 \text{ N} + 1.80 \cdot 10^3 \text{ N} - \mu_s \cdot F_n = m \cdot (0 \text{ m/s}^2)$$

Note: the 0 N forces are forces that are only in the y direction (gravity and normal). Add $\mu_s \cdot F_n$ to both sides, and take out the terms that are zero:

$$\mu_s \cdot F_n = 1.80 \cdot 10^3 \text{ N}$$

Divide both sides by μ_s: $F_n = (1.80 \cdot 10^3 \text{ N})/\mu_s = (1.80 \cdot 10^3 \text{ N})/$
$(0.500) = 3.60 \cdot 10^3 \text{ N}$

Now, to go from F_n to the mass, apply Newton's second law ($F_{net} = m \cdot a$) to the y direction:

$$-(\text{gravity}) + F_n + 0 \text{ N} + 0 \text{ N} = m \cdot (0 \text{ m/s}^2)$$

The 0 N forces are only in the x direction. For gravity, let's put in the amount as m·g, from Equation 5.4.1 (Magnitude of gravity force: m·g, with g = 9.80 m/s²):

$$-m \cdot g + F_n = 0$$

Subtract F_n from both sides: $-m \cdot g = -F_n$

Divide both sides by $-g$: $m = -F_n/(-g)$

Put in numbers: $m = -3.60 \cdot 10^3 \text{ N}/(-9.80 \text{ m/s}^2) = \underline{367 \text{ kg}}$

Section 5.7 (Spring Forces)

18. An object sits on top of a spring. The object is not moving, and has two forces on it: gravity (downward) and a spring force (upward). If the object's mass is 11.0 kg, and the spring compresses by 5.00 cm, what is the spring constant? Note: the length is given in cm; you must answer in N/m.

Solution:

Apply Newton's second law ($F_{net} = m \cdot a$, Equation 4.2.1) to the y direction:

$$F_{spring} - (gravity) = (11.0 \text{ kg}) \cdot (0 \text{ m/s}^2)$$

The acceleration is zero because the object is not moving. We can substitute in for gravity and the spring force using Equations 5.7.1 and 5.4.1:

$$F_{spring} = k \cdot \Delta x \qquad (5.7.1)$$

Magnitude of gravity force: $m \cdot g$, where $g = 9.80 \text{ m/s}^2$ \qquad (5.4.1)

Putting these in gives: $k \cdot \Delta x - m \cdot g = (11.0 \text{ kg}) \cdot (0 \text{ m/s}^2)$

Add $m \cdot g$ to both sides, and multiply out 11·0 to be 0: $k \cdot \Delta x = m \cdot g$

Divide both sides by Δx: $k = m \cdot g / \Delta x$

Put in numbers: $k = (11.0 \text{ kg}) \cdot (9.80 \text{ m/s}^2)/(0.0500 \text{ m}) = \underline{2.16 \cdot 10^3 \text{ N/m}}$

Note: 5.00 cm = 0.0500 m, since 1 m = 100 cm. (See Chapter 1.)

19. An object sits on top of a spring. The object is not moving, and has two forces on it: gravity (downward) and a spring force (upward). If the object's mass is 8.00 kg, and the spring constant is 2000. N/m, how far does the spring compress by?

Solution:

Apply Newton's second law ($F_{net} = m \cdot a$, Equation 4.2.1) to the y direction:

$$F_{spring} - (gravity) = (8.00 \text{ kg}) \cdot (0 \text{ m/s}^2)$$

The acceleration is zero because the object is not moving. We can substitute in for gravity and the spring force using Equations 5.7.1 and 5.4.1:

$$F_{spring} = k \cdot \Delta x \qquad (5.7.1)$$

Magnitude of gravity force: $m \cdot g$, where $g = 9.80 \text{ m/s}^2$ \qquad (5.4.1)

Putting these in gives: $k \cdot \Delta x - m \cdot g = (8.00 \text{ kg}) \cdot (0 \text{ m/s}^2)$

Add m·g to both sides, and multiply out 8·0 to be 0: $k \cdot \Delta x = m \cdot g$

Divide both sides by k: $\Delta x = m \cdot g / k$

Put in numbers: $\Delta x = (8.00 \text{ kg}) \cdot (9.80 \text{ m/s}^2)/(2000. \text{ N/m}) = \underline{3.92 \cdot 10^{-2} \text{ m}}$

20. Inside of a motor, a spring is pulled outward from its normal length. If the spring has been pulled outward by 0.0750 m, and it pulls inward with 507 N of force, what is the spring constant?

Solution:

$$F_{spring} = k \cdot \Delta x \qquad\qquad (5.7.1)$$

Divide both sides by Δx: $k = F_{spring}/\Delta x$

Put in numbers: $k = (507 \text{ N})/(0.0750 \text{ m}) = \underline{6.76 \cdot 10^3 \text{ N/m}}$

Section 5.8 (Ramp Problems)

21. A child is sliding down a playground slide, which is a ramp with incline 20.0° and a coefficient of kinetic friction of 0.300. The child's mass is 23.0 kg. Define the +x direction as "down the ramp", and the +y direction as "perpendicularly away from the slide". The child stays on the slide during the entire motion.

   A. What is the y component of acceleration?

   B. What is the kinetic friction force on the child? Hint: find the normal force first, and ask yourself whether there is any motion in the y direction.

   C. What is the x component of acceleration?

Solution:

Part A:

The y component of acceleration is <u>zero</u>, because there is no motion in the y direction. The y direction is into/off of the slide, and the child stays right on the slide the entire time.

Part B:

To find F_n, we will apply Newton's second law ($F_{net} = m \cdot a$, Equation 4.2.1) to the y direction. In the y direction, the only two forces are the normal force (+y direction) and the y component of gravity:

y component of gravity: $m \cdot g \cdot \cos(\theta)$, in the –y direction   (5.8.2)

Applying Newton's second law gives:

$F_n - m \cdot g \cdot \cos(\theta) = (23.0 \text{ kg}) \cdot (0 \text{ m/s}^2)$

The y component of acceleration is zero, as explained in Part A.
Add $m \cdot g \cdot \cos(\theta)$ to both sides, and multiply out 23·0:
$F_n = m \cdot g \cdot \cos(\theta) + 0$

For now, we will leave the amount of normal force in an equation:
$F_n = m \cdot g \cdot \cos(\theta)$

The equation for kinetic friction force is:

$$F_{kf} = \mu_k \cdot F_n \qquad\qquad (5.5.1)$$

Putting in $F_n = m \cdot g \cdot \cos(\theta)$, which we found earlier in this problem, we get:

$F_{kf} = \mu_k \cdot m \cdot g \cdot \cos(\theta)$

Put in numbers: $F_{kf} = (0.300) \cdot (23.0 \text{ kg}) \cdot (9.80 \text{ m/s}^2) \cdot \cos(20.0°) = \underline{63.5}$
$\underline{N}$

The direction of the friction force is up the ramp (the –x direction), since the child is going down the ramp.

Part C:

Apply Newton's second law ($F_{net} = m \cdot a$, Equation 4.2.1) to the x direction. In the x direction, the only two forces are the kinetic friction force (–x direction) and the x component of gravity (+x direction):

x component of gravity: $m \cdot g \cdot \sin(\theta)$, in the +x direction   (5.8.2)

Applying Newton's second law gives:

$-63.5 \text{ N} + m \cdot g \cdot \sin(\theta) = (23.0 \text{ kg}) \cdot a$

Divide both sides by 23.0 kg: $a = (-63.5 \text{ N} + m \cdot g \cdot \sin(\theta))/(23.0 \text{ kg})$

Put in numbers:
$a = (-63.5 \text{ N} + 23.0 \text{ kg} \cdot 9.80 \text{ m/s}^2 \cdot \sin(20.0°))/(23.0 \text{ kg}) = \underline{0.590 \text{ m/s}^2}$

22. A tired person sits in a recliner after a long day of work. The recliner slides up so that the bottom part of the person's legs (from the knees down) are tilted at a 30.0° angle. We'll look at this as a 7.50 kg object that is on a ramp and not moving, because of friction. (If you try this, you might feel less pull downward as the angle changes.) What is the amount of static friction? Hint: use the gravity force to figure out what the static friction force must be in order to cancel the forces in the other direction. The coefficient is not given because you will not need the equation for maximum amount of force.

Solution:

Apply Newton's second law ($F_{net} = m \cdot a$, Equation 4.2.1) to the x direction. In the x direction, the only two forces are the static friction force (–x direction) and the x component of gravity (+x direction):

x component of gravity: $m \cdot g \cdot \sin(\theta)$, in the +x direction   (5.8.2)

Applying Newton's second law gives:

$-(\text{static friction}) + m \cdot g \cdot \sin(\theta) = (7.50 \text{ kg}) \cdot (0 \text{ m/s}^2)$

The acceleration is zero because the object (leg) is not moving.

Add (static friction) to both sides and multiply out 7.5·0:
$m \cdot g \cdot \sin(\theta) = (\text{static friction}) + 0$

Put in numbers: $(\text{static friction}) = (7.50 \text{ kg}) \cdot (9.80 \text{ m/s}^2) \cdot \sin(30.0°) = \underline{36.8 \text{ N}}$

Note: if you did a similar calculation in the y direction, you'd find that the normal force is equal to 63.7 N. If the static friction force given above was as large as possible, that would give a coefficient of static friction of 0.577.

23. Professional movers slide a heavy couch down a ramp, out of the moving truck. The couch has a mass of 40.0 kg. The coefficient of kinetic friction between the ramp and the couch is 0.500, and the ramp is making an angle of 20.0°. Define the x direction as "down the ramp" (along the ramp), and the y direction as "above and into the slide".

   A. What is the y component of acceleration?

   B. What is the kinetic friction force on the couch?

   C. What is the x component of acceleration?

Solution:

Part A:

The y component of acceleration is <u>zero,</u> because there is no motion in the y direction. The y direction is into/off of the ramp, and the couch stays right on the ramp the entire time.

Part B:

To find F_n, we will apply Newton's second law ($F_{net} = m \cdot a$, Equation 4.2.1) to the y direction. In the y direction, the only two forces are the normal force (+y direction) and the y component of gravity:

   y component of gravity: $m \cdot g \cdot \cos(\theta)$, in the –y direction   (5.8.2)

Applying Newton's second law gives:

$F_n - m \cdot g \cdot \cos(\theta) = (23.0 \text{ kg}) \cdot (0 \text{ m/s}^2)$

The y component of acceleration is zero, as explained in Part A.

Add $m \cdot g \cdot \cos(\theta)$ to both sides, and multiply out 23·0:
$F_n = m \cdot g \cdot \cos(\theta) + 0$

For now, we will leave the amount of normal force in an equation:
$F_n = m \cdot g \cdot \cos(\theta)$

The equation for kinetic friction force is:

$$F_{kf} = \mu_k \cdot F_n \qquad\qquad (5.5.1)$$

Putting in $F_n = m \cdot g \cdot \cos(\theta)$, which we found earlier in this problem, we get:

$F_{kf} = \mu_k \cdot m \cdot g \cdot \cos(\theta)$

Put in numbers: $F_{kf} = (0.500) \cdot (40.0\,\text{kg}) \cdot (9.80\,\text{m/s}^2) \cdot \cos(20.0°) = \underline{184\,\text{N}}$

The direction of the friction force is up the ramp (the –x direction), since the couch is going down the ramp.

Part C:

Apply Newton's second law ($F_{net} = m \cdot a$, Equation 4.2.1) to the x direction. In the x direction, the only two forces are the kinetic friction force (–x direction) and the x component of gravity (+x direction):

$$\text{x component of gravity: } m \cdot g \cdot \sin(\theta), \text{ in the +x direction} \qquad (5.8.2)$$

Applying Newton's second law gives:

$-184\,\text{N} + m \cdot g \cdot \sin(\theta) = (40.0\,\text{kg}) \cdot a$

Divide both sides by 40.0 kg: $a = (-184\,\text{N} + m \cdot g \cdot \sin(\theta))/(40.0\,\text{kg})$

Put in numbers:
$a = (-184\,\text{N} + 40.0\,\text{kg} \cdot 9.80\,\text{m/s}^2 \cdot \sin(20.0°))/(23.0\,\text{kg}) = \underline{-1.25\,\text{m/s}^2}$

The negative number means that the couch slows down as it goes down the ramp.

Section 5.9 (Problems with Elevators)

24. A person in an elevator has a mass of 75.0 kg. If the elevator goes downward with an acceleration of 2.50 m/s² downward, what is the normal force?

    Solution:

    $$F_n - m \cdot g = m \cdot a \qquad (5.9.1)$$

    Add m·g to both sides: $F_n = m \cdot g + m \cdot a$

    a will be negative, since the acceleration is downward.

Put in numbers: $F_n = (75.0 \text{ kg}) \cdot (9.80 \text{ m/s}^2) + (75.0 \text{ kg}) \cdot (-2.50 \text{ m/s}^2) = \underline{548 \text{ N}}$

25. A person in an elevator has a mass of 63.0 kg. If the elevator goes upward with an acceleration of 1.75 m/s² upward, what is the normal force?

Solution:

$$F_n - m \cdot g = m \cdot a \tag{5.9.1}$$

Add m·g to both sides: $F_n = m \cdot g + m \cdot a$

a will be positive, since the acceleration is upward.

Put in numbers: $F_n = (63.0 \text{ kg}) \cdot (9.80 \text{ m/s}^2) + (63.0 \text{ kg}) \cdot (1.75 \text{ m/s}^2) = \underline{728 \text{ N}}$

26. A person is inside of an elevator that is going upward but slowing down. The elevator's acceleration is 2.00 m/s², downward. If the normal force on the person is 915 Newtons, what is the person's mass?

Solution:

$$F_n - m \cdot g = m \cdot a \tag{5.9.1}$$

Add m·g to both sides: $F_n = m \cdot g + m \cdot a$

We can rewrite the right side, since m is in both terms: $F_n = m \cdot (g + a)$

Divide both sides by (g + a): $m = F_n / (g + a)$

The acceleration a will be negative because it is downward.

Put in numbers: $m = (915 \text{ N}) / (9.80 \text{ m/s}^2 - 2.00 \text{ m/s}^2) = \underline{117 \text{ kg}}$

27. A person is moving inside of an elevator. If the person has mass 115.0 kg and the normal force is 989 Newtons, what is the acceleration?

Solution:

$$F_n - m \cdot g = m \cdot a \qquad (5.9.1)$$

Divide both sides by m: $a = (F_n - m \cdot g)/m$

Put in numbers: a = (989 N – 115.0 kg·9.80 m/s²)/(115.0 kg) = <u>–1.20 m/s²</u>

Section 5.10 (Problems with a Rope and a Pulley)

28. A 2.50 kg mass and a 3.75 kg mass are connected by a rope that goes over a pulley. The 3.75 kg mass is hanging over the edge of a table, and the 2.50 kg mass is on the table. Ignore any friction between the 2.50 kg mass and the table. What is the tension in the rope, and the acceleration of both objects? Hint: the acceleration and tension will be the same for both masses.

Solution:

Because these objects are connected by a rope that is pulled tight, which goes over a pulley, we can say that the acceleration (call it a) and the rope force (call it F_T for magnitude of tension) are the same. Let's look at the forces on each mass. Let's start with m_B, which is simpler.

Looking at the forces on m_B, there are two forces:

1. gravity, $m_B \cdot g$, downward. Remember from the introduction to rope/pulley problems above that downward is the positive direction for m_B.

2. Tension force (F_T), upward. This will be the negative direction here, since downward is positive.

Applying Newton's second law to m_B gives:

$m_B \cdot g - F_T = m_B \cdot a$

Looking at the forces on m_A, there are three forces:

1. gravity, $m_A \cdot g$, downward

2. normal force, upward

3. rope force, F_T, to the right

(Note: if we had included friction, it would have been a fourth force and to the left)

Mass m_A will only move to the right, not up and down. Because there is no motion up and down, we can safely say that the one downward force ($m_A \cdot g$) and the one upward force (normal force) are equal to each other. (We could use Newton's second law applied to the vertical direction here, and we'd get the same result; knowing the normal force would help if we needed to calculate friction.)

For the purpose of this problem, we need to look at the direction that the block is moving, which is the horizontal direction for m_A. Applying Newton's second law, with "to the right" as positive, we get:

$$F_T = m_A \cdot a$$

We now have two equations (the one for m_A and the one for m_B) and we have two unknowns (a and F_T). We can now solve. This problem is similar to the type of math reviewed in Section 13.6. What we want to do first is modify one of the two equations to have either just a and not F_T, or just F_T and not a. One way to do this is to use the equation with m_A ($F_T = m_A \cdot a$), and put in $m_A \cdot a$ where the equation for m_B contains F_T.

The equation with m_B is:

$$m_B \cdot g - F_T = m_B \cdot a$$

If we plug in $m_A \cdot a$ for F_T (using the equation ($F_T = m_A \cdot a$), we get:

$$m_B \cdot g - m_A \cdot a = m_B \cdot a$$

We now want to solve for a. So we get all of the terms with a in it onto one side. Add $m_A \cdot a$ to both sides, to get:

$$m_B \cdot g = m_A \cdot a + m_B \cdot a$$

Next, we can factor out a, because it is in both terms on the right side of the equation. This looks like:

$$m_B \cdot g = (m_A + m_B) \cdot a$$

Next, we can divide by ($m_A + m_B$) to get a by itself:

$$m_B \cdot g / (m_A + m_B) = a$$

Finally, we can put in numbers for a:

$a = m_B \cdot g/(m_A + m_B) = (3.75 \text{ kg}) \cdot (9.80 \text{ m/s}^2)/(2.50 \text{ kg} + 3.75 \text{ kg}) = \underline{5.88 \text{ m/s}^2}$

We now have an answer for a, and just need an answer for F_T. One of our two equations was already in the form $F_T =$ something:

$F_T = m_A \cdot a$

We can put in numbers to solve it now:

$F_T = m_A \cdot a = (2.50 \text{ kg}) \cdot (5.88 \text{ m/s}^2) = \underline{14.7 \text{ Newtons}}$

29. A 4.50 kg mass and a 5.75 kg mass are connected by a rope that goes over a pulley. The 5.75 kg mass is hanging over the edge of a table, and the 4.50 kg mass is on the table. Ignore any friction between the 4.50 kg mass and the table. What is the tension in the rope, and the acceleration of both objects? Hint: the acceleration and tension will be the same for both masses.

Solution:

Because these objects are connected by a rope that is pulled tight, which goes over a pulley, we can say that the acceleration (call it a) and the rope force (call it F_T for magnitude of tension) are the same. Let's look at the forces on each mass. Let's start with m_B, which is simpler.

Looking at the forces on m_B, there are two forces:

1. gravity, $m_B \cdot g$, downward. Remember from the introduction to rope/pulley problems above that downward is the positive direction for m_B.

2. Tension force (F_T), upward. This will be the negative direction here, since downward is positive.

Applying Newton's second law to m_B gives:

$m_B \cdot g - F_T = m_B \cdot a$

Looking at the forces on m_A, there are three forces:

1. gravity, $m_A \cdot g$, downward

2. normal force, upward

3. rope force, F_T, to the right

(Note: if we had included friction, it would have been a fourth force and to the left)

Mass m_A will only move to the right, not up and down. Because there is no motion up and down, we can safely say that the one downward force ($m_A \cdot g$) and the one upward force (normal force) are equal to each other. (We could use Newton's second law applied to the vertical direction here, and we'd get the same result; knowing the normal force would help if we needed to calculate friction.)

For the purpose of this problem, we need to look at the direction that the block is moving, which is the horizontal direction for m_A. Applying Newton's second law, with "to the right" as positive, we get:

$$F_T = m_A \cdot a$$

We now have two equations (the one for m_A and the one for m_B) and we have two unknowns (a and F_T). We can now solve. This problem is similar to the type of math reviewed in Section 13.6. What we want to do first is modify one of the two equations to have either just a and not F_T, or just F_T and not a. One way to do this is to use the equation with m_A ($F_T = m_A \cdot a$), and put in $m_A \cdot a$ where the equation for m_B contains F_T.

The equation with m_B is:

$$m_B \cdot g - F_T = m_B \cdot a$$

If we plug in $m_A \cdot a$ for F_T (using the equation ($F_T = m_A \cdot a$), we get:

$$m_B \cdot g - m_A \cdot a = m_B \cdot a$$

We now want to solve for a. So we get all of the terms with a in it onto one side. Add $m_A \cdot a$ to both sides, to get:

$$m_B \cdot g = m_A \cdot a + m_B \cdot a$$

Next, we can factor out a, because it is in both terms on the right side of the equation. This looks like:

$$m_B \cdot g = (m_A + m_B) \cdot a$$

Next, we can divide by $(m_A + m_B)$ to get a by itself:

$$m_B \cdot g/(m_A + m_B) = a$$

Finally, we can put in numbers for a:

$$a = m_B \cdot g/(m_A + m_B) = (5.75 \text{ kg}) \cdot (9.80 \text{ m/s}^2)/(5.75 \text{ kg} + 4.50 \text{ kg}) = \underline{5.50}$$
$$\underline{\text{m/s}^2}$$

We now have an answer for a, and just need an answer for F_T. One of our two equations was already in the form $F_T = $ something:

$$F_T = m_A \cdot a$$

We can put in numbers to solve it now:

$$F_T = m_A \cdot a = (4.50 \text{ kg}) \cdot (5.50 \text{ m/s}^2) = \underline{24.7 \text{ Newtons}}$$

30. A 2.50 kg mass and a 3.75 kg mass are connected by a rope that goes over a pulley. The 3.75 kg mass is hanging over the edge of a table, and the 2.50 kg mass is on the table. The coefficient of kinetic friction between the 2.50 kg mass and the table is 0.500. What is the tension in the rope, and the acceleration of both objects? Hint: the acceleration and tension will be the same for both masses.

Solution:

Because these objects are connected by a rope that is pulled tight, which goes over a pulley, we can say that the acceleration (call it a) and the rope force (call it F_T for magnitude of tension) are the same. Let's look at the forces on each mass. Let's start with m_B.

Looking at the forces on m_B, there are two forces:

1. gravity, $m_B \cdot g$, downward. Remember from the introduction to rope/pulley problems above that downward is the positive direction for m_B.

2. Tension force (F_T), upward. This will be the negative direction here, since downward is positive.

Applying Newton's second law to m_B gives:

$m_B \cdot g - F_T = m_B \cdot a$

Looking at the forces on m_A, there are four forces:

1. gravity, $m_A \cdot g$, downward

2. normal force, upward

3. rope force, F_T, to the right

4. Friction, F_{kf}, to the left

Mass m_A will only move to the right, not up and down. Because there is no motion up and down, the y component of acceleration is equal to zero and so Newton's second law in the y direction becomes $F_n - m_A \cdot g = m_A \cdot 0$. If we add $m_A g$ to both sides, we get that $F_n = m_A \cdot g$. The amount of kinetic friction is $F_{kf} = \mu_k \cdot F_n = \mu_k \cdot m_A \cdot g$

For the purpose of this problem, we need to look at the direction that the block is moving, which is the horizontal direction for m_A. Applying Newton's second law, with "to the right" as positive, we get:

$F_T - \mu_k \cdot m_A \cdot g = m_A \cdot a$

We now have two equations ($F_T - \mu_k \cdot m_A \cdot g = m_A \cdot a$ and $m_B \cdot g - F_T = m_B \cdot a$) and two unknown variables (a and F_T).

Take $F_T - \mu_k \cdot m_A \cdot g = m_A \cdot a$ and add $\mu_k \cdot m_A \cdot g$ to both sides: $F_T = m_A \cdot a + \mu_k \cdot m_A \cdot g$

Put this in for F_T in the second equation: $m_B \cdot g - F_T = m_B \cdot a$ becomes:

$m_B \cdot g - (m_A \cdot a + \mu_k \cdot m_A \cdot g) = m_B \cdot a$

Multiply out the minus sign: $m_B \cdot g - m_A \cdot a - \mu_k \cdot m_A \cdot g = m_B \cdot a$

Add $m_A \cdot a$ to both sides: $m_B \cdot g - \mu_k \cdot m_A \cdot g = m_B \cdot a + m_A \cdot a$

Since both terms on the right side include a, we can rewrite that side:

$m_B \cdot g - \mu_k \cdot m_A \cdot g = (m_B + m_A) \cdot a$

Divide both sides by $(m_B + m_A)$: $a = (m_B \cdot g - \mu_k \cdot m_A \cdot g)/(m_B + m_A)$

Put in numbers: $a = (3.75 \text{ kg} \cdot 9.80 \text{ m/s}^2 - 0.500 \cdot 2.50 \text{ kg} \cdot 9.80 \text{ m/s}^2)/$
$(3.75 \text{ kg} + 2.50 \text{ kg})$

$a = \underline{3.92 \text{ m/s}^2}$

To solve for F_T, use the equation we had before for $F_T = \ldots$

$F_T = m_A \cdot a + \mu_k \cdot m_A \cdot g$

Put in numbers: $F_T = (2.50 \text{ kg}) \cdot (3.92 \text{ m/s}^2) + (0.500) \cdot (2.50 \text{ kg}) \cdot$
$(9.80 \text{ m/s}^2) = \underline{22.1 \text{ N}}$

Section 5.11 (Combining Force and Kinematics Problems)

31. A car has a mass of 905 kg. It accelerates from 15.0 m/s to 25.0 m/s, in a straight line at a constant acceleration. If it travels 32.0 m during this acceleration,

    A. What is the acceleration?
    B. What is the net force on the car?

Solution:

Part A:

To find the acceleration, our options are Newton's second law ($F_{net} = m \cdot a$, Equation 4.2.1) or one of the constant acceleration equations (Equations 2.6.1–2.6.4). We have no information about forces, so it must be one of Equations 2.6.1–2.6.4. We know $v_i = 15.0$ m/s, $v_f = 25.0$ m/s, and $\Delta x = 32.0$ m. We want to find a (acceleration). The equation with the variable that we want to find and the variables that we already know is Equation 2.6.4:

$$v_f^2 = v_i^2 + 2 \cdot a \cdot \Delta x \qquad (2.6.4)$$

Subtract v_i^2 from both sides: $v_f^2 - v_i^2 = 2 \cdot a \cdot \Delta x$

Divide both sides by $(2 \cdot \Delta x)$: $a = \dfrac{v_f^2 - v_i^2}{2 \cdot \Delta x}$

Put in numbers: $a = \dfrac{\left(25.0 \text{m/s}\right)^2 - \left(15.0 \text{m/s}\right)^2}{2 \cdot \left(32.0 \text{m}\right)} = 6.25 \text{ m/s}^2$

Part B:

To find the net force, we use Equation 4.2.1: $F_{net} = m \cdot a$

Put in numbers: $F_{net} = (905 \text{ kg}) \cdot (6.25 \text{ m/s}^2) = \underline{5.66 \cdot 10^3 \text{ N}}$

32. A kite has three forces on it: gravity (mass 0.100 kg), force from the wind ($1.75 \text{ N } i + 2.00 \text{ N } j$), and force from the string ($-1.50 \text{ N } i - 0.800 \text{ N } j$).

   A. What are the x and y components of acceleration?

   B. If the kite starts at velocity ($1.00 \text{ m/s } i + 0.500 \text{ m/s } j$), what are the x and y coordinates of displacement 1.75 seconds later?

Solution:

Part A:

To find the acceleration, our options are Newton's second law ($F_{net} = m \cdot a$, Equation 4.2.1) or one of the constant acceleration equations (Equations 2.6.1–2.6.4). We have lots of information on forces, so we will use Equation 4.2.1:

$F_{net} = m \cdot a$

Divide both sides by m: $a = F_{net}/m$

We will now find F_{net} in each of the x and y directions separately, and divide each by m to get the x and y components of acceleration.

x direction: $(0 \text{ N} + 1.75 \text{ N} - 1.50 \text{ N})/(0.100 \text{ kg}) = \underline{2.50 \text{ m/s}^2}$

Note: the 0 in the x direction is gravity, which is only in the y direction.

y direction: $(-0.100 \text{ kg} \cdot 9.80 \text{ m/s}^2 + 2.00 \text{ N} - 0.800 \text{ N})/$ $(0.100 \text{ kg}) = \underline{2.20 \text{ m/s}^2}$

Note: the $-0.100 \text{ kg} \cdot 9.80 \text{ m/s}^2$ term is gravity (m·g from Equation 5.4.1, negative because it is downward)

Part B:

Now that we have acceleration, we can use Equations 2.6.1–2.6.4 to find displacement. We will do this twice – once for the x direction and once for the y direction. In both cases (both x and y), we know initial velocity (1.00 m/s x, 0.500 m/s y), time (1.75 s for both), and acceleration (2.50 m/s² x, 2.20 m/s² y). The equation that has what we want (Δx) as well as the variables that we already know (v$_i$, t, a) is Equation 2.6.2:

$$\Delta x = v_i \cdot \Delta t + \frac{1}{2} \cdot a \cdot (\Delta t)^2 \qquad (2.6.2)$$

Since the equation is already solved for Δx, we can put in numbers.

x component:

$$\Delta x = (1.00\,m/s) \cdot (1.75s) + \frac{1}{2} \cdot (2.50m/s^2) \cdot (1.75s)^2 = 5.58\,m$$

y component:

$$\Delta x = (0.500\,m/s) \cdot (1.75s) + \frac{1}{2} \cdot (2.20m/s^2) \cdot (1.75s)^2 = 4.24\,m$$

33. A 1805 kg bus is initially traveling 12.0 m/s, but it slows down as it reaches a stop sign. Define the direction the bus is traveling as the positive direction. It comes to a stop after traveling 40.0 m. If the acceleration was constant, and the bus traveled in a straight line,

A. What was the acceleration?

B. What was the net force?

Solution:

Part A:

To find the acceleration, our options are Newton's second law (F$_{net}$ = m·a, Equation 4.2.1) or one of the constant acceleration equations (Equations 2.6.1–2.6.4). We have no information about forces, so it must be one of Equations 2.6.1–2.6.4. We know v$_i$ = 12.0 m/s, v$_f$ = 0 m/s, and Δx = 40.0 m. We want to find a (acceleration). The equation with the variables that we know already and the variable that we want to find is Equation 2.6.4:

$$v_f^2 = v_i^2 + 2 \cdot a \cdot \Delta x \qquad (2.6.4)$$

Subtract v$_i$² from both sides: $v_f^2 - v_i^2 = 2 \cdot a \cdot \Delta x$

Divide both sides by (2·Δx): $a = \dfrac{v_f^2 - v_i^2}{2 \cdot \Delta x}$

Put in numbers: $a = \dfrac{(0\,\text{m}/\text{s})^2 - (12.0\,\text{m}/\text{s})^2}{2 \cdot (40.0\,\text{m})} = -1.80\,\text{m}/\text{s}^2$

Part B:

To find the net force, we use Equation 4.2.1: $F_{net} = m \cdot a$

Put in numbers: $F_{net} = (1805\,\text{kg}) \cdot (-1.80\,\text{m/s}^2) = \underline{-3.25 \cdot 10^3\,\text{N}}$

34. You carry a heavy textbook to class, while walking in a straight line. The book has mass 4.00 kg. There are two forces on the book: gravity, and the force from you. The force from you is 3.00 N i + 39.2 N j. If the book starts from rest, what are the x and y coordinates of the book's velocity 3.00 s later? Define "up" as the positive y direction.

Solution:

We will use the information given about forces to find acceleration, and then use acceleration along with other information in the problem to find the final velocity.

To find the acceleration, we will use Newton's second law ($F_{net} = m \cdot a$, Equation 4.2.1):

$F_{net} = m \cdot a$

Divide both sides by m: $a = F_{net}/m$

We will now find F_{net} in each of the x and y directions separately, and divide each by m to get the x and y components of acceleration.

x direction: $(0\,\text{N} + 3.00\,\text{N})/(4.00\,\text{kg}) = \underline{0.750\,\text{m/s}^2}$

Note: the 0 in the x direction is gravity, which is only in the y direction.

y direction: $(-4.00\,\text{kg} \cdot 9.80\,\text{m/s}^2 + 39.2\,\text{N})/(4.00\,\text{kg}) = \underline{0\,\text{m/s}^2}$

Note: the $-4.00\,\text{kg} \cdot 9.80\,\text{m/s}^2$ term is gravity (m·g from Equation 5.4.1, negative because it is downward)

Now that we have acceleration, we can use Equations 2.6.1–2.6.4 to find displacement. We will do this twice – once for the x direction and once for the y direction. In both cases (both x and y), we know initial velocity (0 m/s in both), time (3.00 s for both), and acceleration

(0.750 m/s² x, 0 m/s² y). We want to find v_f. The equation that has what we want (v_f) as well as the variables that we already know (v_i, t, a) is Equation 2.6.1:

$$v_f = v_i + a \cdot \Delta t \tag{2.6.1}$$

The equation is already solved for v_f, so we can put in numbers.

x component: $v_f = 0 \, m/s + (0.750 \, m/s^2) \cdot (3.00 \, s) = 2.25 \, m/s$

y component: $v_f = 0 \, m/s + (0.600 \, m/s^2) \cdot (3.00 \, s) = 0 \, m/s$

35. An object flies through the air, and gravity is the only force on it.

   A.   What are the x and y components of acceleration?

   B.   What type of problem is this? Hint: it's one that was covered in this textbook, in a different chapter.

   C.   If the object is launched from the ground with an initial velocity of 5.00 m/s, at an angle of 40.0°, how much time does it take for it to land back on the ground?

Solution:

Part A:

$$F_{net} = m \cdot a \tag{4.2.1}$$

Divide both sides by m: $a = F_{net}/m$

Put in numbers for the x and y components separately:

x component: $a = (0 \, N)/(m) = \underline{0 \, m/s^2}$

Note: gravity is only in the y direction, so the x component of the only force is zero.

y component: $a = (m \cdot -9.80 \, m/s^2)/m = \underline{-9.80 \, m/s^2}$

Note: gravity is $-m \cdot g$, where $g = 9.80$ m/s² and m is mass, based on Equation 5.4.1.

Part B:

This is a projectile motion problem. In Chapter 3, we said that if an object is flying through the air with gravity as the only force, the x component of acceleration was 0 m/s² and the y component of acceleration was −9.80 m/s² (negative meaning downward).

Part C:

Look at the y direction. The acceleration is −9.80 m/s², the initial velocity is 5.00 m/s·sin(40.0°) = 3.21 m/s (using Equation 3.3.5), and Δx=0 (since it starts and lands at the same height, the ground). We want to find time (Δt). The equation with the variable that we want to find and the variables that we already know is Equation 2.6.2:

$$\Delta x = v_i \cdot \Delta t + \frac{1}{2} \cdot a \cdot (\Delta t)^2 \qquad (2.6.2)$$

If none of the terms were zero, we would have to use the quadratic formula. But we don't have to because Δx=0.

Putting this in: $0 = v_i \cdot \Delta t + \frac{1}{2} \cdot a \cdot (\Delta t)^2$

Both terms on the right side have a factor of Δt, so we can rewrite it as:

$$0 = \Delta t \cdot \left(v_i + \frac{1}{2} \cdot a \cdot \Delta t \right)$$

This gives two answers: $\Delta t = 0$ and $v_i + \frac{1}{2} \cdot a \cdot \Delta t = 0$

The first answer (Δt = 0) basically tells us that the ball is on the ground at the start. The second answer ($v_i + \frac{1}{2} \cdot a \cdot \Delta t = 0$) tells us how much time passes before it is on the ground again – this is the answer we want. This means that our next step is to solve $v_i + \frac{1}{2} \cdot a \cdot \Delta t = 0$ for Δt.

Subtract v_i from both sides: $\frac{1}{2} \cdot a \cdot \Delta t = -v_i$

Divide both sides by $\frac{1}{2} \cdot a$: $\Delta t = \dfrac{-v_i}{\frac{1}{2} \cdot a}$

Put in numbers: $\Delta t = \dfrac{-(3.21 \text{m/s})}{\frac{1}{2} \cdot (-9.80 \text{m/s}^2)} = 0.656 \text{s}$

Energy, and Work

6.1 INTRODUCTION: MOTION HAS ENERGY

So far, we've looked at motion by describing the way it happens (Chapters 2–3) and by looking at the forces that cause it (Chapters 4–5). We'll now look at motion by talking about energy – something that can go into and out of an object as it moves.

There are many types of energy in the world – for example, the electrical energy that powers computers, phones, lights, cars, etc. In this chapter we will talk specifically about three types of energy that relate to motion: kinetic energy, potential energy, and mechanical energy. But first, we'll start by talking about the physics concept of work.

6.2 WORK BY A CONSTANT FORCE

In physics, we say that a force does **work** when it helps to move an object.

In this section, we'll do a simple case – a constant (non-changing) two-dimensional force **F** on an object that moves by a two-dimensional displacement Δ**x**. (**F** and Δ**x** are bolded because they are vectors; remember from Chapters 2 and 3 that displacement is a change in position.) In this case, the amount of work done by the force **F** is:

$$W = F \cdot \Delta x \cdot \cos(\theta) \tag{6.2.1}$$

In this equation, W is the work, F is the magnitude of force **F**, Δx is the magnitude of vector Δ**x**, and the angle θ is the angle between the direction of **F** and the direction of Δ**x**.

DOI: 10.1201/9781003005049-6

Note that the work is not a vector. None of the work or energy quantities coming up in this chapter are vectors. The units of work are N·m, usually said as "Newton meters". In the equation, one term (force) has units of Newtons, and it is multiplied by a term with units of meters (Δx). The cosine term has no units.

The cosine term in Equation 6.2.1 can make the work positive, negative, or zero.

- $\cos(\theta)$ is a negative number if the angle θ is between 90 and 180 degrees. When the force is opposite to the motion (like with kinetic friction), $\theta = 180$ degrees and $\cos(180°) = -1$.

- $\cos(\theta) = 0$ when $\theta = 90$ degrees. This means that when a force is 90 degrees from the motion (also called perpendicular to it) it is not doing work on an object.

- $\cos(\theta)$ is a positive number when θ is between 0 and 90 degrees.

This equation is written for two dimensions, but it can work just fine for one dimension. In that case, the only options for angle θ are 0 degrees (same direction) or 180 degrees (opposite direction).

Example 6.2.1

You move a couch, so that you can vacuum under it. It takes a force of 154 Newtons, at an angle of 20.0° above "to the right". The couch moves 2.00 m to the right. How much work did you do?

Solution:

$$W = F \cdot \Delta x \cdot \cos(\theta) \qquad (6.2.1)$$

Put in numbers: $W = (154 \text{ N}) \cdot (2.00 \text{ m}) \cdot \cos(20.0°) = \underline{289 \text{ N·m}}$

Problem to Try Yourself

During the same couch motion as Example 6.2.1 above, kinetic friction is 122 Newtons, to the left. How much work is done by the kinetic friction? Hint: be careful about the angle.

Solution:

$$W = F \cdot \Delta x \cdot \cos(\theta) \qquad (6.2.1)$$

The angle is 180°, because the direction of the force ("left") is exactly opposite of the direction of the displacement ("right"). Opposite directions are 180° apart.

Put in numbers: $W = (122 \text{ N}) \cdot (2.00 \text{ m}) \cdot \cos(180°) = \underline{-244 \text{ N·m}}$

6.3 WORK DONE BY SPRINGS, WHERE FORCE CHANGES WITH POSITION

Section 6.2 covers work done by forces that do not change. So far, we have seen exactly one force that changes with position – the spring force discussed in Section 5.7. As a reminder, springs have a length that they would like to be when no other forces are applied (the equilibrium length). Between some initial time point (i) and some final time point (f), the work done by a spring is equal to:

$$W_{spring} = \tfrac{1}{2} \cdot k \cdot x_i^2 - \tfrac{1}{2} \cdot k \cdot x_f^2 \qquad (6.3.1)$$

In this equation, k is the spring constant, x_f is the distance of stretching or compressing (from the equilibrium point) at the final time point, and x_i is the distance of stretching or compressing at the initial time point.

Side point: this equation comes from calculating the area under a graph of force versus distance.

Example 6.3.1 Work Done By a Spring

A spring with spring constant 200 N/m is initially stretched by 0.100 m. Later, it is stretched by 0.500 m. What is the work done between these two time points?

Solution:

$W_{spring} = \tfrac{1}{2} \cdot k \cdot x_f^2 - \tfrac{1}{2} \cdot k \cdot x_i^2 = \tfrac{1}{2} \cdot (200 \text{ N/m}) \cdot (0.100 \text{ m})^2$
$\qquad - \tfrac{1}{2} \cdot (200 \text{ N/m}) \cdot (0.500 \text{ m})^2 = \underline{-24 \text{ N·m}}$

Problem to Try Yourself

In Chapter 5, we approximated the little springs that hold in a battery as being compressed by 0.00500 m and having a spring constant of 40.0 N/m. How much work by the spring would go into compressing one of these springs? Hint: treat the initial time point as not stretched or compressed.

Solution:

$W_{spring} = ½·k·x_i^2 - ½·k·x_f^2 = 0 - ½·(40.0 \text{ N/m})·(0.00500 \text{ m})^2 = \underline{-5.00·10^{-4}}$ N·m

6.4 NET WORK

The **net work** is the total amount of work done on the object, considering all of the forces. There are two ways to calculate this:

1. Calculate the net work done for each force, and add everything up. This is usually the easiest way to calculate net work. In equation form, we can express this using the Greek letter sigma (Σ), which in math means to add everything:

$$W_{net} = Σ \, W_{each force} \quad\quad (6.4.1)$$

   In this equation, W_{net} is the net force

2. If all of the forces are constant during the problem (no springs), then the net work can be expressed in a similar equation to Equation 6.2.1, but now using net force:

$$W_{net} = F_{net}·Δx·\cos(θ) \quad\quad (6.4.2)$$

This equation is just like Equation 6.2.1, but now instead of looking at each force, you look at the net force. The angle θ is now between the direction of the net force, and the direction of the displacement ($Δx$). Remember from Chapter 4 that the net force involves adding all of the forces, taking into account that they are vectors and sometimes have two-dimensional components

Example 6.4.1

In this problem, we will use the same couch moving situation as before. The couch has four forces on it: gravity (down), normal force (up), pull force (154 N, 20.0° above "to the right"), friction (122 N, left). Compute the net work done.

Solution:

If we look at the work done by each of the four forces, we can add them. For gravity and the normal force, the angle between the direction of displacement and force is 90 degrees, and cos(90) = 0.

$$W = F·Δx·\cos(θ)$$

Pull force: $W = (154 \text{ N}) \cdot (2.00 \text{ m}) \cdot \cos(20.0°) = 289 \text{ N·m}$

Friction: $W = (122 \text{ N}) \cdot (2.00 \text{ m}) \cdot \cos(180°) = -244 \text{ N·m}$

Gravity: $W = (m \cdot g) \cdot (2.00 \text{ m}) \cdot \cos(90°) = 0$

Normal: $W = (F_n) \cdot (2.00 \text{ m}) \cdot \cos(90°) = 0$

Add the results: $W_{net} = 289 \text{ N·m} - 244 \text{ N·m} + 0 + 0 = \underline{45 \text{ N·m}}$

Note 1: it doesn't matter what the magnitude of m·g or F_n are, because they are multiplied by zero and become zero.

Note 2: you could compute the net force, in a way very similar to Example 5.5.1. You would find that it is 22.7 N, to the right. You could then use $W_{net} = F_{net} \cdot \Delta x \cdot \cos(\theta)$ (Equation 6.4.2) with $\theta = 0$ since the displacement and net force are both the right. This would give the same answer of 45 N·m, to within two significant figures.

Problem to Try Yourself

Note: this is the same basic situation as the "problem to try yourself" from Section 5.11. You pull a 4.50 kg dinner platter across a table, to get it closer to you. The platter has four forces on it: normal force (upward), gravity (downward), kinetic friction (22.1 N to the left), and the pull (23.0 N, to the right). The platter moves 0.400 m to the right. What is the net work done?

Solution:

$$W = F \cdot \Delta x \cdot \cos(\theta)$$

Pull force: $W = (23.0 \text{ N}) \cdot (0.400 \text{ m}) \cdot \cos(0°) = 9.20 \text{ N·m}$

Friction: $W = (22.1 \text{ N}) \cdot (0.400 \text{ m}) \cdot \cos(180°) = -8.84 \text{ N·m}$

Gravity: $W = (m \cdot g) \cdot (0.400 \text{ m}) \cdot \cos(90°) = 0$

Normal: $W = (F_n) \cdot (0.400 \text{ m}) \cdot \cos(90°) = 0$

$W_{net} = 9.20 \text{ N·m} - 8.84 \text{ N·m} + 0 + 0 = \underline{0.36 \text{ N·m}}$

Note 1: no need to calculate gravity and normal force here, since the cos(90°) makes those terms zero.

Note 2: in that example, it was found that F_{net} was 0.9 N, to the right. We could then use $W_{net} = F_{net} \cdot \Delta x \cdot \cos(\theta)$ (Equation 6.4.2) with $\theta = 0$ since the displacement and net force are both the right. This would give the same answer of 0.36 N·m.

6.5 KINETIC ENERGY

Kinetic energy is a type of energy that moving objects have. The equation for it is:

$$KE = \tfrac{1}{2} \cdot m \cdot v^2 \qquad (6.5.1)$$

In this equation, KE is kinetic energy, m is mass, and v is velocity (more specifically, the magnitude of velocity).

If you are looking at two time points, you can look at the change in kinetic energy:

$$\Delta KE = \tfrac{1}{2} \cdot m \cdot v_f^2 - \tfrac{1}{2} \cdot m \cdot v_i^2 \qquad (6.5.2)$$

In this equation, ΔKE is the change in kinetic energy from some initial time point to some final time point, m is mass, v_f is the velocity (magnitude) at the final time point, and v_i is the velocity (magnitude) at the initial time point.

The unit of kinetic energy is Joules. All energies have units of Joules. Joules are the same as N·m ("Newton meters"), since they have the same base units ($kg \cdot m^2/s^2$), and you can add quantities with N·m to quantities with Joules.

Example 6.5.1 Kinetic Energy Changes

A person is traveling on a bus. At first, the bus is on a local road and the bus is traveling at 12.0 m/s. Later, the bus is on a highway and traveling at 25.0 m/s. If the person has mass 80.0 kg, what is the change in the person's kinetic energy between the local road and the highway?

Solution:

$\Delta KE = \tfrac{1}{2} \cdot m \cdot v_f^2 - \tfrac{1}{2} \cdot m \cdot v_i^2 = \tfrac{1}{2} \cdot (80.0 \text{ kg}) \cdot (25.0 \text{ m/s})^2 - \tfrac{1}{2} \cdot (80.0 \text{ kg}) \cdot (12.0 \text{ m/s})^2 = \underline{1.92 \cdot 10^4 \text{ Joules}}$

Problem to Try Yourself

A 60.0 kg person is walking to class at 1.10 m/s. Suddenly, the person sees the time and starts walking at 1.50 m/s. What is the change in kinetic energy?

Solution:

$\Delta KE = \frac{1}{2}\cdot m\cdot v_f^2 - \frac{1}{2}\cdot m\cdot v_i^2 = \frac{1}{2}\cdot(60.0 \text{ kg})\cdot(1.50 \text{ m/s})^2 - \frac{1}{2}\cdot(60.0 \text{ kg})\cdot(1.10 \text{ m/s})^2 = \underline{31.2 \text{ Joules}}$

6.6 THE WORK-KINETIC ENERGY THEOREM

The work-kinetic energy theorem is that the amount of net work done on an object is equal to its change in kinetic energy. Putting this into an equation:

$$W_{net} = \Delta KE \qquad (6.6.1)$$

We have a definition for ΔKE, which is Equation 6.5.2. If we put this in:

$$W_{net} = \frac{1}{2}\cdot m\cdot v_f^2 - \frac{1}{2}\cdot m\cdot v_i^2 \qquad (6.6.2)$$

Side point: one way of thinking about where this equation comes from is to approximate a one-dimensional equation for work as $W = F\cdot\Delta x$, then combine it with Newton's second law ($F = m\cdot a$) and $v_f^2 = v_i^2 + 2\cdot a\cdot\Delta x$ (Equation 2.6.4) to end up with Equation 6.6.2.

Example 6.6.1

Let's use the moving couch example one last time. We found in Example 6.4.1 that the net work was 45 N·m. If the couch has mass 30.0 kg and starts at rest, how fast is it going after it was moved the 2.00 m distance?

Solution:

$$W_{net} = \frac{1}{2}\cdot m\cdot v_f^2 - \frac{1}{2}\cdot m\cdot v_i^2$$

We are given that $v_i = 0$. Putting this in takes out the term on the right:

$$W_{net} = \frac{1}{2}\cdot m\cdot v_f^2$$

Divide both sides by $\frac{1}{2}\cdot m$: $v_f^2 = W_{net}/(\frac{1}{2}\cdot m)$

Take the square root: $v_f = \sqrt{W_{net}/(\frac{1}{2}\cdot m)}$

Put in numbers: $v_f = \sqrt{(45\,\text{N}\cdot\text{m})/(\frac{1}{2}\cdot 30.0\,\text{kg})} = 1.73\,\text{m/s}$

Note: putting together examples 6.2.1, 6.4.1 and 6.6.1 gives an idea of a longer work problem: calculate the work by a given force, calculate net work, find velocity using the work-kinetic energy theorem.

Problem to Try Yourself

The example with a dinner platter on a table had a 4.5 kg object traveling 0.400 m, and the work needed for this being 0.36 N·m. If it starts at rest, how fast is it going?

Solution:

$$W_{net} = \frac{1}{2}\cdot m\cdot v_f{}^2 - \frac{1}{2}\cdot m\cdot v_i{}^2$$

We are given that $v_i = 0$. Putting this in takes out the term on the right:

$$W_{net} = \frac{1}{2}\cdot m\cdot v_f{}^2$$

Divide both sides by $\frac{1}{2}\cdot m$: $v_f{}^2 = W_{net}/(\frac{1}{2}\cdot m)$

Take the square root: $v_f = \sqrt{W_{net}/(\frac{1}{2}\cdot m)}$

Put in numbers: $v_f = \sqrt{(0.36\,\text{N}\cdot\text{m})/(\frac{1}{2}\cdot 4.50\,\text{kg})} = 0.40\,\text{m/s}$

6.7 POTENTIAL ENERGY

Potential energy is another form of energy. Unlike kinetic energy, potential energy doesn't require motion. *Physics for Scientists and Engineers* by Tipler and Mosca (cited in the acknowledgements section) describes potential energy as something that is related to position.

Only certain forces can be connected to potential energy. These forces are called **conservative forces**. The definition of conservative forces includes the fact that any path that starts and ends at the same place will have zero net work done by that force.

In a typical introductory physics course, you will only see 2–3 forces that are conservative:

- Gravity

- Springs

- Electric force (not covered in this textbook)

In physics theory, the change in potential energy is related to the area under the curve of force versus distance. For springs and gravity, the equations for change in potential energy are:

$$\text{Springs: } \Delta PE = \tfrac{1}{2} \cdot k \cdot x_f^2 - \tfrac{1}{2} \cdot k \cdot x_i^2 \qquad (6.7.1)$$

$$\text{Gravity: } \Delta PE = m \cdot g \cdot \Delta h \qquad (6.7.2)$$

ΔPE is change in potential energy, k is the spring constant, x is the distance that the spring is compressed or stretched from its normal (equilibrium) length, f means final time point, i means initial, m is mass, g is acceleration due to gravity (9.80 m/s^2), and Δh means change in height.

Example 6.7.1

You drop something from your hands, a distance of 1.50 m from the ground. If the object has mass 1.30 kg, what is the change in gravitational potential energy?

Solution:

There is no spring in this problem, so any potential energy change is from gravity.

$$\text{Gravity: } \Delta PE = m \cdot g \cdot \Delta h \qquad (6.7.2)$$

Put in numbers: $\Delta PE = (1.30 \text{ kg}) \cdot (9.80 \text{ m/s}^2) \cdot (-1.50 \text{ m}) = \underline{-19.1 \text{ Joules}}$

Example 6.7.2

You take a spring with spring constant 50.0 N/m, and squeeze it by 0.0150 m (1.50 cm) starting from the equilibrium length. How much potential energy was added?

Solution:

$$\text{Springs: } \Delta PE = \tfrac{1}{2} \cdot k \cdot x_f^2 - \tfrac{1}{2} \cdot k \cdot x_i^2 \qquad (6.7.1)$$

x_i is zero because it starts at the equilibrium length.

Put in numbers: $\Delta PE = \tfrac{1}{2} \cdot (50.0 \text{ N/m}) \cdot (0.0150 \text{ m})^2 - \tfrac{1}{2} \cdot (50.0 \text{ N/m}) \cdot (0)^2 = \underline{5.63 \cdot 10^{-3} \text{ Joules}}$

Note on units: 1 Joule and 1 N·m are the same thing. Think about the work-kinetic energy theorem, where work (units of N·m) is equal to (changes in) kinetic energy (units of Joules).

Problem to Try Yourself

A kite with mass 0.200 kg goes from the ground to a height of 15.0 m. What is the change in potential energy?

Solution:
There is no spring in this problem, so any potential energy change is from gravity.

$$\text{Gravity: } \Delta PE = m \cdot g \cdot \Delta h \qquad (6.7.2)$$

Put in numbers: $\Delta PE = (0.200 \text{ kg}) \cdot (9.80 \text{ m/s}^2) \cdot (15.0 \text{ m}) = \underline{29.4 \text{ Joules}}$

6.8 MECHANICAL ENERGY

Mechanical energy is equal to kinetic energy plus potential energy:

$$ME = KE + PE \qquad (6.8.1)$$

Similarly, the change in mechanical energy is equal to:

$$\Delta ME = \Delta KE + \Delta PE \qquad (6.8.2)$$

Mechanical energy is important because there are physics situations where the total amount of it stays the same, which is called a conservation law in physics. This will be covered in the next section.

Example 6.8.1

Note: this problem is similar to a problem in *Physics for Scientists and Engineers* by Tipler and Mosca, which is cited in the acknowledgements section.

An 80.0 kg person is walking home. At the start, the person is walking 1.50 m/s. At the end, the person is walking 1.20 m/s from being tired. The person has also gone up a hill and changed their vertical position by 106 m. What is the change in mechanical energy?

Solution:
$\Delta ME = \Delta KE + \Delta PE$

The change in potential energy is only from gravity, since there are no springs:

$$\text{Gravity: } \Delta PE = m \cdot g \cdot \Delta h \qquad (6.7.2)$$

The change in kinetic energy is always equal to:

$$\Delta KE = \tfrac{1}{2} \cdot m \cdot v_f^2 - \tfrac{1}{2} \cdot m \cdot v_i^2 \qquad (6.5.2)$$

Add these to get the total change in mechanical energy:

$\Delta ME = \Delta KE + \Delta PE$ becomes

$\Delta ME = \tfrac{1}{2} \cdot m \cdot v_f^2 - \tfrac{1}{2} \cdot m \cdot v_i^2 + m \cdot g \cdot \Delta h$

Put in numbers:

$\Delta ME = \tfrac{1}{2} \cdot (80.0 \text{ kg}) \cdot (1.20 \text{ m/s})^2 - \tfrac{1}{2} \cdot (80.0 \text{ kg}) \cdot (1.50 \text{ m/s})^2 + (80.0 \text{ kg}) \cdot (9.80 \text{ m/s}^2) \cdot (106 \text{ m})$

$\Delta ME = \underline{8.31 \cdot 10^4 \text{ Joules}}$

Problem to Try Yourself

Note: this problem is similar to a problem in *Physics for Scientists and Engineers* by Tipler and Mosca.

A 912 kg airplane is initially at rest, on the ground. It then takes off, and flies at a velocity of 50.0 m/s, at a height of 805 m above the ground. What is the change in mechanical energy?

Solution:

$\Delta ME = \Delta KE + \Delta PE$

The change in potential energy is only from gravity, since there are no springs:

$$\text{Gravity: } \Delta PE = m \cdot g \cdot \Delta h \qquad (6.7.2)$$

The change in kinetic energy is always equal to:

$$\Delta KE = \tfrac{1}{2} \cdot m \cdot v_f^2 - \tfrac{1}{2} \cdot m \cdot v_i^2 \qquad (6.5.2)$$

Add these to get the total change in mechanical energy:

$\Delta ME = \Delta KE + \Delta PE$ becomes

$$\Delta ME = \frac{1}{2} \cdot m \cdot v_f^2 - \frac{1}{2} \cdot m \cdot v_i^2 + m \cdot g \cdot \Delta h$$

Put in numbers:

$$\Delta ME = \frac{1}{2} \cdot (912 \text{ kg}) \cdot (50.0 \text{ m/s})^2 - \frac{1}{2} \cdot (912 \text{ kg}) \cdot (0 \text{ m/s})^2 + (912 \text{ kg}) \cdot (9.80 \text{ m/s}^2) \cdot (805 \text{ m})$$

$$\Delta ME = \underline{8.33 \cdot 10^6 \text{ Joules}}$$

6.9 CONSERVATION OF MECHANICAL ENERGY

In physics, when something is **conserved**, it means that the amount of it stays the same. If you look at a group of objects, called a **system**, the total mechanical energy of the group can be written as:

$$\Delta KE + \Delta PE = W_{nc} \tag{6.9.1}$$

$\Delta KE + \Delta PE$ is the total amount of change in mechanical energy, and W_{nc} is equal to the net work done by non-conservative forces. Non-conservative forces are any forces that are not conservative forces – in this textbook, any forces besides gravity and springs are non-conservative.

Where does this equation come from? You combine the following physics equations: $\Delta KE = W_{net}$ (the work-kinetic energy theorem), $\Delta PE = -W_c$ (a relationship between potential energy and the work done by conservative forces, which called W_c), and $W_{net} = W_{nc} + W_c$ (since all forces are either non-conservative or conservative, the total work is the work done by both put together).

We can make a more specific equation if we combine the following equations:

$$\Delta KE = \frac{1}{2} \cdot m \cdot v_f^2 - \frac{1}{2} \cdot m \cdot v_i^2 \tag{6.5.2}$$

$$\text{Springs: } \Delta PE = \frac{1}{2} \cdot k \cdot x_f^2 - \frac{1}{2} \cdot k \cdot x_i^2 \tag{6.7.1}$$

$$\text{Gravity: } \Delta PE = m \cdot g \cdot \Delta h \tag{6.7.2}$$

$$\Delta KE + \Delta PE = W_{nc} \tag{6.9.1}$$

Putting the first three equations (6.5.2, 6.7.1, 6.7.2) into Equation 6.9.1 gives:

$$\frac{1}{2} \cdot m \cdot v_f^2 - \frac{1}{2} \cdot m \cdot v_i^2 + \frac{1}{2} \cdot k \cdot x_f^2 - \frac{1}{2} \cdot k \cdot x_i^2 + m \cdot g \cdot \Delta h = W_{nc} \tag{6.9.2}$$

In this equation, some terms will often be zero. For example, if there is no spring in the problem, any term with x_f or x_i in it will be zero.

Equations 6.9.1 and 6.9.2 tend to be used in two different ways in order to solve physics problems.

1. When the net work done by non-conservative forces is zero, then the total amount of mechanical energy stays the same between some initial and final time point. In physics, this is called the **conservation of mechanical energy**.

2. If the work done by non-conservative forces is not zero, and the non-conservative forces are constant, Equation 6.2.1 can be used to calculate the term W_{nc}.

One way to think about mechanical energy being conserved is that energy can switch between potential energy and kinetic energy. Picture a person jumping up and down. At the highest point, they are not moving – kinetic energy is zero and potential energy is higher. At lower points, the kinetic energy is higher and the potential energy is lower.

Example 6.9.1

On a nice winter day, you and a friend go tubing. (Tubing is riding a large piece of rubber shaped like a donut down a hill.) Your friend has a mass of 112 kg. The top of the hill is 7.50 m higher than the bottom of the hill. We will ignore any friction, so that only conservative forces are doing work. Your friend is traveling at 1.00 m/s at the top of the hill.

A. How fast is your friend going at the bottom of the hill?
B. After reaching the bottom of the hill, the tube goes onto flat ground and comes to a stop as it bumps into a barrier. The tube compresses as this happens, and we'll model it as if it's a spring compressing. If the tube compresses by 0.250 m, what is the spring constant?

Solution:
Part A:

$$\tfrac{1}{2} \cdot m \cdot v_f^2 - \tfrac{1}{2} \cdot m \cdot v_i^2 + \tfrac{1}{2} \cdot k \cdot x_f^2 - \tfrac{1}{2} \cdot k \cdot x_i^2 + m \cdot g \cdot \Delta h = W_{nc} \quad (6.9.2)$$

The top of the hill is the initial time point, and the bottom of the hill is the final time point. We know m = 112 kg, v_i = 1.00 m/s, g = 9.80

m/s², and $\Delta h = -7.50$ m (going from top to bottom of the hill). $W_{nc} = 0$ because only conservative forces are doing work. The spring terms $(\frac{1}{2} \cdot k \cdot x_f^2 - \frac{1}{2} \cdot k \cdot x_i^2)$ are 0 because there is no spring.

If we take out the terms that are 0, we're left with:

$$\frac{1}{2} \cdot m \cdot v_f^2 - \frac{1}{2} \cdot m \cdot v_i^2 + m \cdot g \cdot \Delta h = 0$$

We want to solve for v_f. To do this, add $\frac{1}{2} \cdot m \cdot v_i^2$ to both sides and subtract $m \cdot g \cdot \Delta h$ from both sides:

$$\frac{1}{2} \cdot m \cdot v_f^2 = \frac{1}{2} \cdot m \cdot v_i^2 - m \cdot g \cdot \Delta h$$

Divide both sides by $\frac{1}{2} \cdot m$: $v_f^2 = \dfrac{\frac{1}{2} \cdot m \cdot v_i^2 - m \cdot g \cdot \Delta h}{\frac{1}{2} \cdot m}$

Take the square root: $v_f = \pm \sqrt{\dfrac{\frac{1}{2} \cdot m \cdot v_i^2 - m \cdot g \cdot \Delta h}{\frac{1}{2} \cdot m}}$

We will take the positive root here. The problem asks "how fast", which we will take to mean the magnitude of velocity. (We also didn't need to specify direction in any other part of this problem.)

Put in numbers:

$$v_f = \sqrt{\dfrac{\frac{1}{2} \cdot (112\,kg) \cdot (1.00\,m/s)^2 - (112\,kg) \cdot (9.80\,m/s^2) \cdot (-7.50\,m)}{\frac{1}{2} \cdot (112\,kg)}}$$

$= \underline{12.2 \text{ m/s}}$

Part B:
Now, the initial time is just before the spring (with $v = 12.2$ m/s), and the final time is when the spring has stopped the motion ($v = 0$).

$$\frac{1}{2} \cdot m \cdot v_f^2 - \frac{1}{2} \cdot m \cdot v_i^2 + \frac{1}{2} \cdot k \cdot x_f^2 - \frac{1}{2} \cdot k \cdot x_i^2 + m \cdot g \cdot \Delta h = W_{nc} \qquad (6.9.2)$$

Here, we know $m = 112$ kg, $v_i = 12.2$ m/s (from Part A), $v_f = 0$ (it stops), $x_i = 0$ (not compressed at the start), $x_f = 0.250$ m (from the problem), and we want to find k. W_{nc} is zero since only conservative forces are doing work, and Δh is zero because this part happens on flat ground.

If we take out the zero terms:

$$- \frac{1}{2} \cdot m \cdot v_i^2 + \frac{1}{2} \cdot k \cdot x_f^2 = 0$$

To solve for k, first add $\frac{1}{2} \cdot m \cdot v_i^2$ to both sides: $\frac{1}{2} \cdot k \cdot x_f^2 = \frac{1}{2} \cdot m \cdot v_i^2$

Divide both sides by ($\frac{1}{2} \cdot x_f^2$): $k = \dfrac{\frac{1}{2} \cdot m \cdot v_i^2}{\frac{1}{2} \cdot x_f^2}$

Put in numbers: $k = \dfrac{\frac{1}{2} \cdot (112\,\text{kg}) \cdot (12.2\,\text{m/s})^2}{\frac{1}{2} \cdot (0.250\,\text{m})^2} = \underline{2.65 \cdot 10^5\ \text{N/m}}$

Example 6.9.2

It is the same situation as Example 6.9.1 – a 112 kg person riding a tube down a hill that is 7.50 m higher at the top than at the bottom, starting at 1.00 m/s at the top. In Part A of Example 6.9.1, we found that the person is traveling 12.2 m/s. You now try this in real life, and find that the person is only actually going 9.00 m/s. For this to be true, how much work must have been done by non-conservative forces? (This could include friction and air resistance.)

Solution:

$$\tfrac{1}{2} \cdot m \cdot v_f^2 - \tfrac{1}{2} \cdot m \cdot v_i^2 + \tfrac{1}{2} \cdot k \cdot x_f^2 - \tfrac{1}{2} \cdot k \cdot x_i^2 + m \cdot g \cdot \Delta h = W_{nc} \quad (6.9.2)$$

This time, we are solving for W_{nc}, which is now not zero. We're given that m = 112 kg, v_i = 1.00 m/s, v_f = 9.00 m/s, Δh = –7.50 m, and both spring terms ($\frac{1}{2} \cdot k \cdot x_f^2 - \frac{1}{2} \cdot k \cdot x_i^2$) are zero because there is no spring. Taking out the zero terms, this leaves:

$$\tfrac{1}{2} \cdot m \cdot v_f^2 - \tfrac{1}{2} \cdot m \cdot v_i^2 + m \cdot g \cdot \Delta h = W_{nc}$$

This equation is already solved for W_{nc}, so we just put in numbers:
$W_{nc} = \frac{1}{2} \cdot (112\,\text{kg}) \cdot (9.00\,\text{m/s})^2 - \frac{1}{2} \cdot (112\,\text{kg}) \cdot (1.00\,\text{m/s})^2 + (112\,\text{kg}) \cdot (9.80$ $\text{m/s}^2) \cdot (-7.50\,\text{m}) = \underline{-3.75 \cdot 10^{-3}\ \text{N·m}}$

Note 1: the work is negative, which makes sense because friction is always in the opposite direction of motion (think about our example in Section 6.2).

Note 2: N·m and Joules are the same basic units. Most books seem to use N·m for work and Joules for energy.

Note 3: this problem could be made harder by combining the work done with Equation 6.2.1 ($W = F \cdot \Delta x \cdot \cos(\theta)$) and asking for something like the amount of force.

Problem to Try Yourself

Inside of a pinball machine, a compressed spring suddenly goes back to its equilibrium length as it sends the pinball up a ramp. From an energy point of view, think of this as the potential energy in the spring being turned into both gravitational potential energy (as it goes up the ramp) and kinetic energy (the ball moves). The spring has spring constant 60.0 N/m, and was initially compressed by 0.0250 m. The ball has mass 0.300 kg, and starts at rest. Assume that friction is small enough to ignore. At the instant where the spring is back at equilibrium, the height has increased by 0.00350 m. At this point, what is the magnitude of the velocity of the ball?

Solution:

$$\tfrac{1}{2}\cdot m\cdot v_f^2 - \tfrac{1}{2}\cdot m\cdot v_i^2 + \tfrac{1}{2}\cdot k\cdot x_f^2 - \tfrac{1}{2}\cdot k\cdot x_i^2 + m\cdot g\cdot \Delta h = W_{nc} \quad (6.9.2)$$

We are given $m = 0.300$ kg, $v_i = 0$ (starts at rest), $x_f = 0$ (spring is back at equilibrium), $x_i = 0.0250$ m, $k = 60.0$ N/m, $g = 9.80$ m/s², $\Delta h = +0.00350$ m (height increases), and $W_{nc} = 0$ (friction is small enough to ignore). Taking out the zero terms, we are left with:

$$\tfrac{1}{2}\cdot m\cdot v_f^2 - \tfrac{1}{2}\cdot k\cdot x_i^2 + m\cdot g\cdot \Delta h = 0$$

We want to solve for v_f. Add $\tfrac{1}{2}\cdot k\cdot x_i^2$ to both sides, and subtract $m\cdot g\cdot \Delta h$ from both sides:

$$\tfrac{1}{2}\cdot m\cdot v_f^2 = \tfrac{1}{2}\cdot k\cdot x_i^2 - m\cdot g\cdot \Delta h$$

Divide both sides by $\tfrac{1}{2}\cdot m$: $v_f^2 = \dfrac{\tfrac{1}{2}\cdot k\cdot x_i^2 - m\cdot g\cdot \Delta h}{\tfrac{1}{2}\cdot m}$

Take the square root: $v_f = \pm\sqrt{\dfrac{\tfrac{1}{2}\cdot k\cdot x_i^2 - m\cdot g\cdot \Delta h}{\tfrac{1}{2}\cdot m}}$

We will take the positive root here, because the problem asks for the magnitude.

Put in numbers:

$$v_f = \sqrt{\dfrac{\tfrac{1}{2}\cdot(60.0\,\mathrm{N/m})\cdot(0.0250\,\mathrm{m})^2 - (0.300\,\mathrm{kg})\cdot(9.80\,\mathrm{m/s^2})\cdot(0.00350\,\mathrm{m})}{\tfrac{1}{2}\cdot(0.300\,\mathrm{kg})}}$$

$$= \underline{0.237\ \mathrm{m/s}}$$

6.10 POWER

Power is a physics quantity that appears often in everyday life – for example, the power output by a battery or the power of an engine.

You can calculate power in three different ways:

$$\text{Power} = (\text{work done})/(\text{time}) \qquad (6.10.1)$$

$$\text{Power} = (\text{energy change})/(\text{time}) \qquad (6.10.2)$$

$$\text{Power} = F \cdot v \cdot \cos(\theta) \qquad (6.10.3)$$

In Equation 6.10.3, F is the magnitude of force, v is magnitude of velocity, and θ is the angle between the directions of force and velocity. This equation is similar to Equation 6.2.1, divided by time. The unit of power is a Watt, which is the same thing as 1 $kg \cdot m^2/s^3$, which is the same thing as 1 N·m/s or 1 Joule/s.

Example 6.10.1

A person throws a baseball at 20.0 m/s, with a force of 1.25 N, with force and velocity in the same direction. What is the power?

Solution:

Here, we have force and velocity, so we use:

$$\text{Power} = F \cdot v \cdot \cos(\theta) \qquad (6.10.3)$$

Put in numbers: Power = (1.25 N)·(20.0 m/s)·cos(0) = <u>25.0 Watts</u>

Note: a harder power problem might ask you to calculate work done (Sections 6.2–6.4) or energy change (Sections 6.5–6.9) first, and then divide by change in time.

Problem to Try Yourself

You bowl a bowling ball at 5.00 m/s, with a force of 80.5 N in the same direction as the velocity. What is the power?

Solution:

Here, we have force and velocity, so we use:

$$\text{Power} = F \cdot v \cdot \cos(\theta) \qquad (6.10.3)$$

Put in numbers: Power = (80.5 N)·(5.00 m/s)·cos(0) = <u>403 Watts</u>

6.11 CHAPTER 6 SUMMARY

Energy is involved in motion.

Work done by a constant force (that stays the same during the problem) is:

$$W = F \cdot \Delta x \cdot \cos(\theta) \qquad (6.2.1)$$

In this equation, W is the work, F is the magnitude of force **F**, Δx is the magnitude of vector $\mathbf{\Delta x}$, and the angle θ is the angle between the direction of **F** and the direction of $\mathbf{\Delta x}$.

Work done by a spring force is:

$$W_{spring} = \tfrac{1}{2} \cdot k \cdot x_i^2 - \tfrac{1}{2} \cdot k \cdot x_f^2 \qquad (6.3.1)$$

In this equation, k is the spring constant, x_f is the distance of stretching or compressing at the final time point, and x_i is the distance of stretching or compressing at the initial time point.

The **net (total) work** on an object can be calculated in two different ways:

$$W_{net} = \Sigma \, W_{each \, force} \qquad (6.4.1)$$

$$W_{net} = F_{net} \cdot \Delta x \cdot \cos(\theta) \qquad (6.4.2)$$

The first equation means to add the work from each force, and the second equation is the same as Equation 6.2.1 except for using net force instead of one force at a time. Equation 6.4.2 is for constant forces only.

Kinetic energy happens when things are moving:

$$KE = \tfrac{1}{2} \cdot m \cdot v^2 \qquad (6.5.1)$$

In this equation, KE is kinetic energy, m is mass, and v is velocity (more specifically, the magnitude of velocity).

If you are looking at two time points, you can look at the change in kinetic energy:

$$\Delta KE = \tfrac{1}{2} \cdot m \cdot v_f^2 - \tfrac{1}{2} \cdot m \cdot v_i^2 \qquad (6.5.2)$$

Kinetic energy and net work are related through the **work-kinetic energy theorem**:

$$W_{net} = \Delta KE \qquad (6.6.1)$$

$$W_{net} = \tfrac{1}{2} \cdot m \cdot v_f^2 - \tfrac{1}{2} \cdot m \cdot v_i^2 \qquad (6.6.2)$$

Equation 6.6.2 comes from putting the equation for ΔKE (Equation 6.5.2) into Equation 6.6.1.

Potential energy comes from conservative forces, which include gravity and springs.

$$\text{Springs: } \Delta PE = \tfrac{1}{2} \cdot k \cdot x_f^2 - \tfrac{1}{2} \cdot k \cdot x_i^2 \qquad (6.7.1)$$

$$\text{Gravity: } \Delta PE = m \cdot g \cdot \Delta h \qquad (6.7.2)$$

Mechanical energy is equal to kinetic energy plus potential energy:

$$ME = KE + PE \qquad (6.8.1)$$

Similarly, the change in mechanical energy is equal to:

$$\Delta ME = \Delta KE + \Delta PE \qquad (6.8.2)$$

For a group of objects, the change in mechanical energy and the work done by non-conservative forces (W_{nc}) are related by:

$$\Delta KE + \Delta PE = W_{nc} \qquad (6.9.1)$$

$$\tfrac{1}{2} \cdot m \cdot v_f^2 - \tfrac{1}{2} \cdot m \cdot v_i^2 + \tfrac{1}{2} \cdot k \cdot x_f^2 - \tfrac{1}{2} \cdot k \cdot x_i^2 + m \cdot g \cdot \Delta h = W_{nc} \qquad (6.9.2)$$

When W_{nc} is zero, the total amount of mechanical energy stays the same. We call this the **conservation of mechanical energy.**

You can calculate power in three different ways:

$$\text{Power} = (\text{work done})/(\text{time}) \qquad (6.10.1)$$

$$\text{Power} = (\text{energy change})/(\text{time}) \qquad (6.10.2)$$

$$\text{Power} = F \cdot v \cdot \cos(\theta) \qquad (6.10.3)$$

Chapter 6 problem types:

Calculate work, for constant forces (Section 6.2)

Calculate work for a spring (Section 6.3)

Calculate net work (Section 6.4)

Calculate the amount (or change in the amount) of kinetic energy (Section 6.5)

Problems using the work-kinetic energy theorem (Section 6.6)

Calculate potential energy changes due to gravity or springs (Section 6.7)

Calculate mechanical energy (Section 6.8)

Problems using the conservation of mechanical energy, or that the change in mechanical energy equals the work done by non-conservative forces (Section 6.9)

Problems calculating with power (Section 6.10)

Practice Problems
Section 6.2 (Work Done by Constant Forces), Section 6.3 (Work Done by Springs) and Section 6.4 (Net Work)

1. At a picnic, you drag a blanket and lunch basket across the ground. There are four forces on them: gravity (mass 3.00 kg), normal force from the ground (23.0 N, upward), force from you (14.0 N, at an angle of 27.0° above "to the right"), and friction (to the left; coefficient of kinetic friction 0.500). You drag the blanket 1.00 m, to the right.

   A. Calculate the work done by the force from you

   B. Calculate the magnitude of the force of friction.

   C. Calculate the work done by friction

   D. Explain why the work done by gravity and the normal force are both zero.

   E. Calculate the net work

Solution:

Part A:

$$W = F \cdot \Delta x \cdot \cos(\theta)$$
(6.2.1)

Put in numbers: $W = (14.0 \text{ N}) \cdot (1.00 \text{ m}) \cdot \cos(27.0°) = \underline{12.5 \text{ N·m}}$

Part B:

$$F_{kf} = \mu_k \cdot F_n$$
(5.5.1)

Put in numbers: $F_{kf} = (0.500) \cdot (23.0 \text{ N}) = \underline{11.5 \text{ N}}$

Part C:

$$W = F \cdot \Delta x \cdot \cos(\theta)$$
(6.2.1)

Put in numbers: $W = (11.5 \text{ N}) \cdot (1.00 \text{ m}) \cdot \cos(180°) = \underline{-11.5 \text{ N·m}}$

Note: the angle is 180° because the direction of the displacement (to the right) and the direction of friction (to the left) are 180° apart.

Part D:

The directions of gravity (downward) and normal force (upward) are each 90° away from the direction of the displacement (to the right). That would make the angle in

$$W = F \cdot \Delta x \cdot \cos(\theta)$$
(6.2.1)

equal to 90°, and $\cos(90°) = 0$.

Part E:

Add the work done by each force: 12.5 N·m – 11.5 N·m + 0 N·m + 0 N·m = $\underline{1.0 \text{ N·m}}$

2. A child sits on a toy horse that is on top of a spring. The spring compresses downward by 0.130 m when the child gets onto the toy horse. The spring has spring constant $1.51 \cdot 10^3$ N/m. How much work was done by the spring during the compressing?

Solution:

$$W_{spring} = \tfrac{1}{2} \cdot k \cdot x_i^2 - \tfrac{1}{2} \cdot k \cdot x_f^2$$
(6.3.1)

Put in numbers: $W_{spring} = \tfrac{1}{2} \cdot (1.51 \cdot 10^3 \text{ N/m}) \cdot (0 \text{ m})^2 - \tfrac{1}{2} \cdot (1.51 \cdot 10^3 \text{ N/m}) \cdot (0.130 \text{ m})^2$

$W_{spring} = \underline{-12.8 \text{ N·m}}$

Note on the sign: a negative sign makes sense because the spring is moving inward as it compresses, but the spring force is outward (towards its equilibrium point). That would be like an angle of 180° in Equation 6.2.1 ($W = F \cdot \Delta x \cdot \cos(\theta)$), which would mean a negative number.

3. A ball with mass 0.100 kg is kicked straight upward, from the ground. It is caught by a goalie at a height of 1.50 m above the ground. If gravity is the only force on the ball as it travels through the air, what is the work done?

Solution:

$$W = F \cdot \Delta x \cdot \cos(\theta) \tag{6.2.1}$$

Here, F is gravity, equal to m·g using Equation 5.4.1. Gravity is downward and the displacement is upward.

Put in numbers: $W = (0.100 \text{ kg·}9.80 \text{ m/s}^2) \cdot (1.50 \text{ m}) \cdot \cos(180°) = \underline{-1.47 \text{ N·m}}$

4. At a summer barbeque, you drag a chair across some grass. The chair has four forces on it: gravity (mass 2.00 kg), normal force from the ground (19.6 N, upward), force from you (11.0 N, to the right), and friction (coefficient of kinetic friction 0.500). You drag the chair 1.50 m, to the right.

A. Calculate the work done by each force

B. Calculate the net work

Solution:

Part A:

$$W = F \cdot \Delta x \cdot \cos(\theta) \tag{6.2.1}$$

Force from you: $W = (11.0 \text{ N}) \cdot (1.50 \text{ m}) \cdot \cos(0°) = \underline{16.5 \text{ N·m}}$

Friction force is $F_{kf} = \mu_k \cdot F_n$, from Equation 5.5.1, and the direction of the friction force (left) is 180° from the direction of the displacement (right).

Friction: W = (0.500·19.6 N)·(1.50 m)·cos(180°) = −14.7 N·m

Normal force and gravity both have work=0, since they are 90° different in direction from the displacement and cos(90°) = 0.

Part B:

Add the work from each force: 16.5 N·m − 14.7 N·m + 0 N·m + 0 N·m = 1.8 N·m

5. A spring with spring constant $1.51 \cdot 10^3$ N/m, is compressed by 0.185 m, with a child (and toy horse) sitting on it. Then, the child gets off and the spring decompresses. What is the work done by the spring during the decompression?

Solution:

$$W_{spring} = \tfrac{1}{2} \cdot k \cdot x_i^2 - \tfrac{1}{2} \cdot k \cdot x_f^2 \qquad (6.3.1)$$

Put in numbers: $W_{spring} = \tfrac{1}{2} \cdot (1.92 \cdot 10^3 \text{ N/m}) \cdot (0.185 \text{ m})^2 - \tfrac{1}{2} \cdot (1.92 \cdot 10^3 \text{ N/m}) \cdot (0 \text{ m})^2$

$W_{spring} = 32.9$ N·m

Note on the sign: a positive sign makes sense because the spring is moving outward as it decompresses, and the spring force is also outward. That would be like an angle of 0° in Equation 6.2.1 ($W = F \cdot \Delta x \cdot \cos(\theta)$)

6. You carry a heavy textbook to class, while walking in a straight line. The book has mass 4.00 kg. There are two forces on the book: gravity, and the force from you. The force from you is 39.3 N, at an angle of 85.6° above "to the right". If you move the book 3.38 meters to the right (+x direction), what is the net work done?

Solution:

$$W = F \cdot \Delta x \cdot \cos(\theta) \qquad (6.2.1)$$

The work done by gravity is zero, since it is 90° different in direction from the displacement and cos(90°) = 0.

Force from you: W = (39.3 N)·(3.38 m)·cos(85.6°) = 10.2 N·m

Net work: 0 N·m + 10.2 N·m = 10.2 N·m

Section 6.5 (Kinetic Energy)

7. A 80.0 kg hockey player skates at 5.00 m/s. What is the player's kinetic energy?

Solution:

$$KE = \tfrac{1}{2} \cdot m \cdot v^2 \qquad (6.5.1)$$

Put in numbers: $KE = \tfrac{1}{2} \cdot (80.0 \text{ kg}) \cdot (5.00 \text{ m/s})^2 = \underline{1.00 \cdot 10^3 \text{ Joules}}$

8. A 905 kg car is driving on the highway, at 20.0 m/s, when a traffic light turns red. The car slows down to a stop. What is the change in kinetic energy?

Solution:

$$\Delta KE = \tfrac{1}{2} \cdot m \cdot v_f^2 - \tfrac{1}{2} \cdot m \cdot v_i^2 \qquad (6.5.2)$$

Put in numbers: $\Delta KE = \tfrac{1}{2} \cdot (905 \text{ kg}) \cdot (0 \text{ m/s})^2 - \tfrac{1}{2} \cdot (905 \text{ kg}) \cdot (20.0 \text{ m/s})^2 = \underline{-1.81 \cdot 10^5 \text{ Joules}}$

9. An 1811 kg bus speeds up from 13.0 m/s to 22.0 m/s. What is the change in kinetic energy?

Solution:

$$\Delta KE = \tfrac{1}{2} \cdot m \cdot v_f^2 - \tfrac{1}{2} \cdot m \cdot v_i^2 \qquad (6.5.2)$$

Put in numbers: $\Delta KE = \tfrac{1}{2} \cdot (1811 \text{ kg}) \cdot (22.0 \text{ m/s})^2 - \tfrac{1}{2} \cdot (1811 \text{ kg}) \cdot (13.0 \text{ m/s})^2 = \underline{2.85 \cdot 10^5 \text{ Joules}}$

Section 6.6 (Work-Kinetic Energy Theorem)

10. At a picnic, you drag a blanket and lunch basket across the ground. There are four forces on them: gravity (mass 3.00 kg), normal force from the ground (23.0 N, upward), force from you (14.0 N, at an angle of 27.0° above "to the right"), and friction (to the left; coefficient of kinetic friction 0.500). You drag the blanket 1.00 m, to the right. If the blanket and lunch basket were initially not moving, what is the magnitude of the velocity after moving 1.00 m? Note: this setup is the same as problem 1, with the same numbers, so if you have the net work from problem 1 you can start from there.

Solution:

Problem 1 finds the net work in this situation to be 1.0 N·m.

$$W_{net} = \tfrac{1}{2} \cdot m \cdot v_f^2 - \tfrac{1}{2} \cdot m \cdot v_i^2 \qquad (6.6.2)$$

Since v_i is given to be zero in this problem, this becomes: $W_{net} = \tfrac{1}{2} \cdot m \cdot v_f^2$

Divide both sides by ($\tfrac{1}{2} \cdot m$): $v_f^2 = W_{net}/(\tfrac{1}{2} \cdot m)$

Take the square root: $v_f = \pm \sqrt{W_{net}/(\tfrac{1}{2} \cdot m)}$

Take the positive square root (the + in the ± sign) because we're asked to find the magnitude and the magnitude is positive.

Put in numbers: $v_f = +\sqrt{(1.0 \, N \cdot m)/(\tfrac{1}{2} \cdot 3.00 \, kg)} = \underline{0.82 \text{ m/s}}$

11. A ball with mass 0.100 kg is kicked straight upward, from the ground. It is caught by a goalie at a height of 1.50 m above the ground. If the initial velocity is 10.0 m/s, what is the magnitude of velocity just before it is caught? Notes: this is the same setup as problem 3 above.

    Solution:

    In problem 3, we found that the work was –1.47 N·m.

$$W_{net} = \tfrac{1}{2} \cdot m \cdot v_f^2 - \tfrac{1}{2} \cdot m \cdot v_i^2 \qquad (6.6.2)$$

v_i is given to be 10.0 m/s.

Add $\tfrac{1}{2} \cdot m \cdot v_i^2$ to both sides: $W_{net} + \tfrac{1}{2} \cdot m \cdot v_i^2 = \tfrac{1}{2} \cdot m \cdot v_f^2$

Divide both sides by ($\tfrac{1}{2} \cdot m$): $v_f^2 = (W_{net} + \tfrac{1}{2} \cdot m \cdot v_i^2)/(\tfrac{1}{2} \cdot m)$

Take the square root: $v_f = \pm \sqrt{(W_{net} + \tfrac{1}{2} \cdot m \cdot v_i^2)/(\tfrac{1}{2} \cdot m)}$

Take the positive square root (the + in the ± sign) because we're asked to find the magnitude and the magnitude is positive.

Put in numbers:

$$v_f = +\sqrt{(-1.47 \, N \cdot m + \tfrac{1}{2} \cdot 0.100 \, kg \cdot (10.0 \, m/s)^2)/(\tfrac{1}{2} \cdot (0.100 \, kg))}$$

$= \underline{8.40 \text{ m/s}}$

12. You carry a heavy textbook to class, while walking in a straight line. The book has mass 4.00 kg. There are two forces on the book: gravity,

and the force from you. The force from you is 39.3 N, at an angle of 85.6° above "to the right". If you move the book 3.38 meters to the right (+x direction), starting from rest, what is the velocity? Note: this is the same setup as problem 6 above.

Solution:

Problem 6 finds the net work in this situation to be 10.2 N·m.

$$W_{net} = \tfrac{1}{2} \cdot m \cdot v_f^2 - \tfrac{1}{2} \cdot m \cdot v_i^2 \qquad (6.6.2)$$

Since v_i is given to be zero in this problem, this becomes: $W_{net} = \tfrac{1}{2} \cdot m \cdot v_f^2$

Divide both sides by $(\tfrac{1}{2} \cdot m)$: $v_f^2 = W_{net}/(\tfrac{1}{2} \cdot m)$

Take the square root: $v_f = \pm \sqrt{W_{net}/(\tfrac{1}{2} \cdot m)}$

Take the positive square root (the $+$ in the $\pm$ sign) because we're asked to find the magnitude and the magnitude is positive.

Put in numbers: $v_f = +\sqrt{(10.2\,N \cdot m)/(\tfrac{1}{2} \cdot 4.00\,kg)} = \underline{2.26\ m/s}$

13. At a summer barbeque, you drag a chair across some grass. The chair has four forces on it: gravity (mass 2.00 kg), normal force from the ground (19.6 N, upward), force from you (11.0 N, to the right), and friction (coefficient of kinetic friction 0.500). You drag the chair 1.50 m, to the right, starting from 0.200 m/s. What is the velocity after moving 1.50 m? Note: this is the same setup as problem 4 above.

Solution:

In problem 4, we found that the work was 1.8 N·m.

$$W_{net} = \tfrac{1}{2} \cdot m \cdot v_f^2 - \tfrac{1}{2} \cdot m \cdot v_i^2 \qquad (6.6.2)$$

v_i is given to be 0.200 m/s.

Add $\tfrac{1}{2} \cdot m \cdot v_i^2$ to both sides: $W_{net} + \tfrac{1}{2} \cdot m \cdot v_i^2 = \tfrac{1}{2} \cdot m \cdot v_f^2$

Divide both sides by $(\tfrac{1}{2} \cdot m)$: $v_f^2 = (W_{net} + \tfrac{1}{2} \cdot m \cdot v_i^2)/(\tfrac{1}{2} \cdot m)$

Take the square root: $v_f = \pm \sqrt{(W_{net} + \tfrac{1}{2} \cdot m \cdot v_i^2)/(\tfrac{1}{2} \cdot m)}$

Take the positive square root (the + in the ± sign) because we're asked to find the magnitude and the magnitude is positive.

Put in numbers:

$$v_f = +\sqrt{\left(1.8\,N\cdot m + \tfrac{1}{2}\cdot 2.00\,kg\cdot(0.200\,m/s)^2\right)/\left(\tfrac{1}{2}\cdot(2.00\,kg)\right)} = \underline{1.36\ m/s}$$

Section 6.7 (Potential Energy)

14. You lift a 10.0 kg grocery bag 1.50 m upwards, and onto a counter. What is the change in potential energy?

Solution:

The two forces that we looked at with potential energy are gravity and springs. Here, gravity is involved because of a change in height.

$$\text{Gravity: } \Delta PE = m\cdot g\cdot \Delta h \qquad (6.7.2)$$

Put in numbers: $(10.0\ kg)\cdot(9.80\ m/s^2)\cdot(1.50\ m) = \underline{147\ Joules}$

Note on units: remember that 1 Joule is equivalent to 1 kg·m²/s² and to 1 N·m.

15. Inside of a motor, a spring with spring constant 915 N/m is compressed by a distance of 0.0500 m, after not being compressed at all. What is the change in potential energy?

Solution:

The two forces that we looked at with potential energy are gravity and springs. Here, the spring force is involved because of a change in a spring's compression.

$$\text{Springs: } \Delta PE = \tfrac{1}{2}\cdot k\cdot x_f^2 - \tfrac{1}{2}\cdot k\cdot x_i^2 \qquad (6.7.1)$$

Put in numbers: $\Delta PE = \tfrac{1}{2}\cdot(915\ N/m)\cdot(0.0500\ m)^2 - \tfrac{1}{2}\cdot(915\ N/m)\cdot(0\ m)^2 = \underline{1.14\ J}$

Note on units: remember that 1 Joule is equivalent to 1 kg·m²/s² and to 1 N·m.

16. At the start of a game, the referee drops a 0.250 kg ball from 1.70 m above the ground to on the ground. What is the change in potential energy?

Solution:

The two forces that we looked at with potential energy are gravity and springs. Here, gravity is involved because of a change in height.

$$\text{Gravity: } \Delta PE = m \cdot g \cdot \Delta h \qquad (6.7.2)$$

Put in numbers: $(0.250 \text{ kg}) \cdot (9.80 \text{ m/s}^2) \cdot (-1.70 \text{ m}) = \underline{-4.17 \text{ Joules}}$

Note on units: remember that 1 Joule is equivalent to 1 kg·m²/s² and to 1 N·m.

Note on sign: the change in height is negative because the ball goes downward.

17. Inside of a motor, a spring with spring constant 758 N/m becomes completely decompressed after being compressed by 0.100 m. What is the change in potential energy?

Solution:

The two forces that we looked at with potential energy are gravity and springs. Here, the spring force is involved because of a change in a spring's compression.

$$\text{Springs: } \Delta PE = \tfrac{1}{2} \cdot k \cdot x_f^2 - \tfrac{1}{2} \cdot k \cdot x_i^2 \qquad (6.7.1)$$

Put in numbers: $\Delta PE = \tfrac{1}{2} \cdot (758 \text{ N/m}) \cdot (0 \text{ m})^2 - \tfrac{1}{2} \cdot (758 \text{ N/m}) \cdot (0.100 \text{ m})^2 = \underline{-3.79 \text{ J}}$

Note on units: remember that 1 Joule is equivalent to 1 kg·m²/s² and to 1 N·m.

Section 6.8 (Mechanical Energy)

18. A person with mass 95.0 kg goes for a stroll, on a road that involves a hill. The person starts at rest, and ends up 5.00 m lower in elevation while walking at a velocity of 1.50 m/s. What is the change in mechanical energy?

Solution:

$$\Delta ME = \Delta KE + \Delta PE \qquad (6.8.2)$$

Since the change in potential energy here is due to a change in height,

$$\text{Gravity: } \Delta PE = m \cdot g \cdot \Delta h \tag{6.7.2}$$

The change in kinetic energy is

$$\Delta KE = \tfrac{1}{2} \cdot m \cdot v_f^2 - \tfrac{1}{2} \cdot m \cdot v_i^2 \tag{6.5.2}$$

Putting these together gives:

$$\Delta ME = \tfrac{1}{2} \cdot m \cdot v_f^2 - \tfrac{1}{2} \cdot m \cdot v_i^2 + m \cdot g \cdot \Delta h$$

Put in numbers: $\Delta ME = \tfrac{1}{2} \cdot (95.0 \text{ kg}) \cdot (1.50 \text{ m/s})^2 - \tfrac{1}{2} \cdot (95.0 \text{ kg}) \cdot (0 \text{ m/s})^2 + (95.0 \text{ kg}) \cdot (9.80 \text{ m/s}^2) \cdot (-5.00 \text{ m}) = \underline{-4.55 \cdot 10^3 \text{ Joules}}$

19. A bus with mass 1800 kg drives on a road that increases in height. It is initially going 18.0 m/s; after a height increase of 25.0 m it has stopped. What is the change in mechanical energy?

Solution:

$$\Delta ME = \Delta KE + \Delta PE \tag{6.8.2}$$

Since the change in potential energy here is due to a change in height,

$$\text{Gravity: } \Delta PE = m \cdot g \cdot \Delta h \tag{6.7.2}$$

The change in kinetic energy is

$$\Delta KE = \tfrac{1}{2} \cdot m \cdot v_f^2 - \tfrac{1}{2} \cdot m \cdot v_i^2 \tag{6.5.2}$$

Putting these together gives:

$$\Delta ME = \tfrac{1}{2} \cdot m \cdot v_f^2 - \tfrac{1}{2} \cdot m \cdot v_i^2 + m \cdot g \cdot \Delta h$$

Put in numbers: $\Delta ME = \tfrac{1}{2} \cdot (1800 \text{ kg}) \cdot (0 \text{ m/s})^2 - \tfrac{1}{2} \cdot (1800 \text{ kg}) \cdot (18.0 \text{ m/s})^2 + (1800 \text{ kg}) \cdot (9.80 \text{ m/s}^2) \cdot (25.0 \text{ m}) = \underline{1.49 \cdot 10^5 \text{ Joules}}$

20. You accidentally drop a book on the floor, starting from rest. The book has mass 1.20 kg, and it falls a distance of 1.10 m. If the book has velocity 4.64 m/s (downward) right before it touches the ground, what is the change in mechanical energy?

Solution:

$$\Delta ME = \Delta KE + \Delta PE \qquad (6.8.2)$$

Since the change in potential energy here is due to a change in height,

$$\text{Gravity: } \Delta PE = m \cdot g \cdot \Delta h \qquad (6.7.2)$$

The change in kinetic energy is

$$\Delta KE = \tfrac{1}{2} \cdot m \cdot v_f^2 - \tfrac{1}{2} \cdot m \cdot v_i^2 \qquad (6.5.2)$$

Putting these together gives:

$$\Delta ME = \tfrac{1}{2} \cdot m \cdot v_f^2 - \tfrac{1}{2} \cdot m \cdot v_i^2 + m \cdot g \cdot \Delta h$$

Put in numbers: $\Delta ME = \tfrac{1}{2} \cdot (1.20 \text{ kg}) \cdot (4.64 \text{ m/s})^2 - \tfrac{1}{2} \cdot (1.20 \text{ kg}) \cdot (0 \text{ m/s})^2 + (1.20 \text{ kg}) \cdot (9.80 \text{ m/s}^2) \cdot (-1.10 \text{ m}) = \underline{-1.82 \cdot 10^{-2} \text{ Joules}}$

Note: this is almost zero. If it is zero, it would be conserved, like Section 6.9.

Section 6.9 (Conservation of Mechanical Energy)

21. A 65.0 kg person rides down an icy hill, starting from a velocity of 1.00 m/s. At the bottom of the hill, the person is traveling at 9.50 m/s. Assume that only conservative forces do work. What is the change in height?

Solution:

$$\tfrac{1}{2} \cdot m \cdot v_f^2 - \tfrac{1}{2} \cdot m \cdot v_i^2 + \tfrac{1}{2} \cdot k \cdot x_f^2 - \tfrac{1}{2} \cdot k \cdot x_i^2 + m \cdot g \cdot \Delta h = W_{nc} \qquad (6.9.2)$$

We are given that $v_i = 1.00$ m/s, $v_f = 9.50$ m/s, m = 65.0 kg, and $W_{nc} = 0$ (work done by non-conservative forces is zero). There is no spring in the problem, so the terms $\frac{1}{2}\cdot k\cdot x_f^2 - \frac{1}{2}\cdot k\cdot x_i^2$ are both zero as well. This leaves us with:

$$\frac{1}{2}\cdot m\cdot v_f^2 - \frac{1}{2}\cdot m\cdot v_i^2 + m\cdot g\cdot \Delta h = 0$$

Add $\frac{1}{2}\cdot m\cdot v_i^2$ to both sides, and subtract $\frac{1}{2}\cdot m\cdot v_f^2$ from both sides:

$$m\cdot g\cdot \Delta h = \frac{1}{2}\cdot m\cdot v_i^2 - \frac{1}{2}\cdot m\cdot v_f^2$$

Divide both sides by (m·g): $\Delta h = (\frac{1}{2}\cdot m\cdot v_i^2 - \frac{1}{2}\cdot m\cdot v_f^2)/(m\cdot g)$

Put in numbers: $\Delta h = (\frac{1}{2}\cdot 65.0\ kg\cdot(1.00\ m/s)^2 - \frac{1}{2}\cdot 65.0\ kg\cdot(9.50\ m/s)^2)/$
$(65.0\ kg\cdot 9.80\ m/s^2)$

$$\Delta h = \underline{-4.55\ m}$$

Note: a negative change in height means that the person went downward.

22. A 79.0 kg person rides down an icy hill, and goes straight into a spring at the bottom that "catches them". At the start, they are moving at 1.50 m/s; at the bottom of the hill they are stopped and have compressed a spring with spring constant $9.00\cdot10^4$ N/m by a distance of 0.250 m. Assume that only conservative forces do work. What is the change in height?

Solution:

$$\frac{1}{2}\cdot m\cdot v_f^2 - \frac{1}{2}\cdot m\cdot v_i^2 + \frac{1}{2}\cdot k\cdot x_f^2 - \frac{1}{2}\cdot k\cdot x_i^2 + m\cdot g\cdot \Delta h = W_{nc} \qquad (6.9.2)$$

We are given $v_f = 0$, $v_i = 1.50$ m/s, m = 79.0 kg, $k = 9.00\cdot10^4$ N/m, $x_f = 0.250$ m, $x_i = 0$ (not compressed at the start), and $W_{nc} = 0$.

Taking out the terms that are zero, we get:

$$- \frac{1}{2}\cdot m\cdot v_i^2 + \frac{1}{2}\cdot k\cdot x_f^2 + m\cdot g\cdot \Delta h = 0$$

Add $\frac{1}{2}\cdot m\cdot v_i^2$ to both sides, and subtract $\frac{1}{2}\cdot k\cdot x_f^2$ from both sides:

$$m\cdot g\cdot \Delta h = \frac{1}{2}\cdot m\cdot v_i^2 - \frac{1}{2}\cdot k\cdot x_f^2$$

Divide both sides by (m·g): $\Delta h = (\frac{1}{2}\cdot m \cdot v_i^2 - \frac{1}{2}\cdot k \cdot x_f^2)/(m \cdot g)$

Put in numbers: $\Delta h = (\frac{1}{2}\cdot 79.0\,\text{kg}\cdot(1.50\,\text{m/s})^2 - \frac{1}{2}\cdot(9.00\cdot 10^4\,\text{N/m})\cdot(0.250\,\text{m})^2)/(79.0\,\text{kg}\cdot 9.80\,\text{m/s}^2) = \underline{-3.52\,\text{m}}$

23. A person rides down an icy hill, starting from a velocity of 1.00 m/s. At the bottom of the hill, their height has decreased by 5.50 m. Assume that only conservative forces do work. What is the magnitude of the velocity at the bottom? Hint: the mass is not given because it cancels out of the calculation.

    Solution:

    $$\frac{1}{2}\cdot m \cdot v_f^2 - \frac{1}{2}\cdot m \cdot v_i^2 + \frac{1}{2}\cdot k \cdot x_f^2 - \frac{1}{2}\cdot k \cdot x_i^2 + m \cdot g \cdot \Delta h = W_{nc} \qquad (6.9.2)$$

    We are given $v_i = 1.00$ m/s, $W_{nc} = 0$, and $\Delta h = -5.50$ m. There are no springs, so the spring terms are zero. Taking out the terms that are zero, we get:

    $$\frac{1}{2}\cdot m \cdot v_f^2 - \frac{1}{2}\cdot m \cdot v_i^2 + m \cdot g \cdot \Delta h = 0$$

    Divide every term by m: $\frac{1}{2}\cdot v_f^2 - \frac{1}{2}\cdot v_i^2 + g \cdot \Delta h = 0$
    Add $\frac{1}{2}\cdot v_i^2$ to both sides, and subtract $g \cdot \Delta h$ from both sides:

    $$\frac{1}{2}\cdot v_f^2 = \frac{1}{2}\cdot v_i^2 - g \cdot \Delta h$$

    Divide both sides by ($\frac{1}{2}$): $v_f^2 = (\frac{1}{2}\cdot v_i^2 - g \cdot \Delta h)/(\frac{1}{2})$

    Take the square root: $v_f = \pm\sqrt{\left(\frac{1}{2}\cdot v_i^2 - g \cdot \Delta h\right)/\left(\frac{1}{2}\right)}$
    We'll take the positive root, since magnitudes are positive.

    Put in numbers: $v_f = +\sqrt{\left(\frac{1}{2}\cdot\left(1.00\frac{\text{m}}{\text{s}}\right)^2 - 9.80\frac{\text{m}}{\text{s}^2}\cdot -5.50\,\text{m}\right)/\left(\frac{1}{2}\right)}$
    $= \underline{10.4\,\text{m/s}}$

24. An 85.0 kg person rides down an icy hill, starting from a velocity of 1.00 m/s. At the bottom of the hill, they have gone downward by 5.00 m. If the person is traveling at 8.50 m/s, how much work was done by non-conservative forces?

    Solution:

    $$\frac{1}{2}\cdot m \cdot v_f^2 - \frac{1}{2}\cdot m \cdot v_i^2 + \frac{1}{2}\cdot k \cdot x_f^2 - \frac{1}{2}\cdot k \cdot x_i^2 + m \cdot g \cdot \Delta h = W_{nc} \qquad (6.9.2)$$

We are given $v_i = 1.00$ m/s, $v_f = 8.50$ m/s, $m = 85.0$ kg, and $\Delta h = -5.00$ m. There are no springs, so the spring terms ($\frac{1}{2} \cdot k \cdot x_f^2 - \frac{1}{2} \cdot k \cdot x_i^2$) are zero. We want to find W_{nc}, and the equation is already solved for it. Putting in numbers:

$W_{nc} = \frac{1}{2} \cdot 85.0$ kg$\cdot(8.50$ m/s$)^2 - \frac{1}{2} \cdot 85.0$ kg$\cdot(1.00$ m/s$)^2 + 0 - 0 + 85.0$ kg$\cdot 9.80$ m/s$^2 \cdot -5.00$ m

$$W_{nc} = \underline{-1.14 \cdot 10^3 \, \text{N} \cdot \text{m}}$$

Notes: remember that the units of work are N·m, and that friction (for example) does negative work.

25. At a workplace, you drop objects from your work position to the person on the floor below you through a small hole in the floor. It lands on a spring that has spring constant 608 N/m. At the initial time point, the velocity is 0.500 m/s. At the second time point, the object is in the air, 2.00 m below where it started from. At the final time point, the velocity is zero and the spring is compressed by a distance of 0.300 m. The object has a mass of 1.00 kg. Assume that only conservative forces do work.

   A. What is the velocity of the object at the second time point?

   B. What is the change in height between the first time point and the third time point?

   Solution:

   Part A:

   Make the first time point the initial time point, and the second time point the final time point.

   $$\frac{1}{2} \cdot m \cdot v_f^2 - \frac{1}{2} \cdot m \cdot v_i^2 + \frac{1}{2} \cdot k \cdot x_f^2 - \frac{1}{2} \cdot k \cdot x_i^2 + m \cdot g \cdot \Delta h = W_{nc} \qquad (6.9.2)$$

   We are given $v_i = 0.500$ m/s, $\Delta h = -2.00$ m, $W_{nc} = 0$, $m = 1.00$ kg, and there are no springs. We want to find v_f. Taking out the terms that are zero, we get:

   $$\frac{1}{2} \cdot m \cdot v_f^2 - \frac{1}{2} \cdot m \cdot v_i^2 + m \cdot g \cdot \Delta h = 0$$

   Divide each term by m: $\frac{1}{2} \cdot v_f^2 - \frac{1}{2} \cdot v_i^2 + g \cdot \Delta h = 0$

Add $\frac{1}{2} \cdot v_i^2$ to both sides, and subtract $g \cdot \Delta h$ from both sides:

$$\tfrac{1}{2} \cdot v_f^2 = \tfrac{1}{2} \cdot v_i^2 - g \cdot \Delta h$$

Divide both sides by ($\frac{1}{2}$): $v_f^2 = (\frac{1}{2} \cdot v_i^2 - g \cdot \Delta h)/(\frac{1}{2})$

Take the square root: $v_f = \pm\sqrt{\left(\tfrac{1}{2} \cdot v_i^2 - g \cdot \Delta h\right)/\left(\tfrac{1}{2}\right)}$

We'll take the positive root, since magnitudes are positive.

Put in numbers: $v_f = +\sqrt{\left(\tfrac{1}{2} \cdot \left(0.500 \dfrac{m}{s}\right)^2 - 9.80\dfrac{m}{s^2} \cdot -2.00\,m\right)/\left(\tfrac{1}{2}\right)}$
$= \underline{6.28\ m/s}$

Note: you didn't actually need the value of the mass here, because it cancels out, but it is necessary for Part B.

Part B:

We now make the first time point the initial time point, and the third time point the final time point.

$$\tfrac{1}{2} \cdot m \cdot v_f^2 - \tfrac{1}{2} \cdot m \cdot v_i^2 + \tfrac{1}{2} \cdot k \cdot x_f^2 - \tfrac{1}{2} \cdot k \cdot x_i^2 + m \cdot g \cdot \Delta h = W_{nc} \qquad (6.9.2)$$

We are given $m = 1.00$ kg, $v_f = 0$, $v_i = 0.500$ m/s, $k = 608$ N/m, $x_i = 0$ (not compressed at the start), $x_f = 0.300$ m, and $W_{nc} = 0$. We want to find Δh. Taking out the terms that are zero, we have:

$$- \tfrac{1}{2} \cdot m \cdot v_i^2 + \tfrac{1}{2} \cdot k \cdot x_f^2 + m \cdot g \cdot \Delta h = 0$$

Add $\frac{1}{2} \cdot m \cdot v_i^2$ to both sides, and subtract $\frac{1}{2} \cdot k \cdot x_f^2$ from both sides:

$$m \cdot g \cdot \Delta h = \tfrac{1}{2} \cdot m \cdot v_i^2 - \tfrac{1}{2} \cdot k \cdot x_f^2$$

Divide both sides by ($m \cdot g$): $\Delta h = (\frac{1}{2} \cdot m \cdot v_i^2 - \frac{1}{2} \cdot k \cdot x_f^2)/(m \cdot g)$

Put in numbers: $\Delta h = (\frac{1}{2} \cdot 1.00$ kg $\cdot (0.500$ m/s$)^2 - \frac{1}{2} \cdot (608$ N/m$) \cdot (0.300$ m$)^2)/(1.00$ kg $\cdot 9.80$ m/s$^2) = \underline{-2.78\ m}$

Note: the object goes downward, so a negative change in height makes sense.

26. At work, you end up needing to send so many things to your coworker that you end up devising a machine to do it. At the start of

the machine, the object (mass 1.50 kg) is placed against a compressed spring, at rest, and then let go. At the end (at your coworker), the spring is uncompressed, the package has increased in height by 1.20 m and the package is stopped again. The spring has spring constant $4.00 \cdot 10^3$ N/m. Assume that only conservative forces do work.

A. How far was the spring compressed at the start? Note: it isn't compressed at all at the end.

B. In reality, it turns out that the object only goes upward by 1.00 m instead of 1.20 m. This is because friction (a non-conservative force) does some work. The spring is compressed initially by the same distance that you got in Part A. How much work does it do?

Solution:

Part A:

$$\tfrac{1}{2} \cdot m \cdot v_f^2 - \tfrac{1}{2} \cdot m \cdot v_i^2 + \tfrac{1}{2} \cdot k \cdot x_f^2 - \tfrac{1}{2} \cdot k \cdot x_i^2 + m \cdot g \cdot \Delta h = W_{nc} \qquad (6.9.2)$$

We are given m = 1.50 kg, $v_i = 0$, $v_f = 0$, $W_{nc} = 0$, $x_f = 0$, $\Delta h = 1.20$ m, and $k = 4.00 \cdot 10^3$ N/m. We want to find x_i. Taking out the terms that are zero:

$$-\tfrac{1}{2} \cdot k \cdot x_i^2 + m \cdot g \cdot \Delta h = 0$$

Subtract $m \cdot g \cdot \Delta h$ from both sides: $-\tfrac{1}{2} \cdot k \cdot x_i^2 = -m \cdot g \cdot \Delta h$

Divide both sides by $-\tfrac{1}{2} \cdot k$: $x_i^2 = (-m \cdot g \cdot \Delta h)/(-\tfrac{1}{2} \cdot k)$

Take the square root: $x_i = \pm \sqrt{(-m \cdot g \cdot \Delta h)/(-\tfrac{1}{2} \cdot k)}$

Take the positive root, since the amount of compression will be a positive number. Put in numbers:

$$x_i = +\sqrt{\left(-1.50 \,\mathrm{kg} \cdot 9.80 \frac{m}{s^2} \cdot 1.20 \,\mathrm{m}\right) / \left(-\tfrac{1}{2} \cdot 4.00 \cdot 10^3 \,\mathrm{N}/\mathrm{m}\right)}$$

$$= \underline{9.39 \cdot 10^{-2} \ \mathrm{m}}$$

Part B:

$$\tfrac{1}{2} \cdot m \cdot v_f^2 - \tfrac{1}{2} \cdot m \cdot v_i^2 + \tfrac{1}{2} \cdot k \cdot x_f^2 - \tfrac{1}{2} \cdot k \cdot x_i^2 + m \cdot g \cdot \Delta h = W_{nc}$$

We are given m = 1.50 kg, $v_f = 0$, $v_i = 0$, $x_f = 0$, $x_i = 9.39 \cdot 10^{-2}$ m, $k = 4.00 \cdot 10^3$ N/m, and $\Delta h = 1.00$ m. We want to find W_{nc}. The equation is already solved for W_{nc}. If we take out the terms that are zero,

$$W_{nc} = -\frac{1}{2} \cdot k \cdot x_i^2 + m \cdot g \cdot \Delta h$$

Put in numbers: $W_{nc} = -\frac{1}{2} \cdot 4.00 \cdot 10^3$ N/m$\cdot(9.39 \cdot 10^{-2}$ m$)^2 + 1.50$ kg$\cdot 9.80$ m/s$^2 \cdot 1.00$ m

$$\underline{Wnc = -2.93 \text{ N·m}}$$

Note: friction does negative work because it is in the opposite direction to the displacement, which makes an angle of 180° in $W = F \cdot \Delta x \cdot \cos(\theta)$ (Equation 6.2.1)

Section 6.10 (Power)

27. A car speeds up from 0 to 15.0 m/s, using a force of 1800. Newtons in the direction it travels. What is the amount of the power, in Watts?

Solution:

Here we are given a force and a velocity. To find power, we use Equation 6.10.3:

$$\text{Power} = F \cdot v \cdot \cos(\theta) \tag{6.10.3}$$

Put in numbers: (1800. N)$\cdot$(15.0 m/s)$\cdot\cos(0) = \underline{2.70 \cdot 10^4 \text{ Watts}}$

Note: $\theta = 0$ because force and velocity are in the same direction.

28. You move a bag of carrots at the store, with a force of 1.20 N, in the same direction as the carrots are moving with. If the power is 1.50 Watts, what is the magnitude of the bag's velocity?

Solution:

Here we are given a force and a power. To find velocity, we use Equation 6.10.3:

$$\text{Power} = F \cdot v \cdot \cos(\theta) \tag{6.10.3}$$

Divide both sides by $F \cdot \cos(\theta)$: $v = \text{Power}/(F \cdot \cos(\theta))$

Put in numbers: $v = (1.50$ W$)/(1.20$ N$\cdot\cos(0)) = \underline{1.25 \text{ m/s}}$

Note on units: 1 Watt is equivalent to 1 N·m/s.

29. You pick up an object using a power of 3.00 Watts. If the velocity is 2.00 m/s, what is the force?

Solution:

Here we are given a power and a velocity. To find force, we use Equation 6.10.3:

$$Power = F \cdot v \qquad\qquad (6.10.3)$$

Divide both sides by v·cos(θ): F = Power/(v·cos(θ))

Put in numbers: F = (3.00 W)/(2.00 m/s·cos(0)) = 1.50 N

Note on units: 1 Watt is equivalent to 1 N·m/s.

Linear Momentum and Collisions

7.1 INTRODUCTION: COLLISIONS AND MOMENTUM

Collisions happen often, whether it is a charging foul in basketball, two people literally bumping into each other, or two bumper cars bumping into each other. Chapter 6 talked about the idea that the amount of something can stay the same during a certain event – this is called a conservation law. In Chapter 6 it was mechanical energy that stayed the same. For collisions, a quantity called **linear momentum** stays the same.

7.2 LINEAR MOMENTUM

Linear momentum is equal to mass times velocity. It is represented by the letter p:

$$p = m \cdot v \qquad (7.2.1)$$

(p is the magnitude of momentum, m is mass, v is the magnitude of velocity)

Linear momentum is a vector. This means:

- In one dimensional problems (with motion in a straight line), you need to have one direction be positive for the velocity, and one direction be negative. This is just like the problems done in Chapter 2, with acceleration, velocity, and displacement.

DOI: 10.1201/9781003005049-7

- When we do two-dimensional collisions later in the chapter, we'll use an equation for each of the x and y components. It might help to think of it as $p_x = m \cdot v_x$ and $p_y = m \cdot v_y$

7.3 LINEAR MOMENTUM PROBLEMS WITHOUT COLLISIONS

Two problem types involve linear momentum but don't involve collisions:

1. Calculate the amount of linear momentum. For these problems, just use $p = m \cdot v$ (Equation 7.2.1). If the direction is requested, the direction of momentum is the same direction as the velocity.

2. Calculate the change in linear momentum, between an initial and final time point. Here, you use Equation 7.2.1 twice – once at the final time point and once at the initial time point. As with many "change" questions, you take the (final amount) – (initial amount). If the problem is one-dimensional, the equation would be:

$$\Delta p = m \cdot v_f - m \cdot v_i \qquad (7.3.1)$$

(Δp is the change in linear momentum, m is the mass of the object, v_f is the magnitude of velocity of the final time point, v_i is the magnitude of velocity at the initial time point)

If the problem is in two dimensions, you need to make everything into x and y components. The following examples will be in one dimension, with one example for each type of problem listed above.

Example 7.3.1

A person has mass 125 kg, and is walking in a straight line. If the person's linear momentum is 190 kg·m/s, what is the person's velocity?

Solution:
 Since $p = m \cdot v$, and we know p and m, we can solve for v.
 Divide both sides by m: $v = p/m$
 Put in numbers: v = (190 kg·m/s)/(125 kg) = 1.52 m/s

Example 7.3.2

A 65.0 kg person is walking at a velocity of 1.75 m/s, to the east. Make east the positive direction. If the person slows down to 1.05 m/s still to the east, what is the change in linear momentum?

Solution:

$\Delta p = m \cdot v_f - m \cdot v_i$

Put in numbers: $\Delta p = (65.0$ kg$) \cdot (1.05$ m/s$) - (65.0$ kg$) \cdot (1.75$ m/s$) = \underline{-45.5 \text{ kg·m/s}}$

Note: the change can be negative.

Problem to Try Yourself

A rowboat has mass 50.0 kg. It starts off traveling at 1.50 m/s, to the north. A little later, it is traveling at 5.75 m/s, still to the north. What is the change in linear momentum between the two time points that are given?

Solution:

$\Delta p = m \cdot v_f - m \cdot v_i$

Put in numbers: $\Delta p = (50.0$ kg$) \cdot (5.75$ m/s$) - (50.0$ kg$) \cdot (1.50$ m/s$) = \underline{213 \text{ kg·m/s}}$

7.4 COLLISIONS AND LINEAR MOMENTUM

Imagine that two objects collide. To start with a simpler case, we're going to imagine a collision in one dimension, meaning that both objects were moving along the same line. The first object has mass m_1, velocity v_{1i} before the collision, and velocity v_{1f} after the collision. ("i" stands for initial and "f" stands for final.) The second object has mass m_2, velocity v_{2i} before the collision, and velocity v_{2f} after the collision.

In a collision, what happens is that total amount of linear momentum is "conserved", meaning that the total amount stays the same. Based on the variables from the last paragraph, the total linear momentum staying the same will look like this:

$$m_1 \cdot v_{1i} + m_2 \cdot v_{2i} = m_1 \cdot v_{1f} + m_2 \cdot v_{2f} \tag{7.4.1}$$

In this equation, the left side of the equation is the total amount of linear momentum before the collision, and the right side of the equation is the total amount of linear momentum after the collision.

Where does this come from? In physics theory, the total amount of linear momentum of a group of objects is conserved if the total (net) force on them from everything outside of the group is zero. For collisions, we can say that this is approximately true. *Physics for Scientists and Engineers* by

Tipler and Mosca (cited in the acknowledgements section) argues that the collision forces are so large that they are the only important ones for the short time that the collision happens.

7.5 COLLISIONS, PROBLEM TYPE 1: PERFECTLY INELASTIC COLLISIONS WITH ONE DIMENSION

We'll now cover two types of collision problems. Each will add an equation to be used along with Equation 7.3.1. The first type is **perfectly inelastic collisions**, which are collisions when two objects stick together. When two objects stick together, their velocity is the same. So if two objects stick together after a collision, this means

$$v_{1f} = v_{2f} \qquad (7.5.1)$$

Every collision problem with motion in a line (1 dimension) where the objects stick together can be solved by using Equations 7.4.1 and 7.5.1. Often, two variables will be asked for since there are two equations.

$$m_1 \cdot v_{1i} + m_2 \cdot v_{2i} = m_1 \cdot v_{1f} + m_2 \cdot v_{2f} \qquad (7.4.1)$$

Example 7.5.1

Two bumper cars bump into each other. The first car has mass 155 kg, including the driver, and before the collision it is traveling at 1.50 m/s to the west. The second car has mass 103 kg, including the driver, and before the collision it is traveling at 1.75 m/s to the east. If the two cars stick together after the collision, what is their velocity?

Solution:
　　Use $m_1 \cdot v_{1i} + m_2 \cdot v_{2i} = m_1 \cdot v_{1f} + m_2 \cdot v_{2f}$ and $v_{1f} = v_{2f}$
　　Set "east" as positive, which makes v_{1i} negative (–1.50 m/s) and v_{2i} positive (1.75 m/s)
　　Note: a review of using two equations at once is included in Section 13.6.
　　Since $v_{1f} = v_{2f}$ we can put in v_{1f} for v_{2f} in the first equation
　　It becomes: $m_1 \cdot v_{1i} + m_2 \cdot v_{2i} = m_1 \cdot v_{1f} + m_2 \cdot v_{1f}$
　　On the right side, v_{1f} is in both terms. We can rewrite the right side to show this:

$$m_1 \cdot v_{1i} + m_2 \cdot v_{2i} = (m_1 + m_2) \cdot v_{1f}$$

Divide both sides by $(m_1 + m_2)$: $v_{1f} = (m_1 \cdot v_{1i} + m_2 \cdot v_{2i})/(m_1 + m_2)$

Put in numbers: $v_{1f} = (155 \text{ kg} \cdot -1.50 \text{ m/s} + 103 \text{ kg} \cdot 1.75 \text{ m/s})/(155 \text{ kg} + 103 \text{ kg}) = \underline{-0.203 \text{ m/s}}$

Since $v_{1f} = v_{2f}$, $v_{2f} = \underline{-0.203 \text{ m/s}}$ as well

Note: the negative sign means that the cars (stuck together) move towards the west.

Problem to Try Yourself

Two people are walking, maybe while looking down at cell phones. The first person has mass 75.0 kg, and is walking north at 1.55 m/s. The second person has mass 95.0 kg, and is walking south at 0.625 m/s. The two people bump into each other, and very briefly stick together after the collision. What is their velocity?

Solution:

Use $m_1 \cdot v_{1i} + m_2 \cdot v_{2i} = m_1 \cdot v_{1f} + m_2 \cdot v_{2f}$ and $v_{1f} = v_{2f}$

Set "north" as positive, which makes v_{1i} positive (1.55 m/s) and v_{2i} negative (−0.625 m/s)

Since $v_{1f} = v_{2f}$ we can put in v_{1f} for v_{2f} in the first equation

It becomes: $m_1 \cdot v_{1i} + m_2 \cdot v_{2i} = m_1 \cdot v_{1f} + m_2 \cdot v_{1f}$

On the right side, v_{1f} is in both terms. We can rewrite the right side to show this:

$$m_1 \cdot v_{1i} + m_2 \cdot v_{2i} = (m_1 + m_2) \cdot v_{1f}$$

Divide both sides by $(m_1 + m_2)$: $v_{1f} = (m_1 \cdot v_{1i} + m_2 \cdot v_{2i})/(m_1 + m_2)$

Put in numbers: $v_{1f} = (75.0 \text{ kg} \cdot 1.55 \text{ m/s} + 95.0 \text{ kg} \cdot -0.625 \text{ m/s})/(75.0 \text{ kg} + 95.0 \text{ kg}) = \underline{0.335 \text{ m/s}}$

Since $v_{1f} = v_{2f}$, $v_{2f} = \underline{0.335 \text{ m/s}}$ as well

7.6 COLLISIONS, PROBLEM TYPE 2: ELASTIC COLLISIONS WITH ONE DIMENSION

The second type of collision problem that we will cover is **elastic collisions**, which are collisions where the total amount of kinetic energy stays the same in addition to the total amount of linear momentum. We could make an equation similar to Equation 7.4.1, just for kinetic energy instead, but it turns out that we can use an easier equation:

$$v_{1i} + v_{1f} = v_{2i} + v_{2f} \tag{7.6.1}$$

This equation comes from combining Equation 7.4.1 and a conservation equation for kinetic energy, and comes from *Physics for Scientists and Engineers* by Tipler and Mosca (cited in the acknowledgements section).

If an equation involves two objects moving in a line (1 dimension) and an elastic collision, you use Equations 7.4.1 and 7.6.1 together.

$$m_1 \cdot v_{1i} + m_2 \cdot v_{2i} = m_1 \cdot v_{1f} + m_2 \cdot v_{2f} \qquad (7.4.1)$$

Side note: *This type of problem involves some of the more difficult math in the entire textbook. Section 13.6 reviews how to solve two equations at the same time.*

Example 7.6.1

Imagine that two objects bump into each other and have a one-dimensional elastic collision. The first object has a mass of 2.50 kg, and the second object has a mass of 3.85 kg. Before the collision, the first object is going to the right at 6.50 m/s, and the second object is going to the left at 7.45 m/s. What are the velocities of the two objects after the collision? Define "to the right" as positive, and be careful about getting the signs correct.

Solution:

We'll use $m_1 \cdot v_{1i} + m_2 \cdot v_{2i} = m_1 \cdot v_{1f} + m_2 \cdot v_{2f}$ and $v_{1i} + v_{1f} = v_{2i} + v_{2f}$

Note: the algebra in this problem is very similar to problem 13.6.1 in the math review chapter.

Any negative velocities mean "to the left"; positive velocities are "to the right".

Use the second equation to relate v_{1f} and v_{2f}. It doesn't matter which one we solve for to start; we'll choose v_{1f}.

Take the second equation, and subtract v_{1i} from both sides:

$$v_{1f} = v_{2i} + v_{2f} - v_{1i}$$

We have now solved one equation for one of the unknown variables (v_{1f}), in terms of the other unknown variable (v_{2f}) and things that we know the values of.

Next, put the equation for v_{1f} in for v_{1f} in the first equation:

$$m_1 \cdot v_{1i} + m_2 \cdot v_{2i} = m_1 \cdot (v_{2i} + v_{2f} - v_{1i}) + m_2 \cdot v_{2f}$$

Multiply out m_1 times the three velocities in parentheses:

$$m_1 \cdot v_{1i} + m_2 \cdot v_{2i} = m_1 \cdot v_{2i} + m_1 \cdot v_{2f} - m_1 \cdot v_{1i} + m_2 \cdot v_{2f}$$

v_{2f} is now the only unknown variable in this equation. We want to get it by itself. Start by moving all terms on the right that do not have v_{2f} in it.

Add $m_1 \cdot v_{1i}$ to both sides, and subtract $m_1 \cdot v_{2i}$ from both sides:

$$m_1 \cdot v_{1i} + m_2 \cdot v_{2i} + m_1 \cdot v_{1i} - m_1 \cdot v_{2i} = m_1 \cdot v_{2f} + m_2 \cdot v_{2f}$$

We can rewrite the right side as $(m_1 + m_2) \cdot v_{2f}$, since both sides have v_{2f} multiplied in it

$$m_1 \cdot v_{1i} + m_2 \cdot v_{2i} + m_1 \cdot v_{1i} - m_1 \cdot v_{2i} = (m_1 + m_2) \cdot v_{2f}$$

Divide both sides by $(m_1 + m_2)$:

$$(m_1 \cdot v_{1i} + m_2 \cdot v_{2i} + m_1 \cdot v_{1i} - m_1 \cdot v_{2i})/(m_1 + m_2) = v_{2f}$$

Put in numbers:
$v_{2f} = (2.50 \text{ kg} \cdot 6.50 \text{ m/s} + 3.85 \text{ kg} \cdot -7.45 \text{ m/s} + 2.50 \text{ kg} \cdot 6.50 \text{ m/s} - 2.50 \text{ kg} \cdot -7.45 \text{ m/s})/(2.50 \text{ kg} + 3.85 \text{ kg}) = \underline{3.53 \text{ m/s}}$

We now have v_{2f}. To find v_{1f}, use the equation for v_{1f} that we solved for earlier in this problem:

$$v_{1f} = v_{2i} + v_{2f} - v_{1i} = -7.45 \text{ m/s} + 3.53 \text{ m/s} - 6.50 \text{ m/s} = \underline{-10.4 \text{ m/s}}$$

Problem to Try Yourself

Two objects go through an elastic collision. The first object has mass 6.70 kg, and velocity 8.00 m/s to the left before the collision. The second object has mass 8.90 kg, and velocity 11.0 m/s to the right before the collision. What are their velocities after the collision?

Solution:

We'll use $m_1 \cdot v_{1i} + m_2 \cdot v_{2i} = m_1 \cdot v_{1f} + m_2 \cdot v_{2f}$ and $v_{1i} + v_{1f} = v_{2i} + v_{2f}$

Any negative velocities mean "to the left"; positive velocities are "to the right".

Use the second equation to relate v_{1f} and v_{2f}. It doesn't matter which one we solve for to start; we'll choose v_{1f}.

Take the second equation, and subtract v_{1i} from both sides:

$$v_{1f} = v_{2i} + v_{2f} - v_{1i}$$

We have now solved one equation for one of the unknown variables (v_{1f}), in terms of the other unknown variable (v_{2f}) and things that we know the values of.

Next, put the equation for v_{1f} in for v_{1f} in the first equation:

$$m_1 \cdot v_{1i} + m_2 \cdot v_{2i} = m_1 \cdot (v_{2i} + v_{2f} - v_{1i}) + m_2 \cdot v_{2f}$$

Multiply out m_1 times the three velocities in parentheses:

$$m_1 \cdot v_{1i} + m_2 \cdot v_{2i} = m_1 \cdot v_{2i} + m_1 \cdot v_{2f} - m_1 \cdot v_{1i} + m_2 \cdot v_{2f}$$

v_{2f} is now the only unknown variable in this equation. We want to get it by itself. Start by moving all terms on the right that do not have v_{2f} in it.

Add $m_1 \cdot v_{1i}$ to both sides, and subtract $m_1 \cdot v_{2i}$ from both sides:

$$m_1 \cdot v_{1i} + m_2 \cdot v_{2i} + m_1 \cdot v_{1i} - m_1 \cdot v_{2i} = m_1 \cdot v_{2f} + m_2 \cdot v_{2f}$$

We can rewrite the right side as $(m_1 + m_2) \cdot v_{2f}$, since both sides have v_{2f} multiplied in it

$$m_1 \cdot v_{1i} + m_2 \cdot v_{2i} + m_1 \cdot v_{1i} - m_1 \cdot v_{2i} = (m_1 + m_2) \cdot v_{2f}$$

Divide both sides by $(m_1 + m_2)$:

$$(m_1 \cdot v_{1i} + m_2 \cdot v_{2i} + m_1 \cdot v_{1i} - m_1 \cdot v_{2i})/(m_1 + m_2) = v_{2f}$$

Put in numbers:
$v_{2f} = (6.70 \text{ kg} \cdot -8.00 \text{ m/s} + 8.90 \text{ kg} \cdot 11.0 \text{ m/s} + 6.70 \text{ kg} \cdot -8.00 \text{ m/s} - 6.70$ $\text{kg} \cdot 11.0 \text{ m/s})/(6.70 \text{ kg} + 8.90 \text{ kg}) = \underline{-5.32 \text{ m/s}}$

We now have v_{2f}. To find v_{1f}, use the equation for v_{1f} that we solved for earlier in this problem:
$v_{1f} = v_{2i} + v_{2f} - v_{1i} = 11.0 \text{ m/s} + -5.32 \text{ m/s} - (-8.00 \text{ m/s}) = \underline{13.7 \text{ m/s}}$

7.7 PERFECTLY INELASTIC COLLISIONS WITH TWO DIMENSIONS

If a problem has a perfectly inelastic collision, but the motion is in two dimensions, you just repeat the one-dimensional version of the problem (Section 7.5) twice: once for the x components, and once for the y components. In addition to this, you may need the standard equations for combining components, from Chapter 3:

$$(\text{magnitude}) = \sqrt{\left(\text{x component}\right)^2 + \left(\text{y component}\right)^2} \qquad (3.3.2)$$

$$\theta = \tan^{-1}(|\text{y component}|/|\text{x component}|) \qquad (3.3.3)$$

$$(\text{x component magnitude}) = (\text{magnitude}) \cdot \cos(\theta) \qquad (3.3.4)$$

$$(\text{y component magnitude}) = (\text{magnitude}) \cdot \sin(\theta) \qquad (3.3.5)$$

Here, these equations could be applied to the velocities, or to the linear momentum values themselves (with the direction of the linear momentum the same as the velocity).

Remember that for the x and y components, you need to choose the sign, and for the angle you need to pick whether it is the positive or negative x direction, and whether it is above or below that direction.

The two equations from Section 7.5, to be used on each component here, are:

$$v_{1f} = v_{2f} \qquad (7.5.1)$$

$$m_1 \cdot v_{1i} + m_2 \cdot v_{2i} = m_1 \cdot v_{1f} + m_2 \cdot v_{2f} \qquad (7.4.1)$$

Example 7.7.1

Two people are walking to class, and not looking carefully because they are in a hurry. The first person has mass 65.0 kg, and is walking at 1.35 m/s at an angle of 35.0° north of west. The second person has mass 97.0 kg, and is walking at 1.20 m/s at an angle of 46.0° south of west. The two people stick together after the collision, just for a short time. What is their shared velocity? Hint: since this is a two-dimensional problem, you will need to give a magnitude and an angle for the velocity. You can define north as the positive y direction and east as the positive x direction.

Solution:

As stated above, we need to solve this problem twice: once for the x components and once for the y components. The x component solution will give us the x component of final velocity; the y component solution will give the y component. We'll then combine those components to get the magnitude and angle.

x components: to do the x component, we'll need the x components of the initial velocities:

(magnitude of v_{1ix}) = $v_{1i}\cdot\cos(\text{angle})$ = 1.35 m/s·cos(35.0°) = 1.11 m/s

Since v_1 is towards the north (+y) and west (–x), the x component is negative: v_{1ix} = –1.11 m/s

Similarly, for the x component of v_{2i}, (magnitude of v_{2ix}) = 1.20 m/s·cos(46.0°) = 0.834 m/s

Since v_2 is towards the south (–y) and west (–x), the x component is negative: v_{2ix} = –0.834 m/s

We now combine $m_1\cdot v_{1i} + m_2\cdot v_{2i} = m_1\cdot v_{1f} + m_2\cdot v_{2f}$ and $v_{1f} = v_{2f}$ for the x components

Since the two final velocities are equal, let's substitute in v_{1f} for v_{2f} in the first equation:

$$m_1\cdot v_{1i} + m_2\cdot v_{2i} = m_1\cdot v_{1f} + m_2\cdot v_{1f}$$

We can rewrite the right side: $m_1\cdot v_{1i} + m_2\cdot v_{2i} = (m_1 + m_2)\cdot v_{1f}$

Divide both sides by $(m_1 + m_2)$: $v_{1f} = (m_1\cdot v_{1i} + m_2\cdot v_{2i})/(m_1 + m_2)$

Put in numbers: v_{1f} = (65.0 kg·–1.11 m/s + 97.0 kg·–0.834 m/s)/(65.0 kg + 97.0 kg) = –0.945 m/s

Since the people stick together, this is the x component of final velocity for both people.

The y component is the same process:

(magnitude of v_{1iy}) = $v_{1i}\cdot\sin(\text{angle})$ = 1.35 m/s·sin(35.0°) = 0.774 m/s

Since v_1 is towards the north (+y) and west (–x), the y component is positive: v_{1iy} = 0.774 m/s

(magnitude of v_{2iy}) = 1.20 m/s·sin(46.0°) = 0.863 m/s

Since v_2 is towards the south (–y) and west (–x), the y component is negative: $v_{2iy} = -0.863$ m/s

We now combine $m_1 \cdot v_{1i} + m_2 \cdot v_{2i} = m_1 \cdot v_{1f} + m_2 \cdot v_{2f}$ and $v_{1f} = v_{2f}$ for the y components

Since the two final velocities are equal, let's substitute in v_{1f} for v_{2f} in the first equation:

$$m_1 \cdot v_{1i} + m_2 \cdot v_{2i} = m_1 \cdot v_{1f} + m_2 \cdot v_{1f}$$

We can rewrite the right side: $m_1 \cdot v_{1i} + m_2 \cdot v_{2i} = (m_1 + m_2) \cdot v_{1f}$

Divide both sides by $(m_1 + m_2)$: $v_{1f} = (m_1 \cdot v_{1i} + m_2 \cdot v_{2i})/(m_1 + m_2)$

Put in numbers: $v_{1f} = (65.0$ kg$\cdot 0.774$ m/s $+ 97.0$ kg$\cdot -0.863$ m/s$)/(65.0$ kg $+ 97.0$ kg$) = -0.206$ m/s

Since the people stick together, this is the y component of final velocity for both people.

We now have the x and y components: x component –0.945 m/s, y component –0.206 m/s.

$$\left(\text{magnitude}\right) = \sqrt{\left(\text{x component}\right)^2 + \left(\text{y component}\right)^2} \quad (3.3.2)$$

Put in numbers: $\left(\text{magnitude}\right) = \sqrt{\left(-0.945\,\text{m/s}\right)^2 + \left(-0.206\,\text{m/s}\right)^2}$ $= 0.967$ m/s

$$\theta = \tan^{-1}(|\text{y component}|/|\text{x component}|) \quad (3.3.3)$$

Put in numbers: $\theta = \tan^{-1}(|-0.206$ m/s$|/|-0.945$ m/s$|) = 12.3°$

Since the x and y components are negative, the velocity is towards the south (–y) and west (–x). We can say that the angle is 12.3° south of west

Final solution: the velocity of the people stuck together is 0.967 m/s, 12.3° south of west

Problem to Try Yourself

Two basketball players are chasing after the ball for a rebound. The first player has mass 85.0 kg, and is running at 2.50 m/s, at 70.0° north of east. The second player has mass 74.0 kg, and is running 3.50 m/s, at 25.0° south of east. The players collide and stick together. What is their velocity after the collision? Define north as the positive y direction, and east as the positive x direction.

Solution:

x components:

initial velocities:

(magnitude of v_{1ix}) = $v_{1i} \cdot \cos(\text{angle})$ = 2.50 m/s·cos(70.0°) = 0.855 m/s

Since v_1 is towards the north (+y) and east (+x), the x component is positive: $v_{1ix} = 0.855$ m/s

(magnitude of v_{2ix}) = 3.50 m/s·cos(25.0°) = 3.17 m/s

Since v_2 is towards the south (–y) and east (+x), the x component is positive: $v_{2ix} = 3.17$ m/s

We now combine $m_1 \cdot v_{1i} + m_2 \cdot v_{2i} = m_1 \cdot v_{1f} + m_2 \cdot v_{2f}$ and $v_{1f} = v_{2f}$ for the x components

Since the two final velocities are equal, let's substitute in v_{1f} for v_{2f} in the first equation:

$$m_1 \cdot v_{1i} + m_2 \cdot v_{2i} = m_1 \cdot v_{1f} + m_2 \cdot v_{1f}$$

We can rewrite the right side: $m_1 \cdot v_{1i} + m_2 \cdot v_{2i} = (m_1 + m_2) \cdot v_{1f}$

Divide both sides by $(m_1 + m_2)$: $v_{1f} = (m_1 \cdot v_{1i} + m_2 \cdot v_{2i})/(m_1 + m_2)$

Put in numbers: v_{1f} = (85.0 kg·0.855 m/s + 74.0 kg·3.17 m/s)/(85.0 kg + 74.0 kg) = 1.93 m/s

Remember that v_{1f} and v_{2f} are the same.

y components:

initial velocities:

(magnitude of v_{1iy}) = $v_{1i} \cdot \sin(\text{angle})$ = 2.50 m/s·sin(70.0°) = 2.35 m/s

Since v_1 is towards the north (+y) and east (+x), the y component is positive: $v_{1iy} = 2.35$ m/s

(magnitude of v_{2iy}) = 3.50 m/s·sin(25.0°) = 1.48 m/s

Since v_2 is towards the south (–y) and east (+x), the y component is negative: $v_{2iy} = -1.48$ m/s

We now combine $m_1 \cdot v_{1i} + m_2 \cdot v_{2i} = m_1 \cdot v_{1f} + m_2 \cdot v_{2f}$ and $v_{1f} = v_{2f}$ for the y components

Since the two final velocities are equal, let's substitute in v_{1f} for v_{2f} in the first equation:

$$m_1 \cdot v_{1i} + m_2 \cdot v_{2i} = m_1 \cdot v_{1f} + m_2 \cdot v_{1f}$$

We can rewrite the right side: $m_1 \cdot v_{1i} + m_2 \cdot v_{2i} = (m_1 + m_2) \cdot v_{1f}$

Divide both sides by $(m_1 + m_2)$: $v_{1f} = (m_1 \cdot v_{1i} + m_2 \cdot v_{2i})/(m_1 + m_2)$

Put in numbers: $v_{1f} = (85.0 \text{ kg} \cdot 2.35 \text{ m/s} + 74.0 \text{ kg} \cdot -1.48 \text{ m/s})/(85.0 \text{ kg} + 74.0 \text{ kg}) = 0.567 \text{ m/s}$

We now have the x and y components: x component 1.93 m/s, y component 0.567 m/s.

$$(\text{magnitude}) = \sqrt{(\text{x component})^2 + (\text{y component})^2} \quad (3.3.2)$$

Put in numbers: $(\text{magnitude}) = \sqrt{(1.93 \text{ m/s})^2 + (0.567 \text{ m/s})^2} = \underline{2.01 \text{ m/s}}$

$$\theta = \tan^{-1}(|\text{y component}|/|\text{x component}|) \quad (3.3.3)$$

Put in numbers: $\theta = \tan^{-1}(|0.567 \text{ m/s}|/|1.93 \text{ m/s}|) = 16.4°$

Since the x and y components are both positive, the velocity is towards the north (+y) and east (+x). We can say that the angle is 12.3° north of east

Final solution: <u>2.01 m/s, 12.3° north of east</u>

7.8 HOW MUCH FORCE HAPPENED?

During a collision that takes an amount of time Δt, the average force is equal to:

$$F_{average} = \Delta p/\Delta t \quad (7.8.1)$$

$F_{average}$ is the average force, Δt is the amount of time, and Δp is the change in momentum:

$$\Delta p = m \cdot v_f - m \cdot v_i \quad (7.3.1)$$

Side note: this equation is a form of Newton's Second Law ($F = m \cdot a$). To find it, you can rewrite Δp as being equal to $m \cdot \Delta v$, and remember that average acceleration equals $\Delta v/\Delta t$.

Example 7.8.1

Walking through a door, you hit your elbow on the doorway. We'll look at the arm separately, as a mass of 5.00 kg that was moving at 1.50 m/s before the collision. During the collision, let's model that it slows down to zero over 0.400 seconds. What is the average force?

Solution:

We'll use $F_{average} = \Delta p / \Delta t$ and $\Delta p = m \cdot v_f - m \cdot v_i$

Put the equation for Δp into the equation for force: $F_{average} = (m \cdot v_f - m \cdot v_i)/\Delta t$

Put in numbers: $F_{average} = (5.00 \text{ kg} \cdot 0 \text{ m/s} - 5.00 \text{ kg} \cdot 1.50 \text{ m/s})/(0.400 \text{ s}) = \underline{-18.8 \text{ Newtons}}$

In this problem, the minus sign means that the force is in the opposite direction of the velocity, since we made the velocity positive.

With units, remember that 1 Newton = 1 kg·m/s^2

Problem to Try Yourself

A baseball player crashes into a fence, while making a great catch. The player, who has mass 80.0 kg, goes from running 7.50 m/s to zero (at the wall) in 0.500 s. What is the average force?

Solution:

We'll use $F_{average} = \Delta p / \Delta t$ and $\Delta p = m \cdot v_f - m \cdot v_i$

Put the equation for Δp into the equation for force: $F_{average} = (m \cdot v_f - m \cdot v_i)/\Delta t$

Put in numbers: $F_{average} = (80.0 \text{ kg} \cdot 0 \text{ m/s} - 80.0 \text{ kg} \cdot 7.50 \text{ m/s})/(0.500 \text{ s}) = \underline{1.20 \cdot 10^3 \text{ Newtons}}$

In this problem, the minus sign means that the force is in the opposite direction of the velocity, since we made the velocity positive.

1 Newton = 1 kg·m/s^2

7.9 EXTRA TOPIC: CENTER OF MASS AND LINEAR MOMENTUM CONSERVATION

For a number of objects, or a number of pieces of a big object, you can calculate the location of the center of mass. For a number of objects that are all on one line (a one-dimensional problem), you can calculate the center of mass using:

$$x_{com} = \frac{\sum_{i=1}^{N} m_i \cdot x_i}{m_{total}} \tag{7.9.1}$$

Putting this equation into words: to calculate the location of the center of mass (x_{com}), add up (mass)·(position) for every single object, and then divide that by the total mass of all the objects (m_{total}).

For two dimensions, use Equation 7.9.1 twice – one with x components of position and once with y components.

You can also do the same thing with velocities, which gives the velocity of the center of mass. If the velocities are all on one line (a one-dimensional problem), you can use Equation 7.9.1 with velocities replacing positions:

$$v_{com} = \frac{\sum_{i=1}^{N} m_i \cdot v_i}{m_{total}} \tag{7.9.2}$$

If the objects are not all on the same line, and are instead a two-dimensional problem, do Equation 7.9.1 or 7.9.2 once for each dimension: one time with x components of position (7.9.1) or velocity (7.9.2), and once with y components.

Why do we include this problem type in linear momentum chapter? Multiply both sides of Equation 7.9.2 by total mass:

$$m_{total} \cdot v_{com} = \sum_{i=1}^{N} m_i \cdot v_i \tag{7.9.3}$$

Both sides of this equation are equal to the total linear momentum of a group of objects. The right side is exactly what we've been doing before – look at either side of Equation 7.4.1. Looking at Equation 7.9.3, if the total amount of linear momentum stays the same, it means that **the velocity of the center of mass has to stay the same.** For example, when two objects collide, we assume that their center of mass keeps moving at the same velocity before and after the collision.

Side note: if the velocity of the center of mass stays the same, then the acceleration of the center of mass must be zero. If acceleration is zero, then net force is zero. Remember that the requirement for linear momentum to be conserved was that the total (external) force on the objects equals zero.

Example 7.9.1

Two people are standing in line at a store's checkout. Person 1 is 0.500 m from the counter, and has mass 50.0 kg. Person 2 is 2.50 m from the counter, and has mass 80.0 kg. If we use the counter as $x = 0$, what is the position of the center of mass of these two people as a group?

Solution:

Equation 7.9.1 tells us to add (mass)·(position) for each object, then divide by total mass.

$x_{com} = (50.0 \text{ kg·} 0.500 \text{ m} + 80.0 \text{ kg·} 2.50 \text{ m})/(50.0 \text{ kg} + 80.0 \text{ kg}) = \underline{1.73}$ m (from the counter)

Note: the center of mass is closer to the person with the higher mass.

Problem to Try Yourself

Two people are walking in line to get on a bus. One person has mass 105 kg, and is walking towards the bus door at 1.05 m/s. The second person has mass 65.0 kg, and is walking towards the bus door at 1.50 m/s. What is the velocity of the center of mass of those two people as a group? Hint: this problem is asking for velocity of the center of mass, whereas the example above was asking for position. You'll do the same process, but use velocity instead of position.

Solution:

Equation 7.9.2 tells us to add (mass)·(velocity) for each object, then divide by total mass.

$v_{com} = (105 \text{ kg·} 1.05 \text{ m/s} + 65.0 \text{ kg·} 1.50 \text{ m/s})/(105 \text{ kg} + 65.0 \text{ kg}) = \underline{1.22}$ m/s (towards the door)

7.10 CHAPTER 7 SUMMARY

Linear momentum is a physics quantity equal to mass times velocity:

$$p = m \cdot v \tag{7.2.1}$$

The amount of change in the linear momentum is equal to:

$$\Delta p = m \cdot v_f - m \cdot v_i \tag{7.3.1}$$

The total amount of linear momentum stays the same during collisions:

$$m_1 \cdot v_{1i} + m_2 \cdot v_{2i} = m_1 \cdot v_{1f} + m_2 \cdot v_{2f} \tag{7.4.1}$$

If the collision is one-dimensional and the objects stick together (perfectly inelastic), combine Equation 7.4.1 with:

$$v_{1f} = v_{2f} \tag{7.5.1}$$

If the collision is one-dimensional and elastic (kinetic energy stays the same), combine Equation 7.4.1 with:

$$v_{1i} + v_{1f} = v_{2i} + v_{2f} \tag{7.6.1}$$

If the collision is two-dimensional and the objects stick together, use Equations 7.4.1 and 7.5.1 twice – once for each set of components (x and y). You might need to use some equations from Chapter 3 when working with components:

$$(\text{magnitude}) = \sqrt{(\text{x component})^2 + (\text{y component})^2} \quad (3.3.2)$$

$$\theta = \tan^{-1}(|\text{y component}|/|\text{x component}|) \quad (3.3.3)$$

$$(\text{x component magnitude}) = (\text{magnitude}) \cdot \cos(\theta) \quad (3.3.4)$$

$$(\text{y component magnitude}) = (\text{magnitude}) \cdot \sin(\theta) \quad (3.3.5)$$

To find the average amount of force in a collision:

$$F_{\text{average}} = \Delta p / \Delta t, \text{ where } \Delta p = m \cdot v_f - m \cdot v_i \quad (7.3.1 \text{ and } 7.8.1)$$

Linear momentum conservation can also be said as the velocity of the center of mass staying the same for a group of objects:

$$m_{\text{total}} \cdot v_{\text{com}} = \sum_{i=1}^{N} m_i \cdot v_i \quad (7.9.3)$$

$$x_{\text{com}} = \frac{\sum_{i=1}^{N} m_i \cdot x_i}{m_{\text{total}}} \quad (7.9.1)$$

Chapter 7 problem types:

Calculate the amount (or the change in) linear momentum (Section 7.3)

Collision problems with one dimension and objects sticking together (Section 7.5)

Collision problems with two dimensions and objects sticking together (Section 7.7)

Collision problems with one dimension and elastic collisions (Section 7.6)

Calculate using the average force in a collision (Section 7.8)

Calculate the position or velocity of the center of mass of a group of objects (Section 7.9)

Practice Problems
Section 7.3 (Linear Momentum Problems without Collisions)

1. What is the magnitude of the linear momentum of a 0.250 kg bird that is flying at 10.0 m/s?

   Solution:

   $$p = m \cdot v \tag{7.2.1}$$

   Put in numbers: $p = (0.250 \text{ kg}) \cdot (10.0 \text{ m/s}) = \underline{2.50 \text{ kg} \cdot \text{m/s}}$

2. A baby stroller is moving at 1.25 m/s, and has a linear momentum of 25.0 kg·m/s, to the right. What is the velocity?

   Solution:

   $$p = m \cdot v \tag{7.2.1}$$

   Divide both sides by m: $v = p/m$

   Put in numbers: $v = (25.0 \text{ kg} \cdot \text{m/s})/(1.25 \text{ m/s}) = 20.0 \text{ m/s}$

   The direction is to the right, since the momentum and the velocity are in the same direction.

   $\underline{v = 20.0 \text{ m/s, to the right}}$

3. Your friend is working on a ladder, and tosses a tool down to the ground. The tool has a mass of 1.50 kg. Right after the toss, the tool is traveling at 1.25 m/s, downward; just before the tool hits the ground it reaches a velocity of 7.00 m/s, downward. What is the change in momentum from the initial toss?

   Solution:

   $$\Delta p = m \cdot v_f - m \cdot v_i \tag{7.3.1}$$

   Put in numbers: $\Delta p = (1.50 \text{ kg}) \cdot (-7.00 \text{ m/s}) - (1.50 \text{ kg}) \cdot (-1.25 \text{ m/s}) = \underline{-8.63 \text{ kg} \cdot \text{m/s}}$

Note: this solution made the velocities negative since they were downward.

4. A baby stroller is moving at 1.15 m/s, to the right, and has a linear momentum of 21.0 kg·m/s, to the right. It then speeds up to 2.90 m/s, as the person behind it begins to jog. What is the change in linear momentum? Hint: use the information about the original time point to find the mass.

Solution:

As mentioned in the hint, we need to find the mass first. Looking at the initial time point, we know the momentum (21.0 kg·m/s, to the right) and the velocity (1.15 m/s, to the right). We'll choose "to the right" as the positive direction. We'll use Equation 7.2.1:

$$p = m \cdot v \tag{7.2.1}$$

Divide both sides by v: $m = p/v$

Put in numbers: $m = (21.0 \text{ kg·m/s})/(1.15 \text{ m/s}) = 18.3 \text{ kg}$

Note that both the momentum (p) and the velocity (v) were positive numbers, since they were to the right and we chose to the right as the positive direction.

Now that we have the mass, we can calculate change in momentum using Equation 7.3.1:

$$\Delta p = m \cdot v_f - m \cdot v_i \tag{7.3.1}$$

Put in numbers: $\Delta p = (18.3 \text{ kg}) \cdot (2.90 \text{ m/s}) - (18.3 \text{ kg}) \cdot (1.15 \text{ m/s}) = \underline{32.0}$ __kg·m/s__

Section 7.5 (Perfectly Inelastic Collisions with One Dimension)

5. A 75.0 kg person is walking at 2.00 m/s, and then walks into a 15.0 kg chair that was not moving. The chair and person stick together. What is their shared velocity after the collision?

Solution:

Use $m_1 \cdot v_{1i} + m_2 \cdot v_{2i} = m_1 \cdot v_{1f} + m_2 \cdot v_{2f}$ and $v_{1f} = v_{2f}$

The person can be number 1 and the chair can be number 2.

Set the direction that the person is walking as positive, which makes v_{1i} positive.

Note: a review of using two equations at once is included in Section 13.6.

Since $v_{1f} = v_{2f}$ we can put in v_{1f} for v_{2f} in the first equation

It becomes: $m_1 \cdot v_{1i} + m_2 \cdot v_{2i} = m_1 \cdot v_{1f} + m_2 \cdot v_{1f}$

On the right side, v_{1f} is in both terms. We can rewrite the right side to show this:

$$m_1 \cdot v_{1i} + m_2 \cdot v_{2i} = (m_1 + m_2) \cdot v_{1f}$$

Divide both sides by $(m_1 + m_2)$: $v_{1f} = (m_1 \cdot v_{1i} + m_2 \cdot v_{2i})/(m_1 + m_2)$

Put in numbers: $v_{1f} = (75.0 \text{ kg} \cdot 2.00 \text{ m/s} + 15.0 \text{ kg} \cdot 0 \text{ m/s})/(75.0 \text{ kg} + 15.0 \text{ kg}) = \underline{1.67 \text{ m/s}}$

Since $v_{1f} = v_{2f}$, $v_{2f} = \underline{1.67 \text{ m/s}}$ as well

Note: the positive sign means that the person and chair move in the direction the person was moving in, since we made that direction the positive direction.

6. A 55.0 kg person and an 85.0 kg person walk into each other in a straight-line hallway. The 55.0 kg person was walking to the right at 1.15 m/s; the 85.0 kg person was walking to the left at 1.60 m/s. The two people stick together. What is their shared velocity after the collision? Hint: you need to specify whether the final velocity is to the left or to the right.

Solution:

Use $m_1 \cdot v_{1i} + m_2 \cdot v_{2i} = m_1 \cdot v_{1f} + m_2 \cdot v_{2f}$ and $v_{1f} = v_{2f}$

Make the 55.0 kg person number 1, and the 85.0 kg person number 2.

Set "to the right" as positive, which makes v_{1i} positive (1.15 m/s) and v_{2i} negative (–1.60 m/s)

Note: a review of using two equations at once is included in Section 13.6.

Since $v_{1f} = v_{2f}$ we can put in v_{1f} for v_{2f} in the first equation

It becomes: $m_1 \cdot v_{1i} + m_2 \cdot v_{2i} = m_1 \cdot v_{1f} + m_2 \cdot v_{1f}$

On the right side, v_{1f} is in both terms. We can rewrite the right side to show this:

$$m_1 \cdot v_{1i} + m_2 \cdot v_{2i} = (m_1 + m_2) \cdot v_{1f}$$

Divide both sides by $(m_1 + m_2)$: $v_{1f} = (m_1 \cdot v_{1i} + m_2 \cdot v_{2i})/(m_1 + m_2)$

Put in numbers: $v_{1f} = (55.0$ kg·1.15 m/s + 85.0 kg·–1.60 m/s)/(55.0 kg + 85.0 kg) = –0.520 m/s

Since $v_{1f} = v_{2f}$, $v_{2f} =$ –0.520 m/s as well

Note: the negative sign means that the people (stuck together) move to the left.

7. A 65.0 kg person is walking in a significant windstorm when a 20.0 kg table moves with the wind and collides with the person. The person was moving at 1.00 m/s to the right, and the table was moving at 0.400 m/s to the left. The person and table stick together. What is the velocity after the collision? Hint: you need to specify whether the final velocity is to the left or to the right.

Solution:

Use $m_1 \cdot v_{1i} + m_2 \cdot v_{2i} = m_1 \cdot v_{1f} + m_2 \cdot v_{2f}$ and $v_{1f} = v_{2f}$

The person can be number 1, and the table can be number 2.

Set "to the right" as positive, which makes v_{1i} positive (1.00 m/s) and v_{2i} negative (–0.400 m/s)

Note: a review of using two equations at once is included in Section 13.6.

Since $v_{1f} = v_{2f}$ we can put in v_{1f} for v_{2f} in the first equation

It becomes: $m_1 \cdot v_{1i} + m_2 \cdot v_{2i} = m_1 \cdot v_{1f} + m_2 \cdot v_{1f}$

On the right side, v_{1f} is in both terms. We can rewrite the right side to show this:

$$m_1 \cdot v_{1i} + m_2 \cdot v_{2i} = (m_1 + m_2) \cdot v_{1f}$$

Divide both sides by $(m_1 + m_2)$: $v_{1f} = (m_1 \cdot v_{1i} + m_2 \cdot v_{2i})/(m_1 + m_2)$

Put in numbers: $v_{1f} = (65.0 \text{ kg·}1.00 \text{ m/s} + 20.0 \text{ kg·}-0.400 \text{ m/s})/(65.0 \text{ kg} + 20.0 \text{ kg}) = \underline{0.671 \text{ m/s}}$

Since $v_{1f} = v_{2f}$, $v_{2f} = \underline{0.671 \text{ m/s}}$ as well

Note: the positive sign means that the person and table move to the right.

8. Two basketball players collide while chasing down a loose ball. (Luckily both are OK.) One player has mass 80.0 kg and is moving to the right at 3.00 m/s; the second player has mass 95.0 kg and is moving to the left at 2.00 m/s. The two players stick together. What is their shared velocity after the collision? Hint: you need to specify whether the final velocity is to the left or to the right.

   Solution:

   Use $m_1 \cdot v_{1i} + m_2 \cdot v_{2i} = m_1 \cdot v_{1f} + m_2 \cdot v_{2f}$ and $v_{1f} = v_{2f}$

   The 80.0 kg person can be number 1, and the 95.0 kg person can be number 2.

   Set "to the right" as positive, which makes v_{1i} positive (3.00 m/s) and v_{2i} negative (–2.00 m/s)

   Note: a review of using two equations at once is included in Section 13.6.

   Since $v_{1f} = v_{2f}$ we can put in v_{1f} for v_{2f} in the first equation

   It becomes: $m_1 \cdot v_{1i} + m_2 \cdot v_{2i} = m_1 \cdot v_{1f} + m_2 \cdot v_{1f}$

   On the right side, v_{1f} is in both terms. We can rewrite the right side to show this:

   $$m_1 \cdot v_{1i} + m_2 \cdot v_{2i} = (m_1 + m_2) \cdot v_{1f}$$

   Divide both sides by $(m_1 + m_2)$: $v_{1f} = (m_1 \cdot v_{1i} + m_2 \cdot v_{2i})/(m_1 + m_2)$

   Put in numbers: $v_{1f} = (80.0 \text{ kg·}3.00 \text{ m/s} + 95.0 \text{ kg·}-2.00 \text{ m/s})/(80.0 \text{ kg} + 95.0 \text{ kg}) = \underline{0.286 \text{ m/s}}$

   Since $v_{1f} = v_{2f}$, $v_{2f} = \underline{0.286 \text{ m/s}}$ as well

   Note: the positive sign means that the two players move to the right.

9. Two friends collide during a game. (Luckily both are OK.) The first person has mass 115 kg and is running at 2.50 m/s to the left; the second person has mass 95.0 kg and is running at 2.00 m/s to the right. The two people stick together. What is their shared velocity after the

collision? Hint: you need to specify whether the final velocity is to the left or to the right.

Solution:

Use $m_1 \cdot v_{1i} + m_2 \cdot v_{2i} = m_1 \cdot v_{1f} + m_2 \cdot v_{2f}$ and $v_{1f} = v_{2f}$

The 115 kg person can be number 1, and the 95.0 kg person can be number 2.

Set "to the right" as positive, which makes v_{1i} negative (–2.50 m/s) and v_{2i} positive (2.00 m/s)

Note: a review of using two equations at once is included in Section 13.6.

Since $v_{1f} = v_{2f}$ we can put in v_{1f} for v_{2f} in the first equation

It becomes: $m_1 \cdot v_{1i} + m_2 \cdot v_{2i} = m_1 \cdot v_{1f} + m_2 \cdot v_{1f}$

On the right side, v_{1f} is in both terms. We can rewrite the right side to show this:

$$m_1 \cdot v_{1i} + m_2 \cdot v_{2i} = (m_1 + m_2) \cdot v_{1f}$$

Divide both sides by $(m_1 + m_2)$: $v_{1f} = (m_1 \cdot v_{1i} + m_2 \cdot v_{2i})/(m_1 + m_2)$

Put in numbers: $v_{1f} = (115 \ kg \cdot -2.50 \ m/s + 95.0 \ kg \cdot 2.00 \ m/s)/(115 \ kg + 95.0 \ kg) = \underline{-0.464 \ m/s}$

Since $v_{1f} = v_{2f}$, $v_{2f} = \underline{-0.464 \ m/s}$ as well

Note: the negative sign means that the two players move to the left.

Section 7.6 (Elastic Collisions with One Dimension)

10. Two objects undergo an elastic collision. The first object has mass 2.00 kg, and is traveling at 2.00 m/s to the right. The second object has mass 1.50 kg, and is traveling at 1.75 m/s to the left. What are the velocities (including left or right) after the collision?

    Solution:

    We'll use $m_1 \cdot v_{1i} + m_2 \cdot v_{2i} = m_1 \cdot v_{1f} + m_2 \cdot v_{2f}$ and $v_{1i} + v_{1f} = v_{2i} + v_{2f}$

    Note: the algebra in this problem is very similar to problem 13.6.1 in the math review chapter.

    Any negative velocities mean "to the left"; positive velocities are "to the right".

Use the second equation to relate v_{1f} and v_{2f}. It doesn't matter which one we solve for to start; we'll choose v_{1f}.

Take the second equation, and subtract v_{1i} from both sides:

$$v_{1f} = v_{2i} + v_{2f} - v_{1i}$$

We have now solved one equation for one of the unknown variables (v_{1f}), in terms of the other unknown variable (v_{2f}) and things that we know the values of.

Next, put the equation for v_{1f} in for v_{1f} in the first equation:

$$m_1 \cdot v_{1i} + m_2 \cdot v_{2i} = m_1 \cdot (v_{2i} + v_{2f} - v_{1i}) + m_2 \cdot v_{2f}$$

Multiply out m_1 times the three velocities in parentheses:

$$m_1 \cdot v_{1i} + m_2 \cdot v_{2i} = m_1 \cdot v_{2i} + m_1 \cdot v_{2f} - m_1 \cdot v_{1i} + m_2 \cdot v_{2f}$$

v_{2f} is now the only unknown variable in this equation. We want to get it by itself. Start by moving all terms on the right that do not have v_{2f} in it.

Add $m_1 \cdot v_{1i}$ to both sides, and subtract $m_1 \cdot v_{2i}$ from both sides:

$$m_1 \cdot v_{1i} + m_2 \cdot v_{2i} + m_1 \cdot v_{1i} - m_1 \cdot v_{2i} = m_1 \cdot v_{2f} + m_2 \cdot v_{2f}$$

We can rewrite the right side as $(m_1 + m_2) \cdot v_{2f}$, since both sides have v_{2f} multiplied in it

$$m_1 \cdot v_{1i} + m_2 \cdot v_{2i} + m_1 \cdot v_{1i} - m_1 \cdot v_{2i} = (m_1 + m_2) \cdot v_{2f}$$

Divide both sides by $(m_1 + m_2)$:

$$(m_1 \cdot v_{1i} + m_2 \cdot v_{2i} + m_1 \cdot v_{1i} - m_1 \cdot v_{2i})/(m_1 + m_2) = v_{2f}$$

Put in numbers:

$v_{2f} = (2.00 \text{ kg} \cdot 2.00 \text{ m/s} + 1.50 \text{ kg} \cdot -1.75 \text{ m/s} + 2.00 \text{ kg} \cdot 2.00 \text{ m/s} - 2.00 \text{ kg} \cdot -1.75 \text{ m/s})/(2.00 \text{ kg} + 1.50 \text{ kg}) = \underline{2.54 \text{ m/s}}$

We now have v_{2f}. To find v_{1f}, use the equation for v_{1f} that we solved for earlier in this problem:

$v_{1f} = v_{2i} + v_{2f} - v_{1i} = -1.75 \text{ m/s} + 2.54 \text{ m/s} - 2.00 \text{ m/s} = \underline{-1.21 \text{ m/s}}$

Note: negative means to the left, positive means to the right.

11. Two objects undergo an elastic collision. The first object has mass 3.00 kg, and is traveling at 1.75 m/s to the left. The second object has mass 4.50 kg, and is traveling at 4.75 m/s to the right. What are the velocities (including left or right) after the collision?

Solution:

We'll use $m_1 \cdot v_{1i} + m_2 \cdot v_{2i} = m_1 \cdot v_{1f} + m_2 \cdot v_{2f}$ and $v_{1i} + v_{1f} = v_{2i} + v_{2f}$

Note: the algebra in this problem is very similar to problem 13.6.1 in the math review chapter.

Any negative velocities mean "to the left"; positive velocities are "to the right".

Use the second equation to relate v_{1f} and v_{2f}. It doesn't matter which one we solve for to start; we'll choose v_{1f}.

Take the second equation, and subtract v_{1i} from both sides:

$$v_{1f} = v_{2i} + v_{2f} - v_{1i}$$

We have now solved one equation for one of the unknown variables (v_{1f}), in terms of the other unknown variable (v_{2f}) and things that we know the values of.

Next, put the equation for v_{1f} in for v_{1f} in the first equation:

$$m_1 \cdot v_{1i} + m_2 \cdot v_{2i} = m_1 \cdot (v_{2i} + v_{2f} - v_{1i}) + m_2 \cdot v_{2f}$$

Multiply out m_1 times the three velocities in parentheses:

$$m_1 \cdot v_{1i} + m_2 \cdot v_{2i} = m_1 \cdot v_{2i} + m_1 \cdot v_{2f} - m_1 \cdot v_{1i} + m_2 \cdot v_{2f}$$

v_{2f} is now the only unknown variable in this equation. We want to get it by itself. Start by moving all terms on the right that do not have v_{2f} in it.

Add $m_1 \cdot v_{1i}$ to both sides, and subtract $m_1 \cdot v_{2i}$ from both sides:

$$m_1 \cdot v_{1i} + m_2 \cdot v_{2i} + m_1 \cdot v_{1i} - m_1 \cdot v_{2i} = m_1 \cdot v_{2f} + m_2 \cdot v_{2f}$$

We can rewrite the right side as $(m_1 + m_2) \cdot v_{2f}$, since both sides have v_{2f} multiplied in it

$$m_1 \cdot v_{1i} + m_2 \cdot v_{2i} + m_1 \cdot v_{1i} - m_1 \cdot v_{2i} = (m_1 + m_2) \cdot v_{2f}$$

Divide both sides by $(m_1 + m_2)$:

$$(m_1 \cdot v_{1i} + m_2 \cdot v_{2i} + m_1 \cdot v_{1i} - m_1 \cdot v_{2i})/(m_1 + m_2) = v_{2f}$$

Put in numbers:

$v_{2f} = (3.00$ kg·-1.75 m/s$ + 4.50$ kg·4.75 m/s$ + 3.00$ kg·-1.75 m/s$ - 3.00$ kg·4.75 m/s$)/(3.00$ kg$ + 4.50$ kg$) = \underline{-0.450}$ m/s

We now have v_{2f}. To find v_{1f}, use the equation for v_{1f} that we solved for earlier in this problem:

$v_{1f} = v_{2i} + v_{2f} - v_{1i} = 4.75$ m/s$ + -0.450$ m/s$ - (-1.75$ m/s$) = \underline{6.05}$ m/s

Note: negative means to the left, positive means to the right.

12. Two objects undergo an elastic collision. The first object has mass 5.00 kg, and is traveling at 3.00 m/s to the right. The second object has mass 6.50 kg, and is traveling at 3.75 m/s to the left. What are the velocities (including left or right) after the collision?

Solution:

We'll use $m_1 \cdot v_{1i} + m_2 \cdot v_{2i} = m_1 \cdot v_{1f} + m_2 \cdot v_{2f}$ and $v_{1i} + v_{1f} = v_{2i} + v_{2f}$

Note: the algebra in this problem is very similar to problem 13.6.1 in the math review chapter.

Any negative velocities mean "to the left"; positive velocities are "to the right".

Use the second equation to relate v_{1f} and v_{2f}. It doesn't matter which one we solve for to start; we'll choose v_{1f}.

Take the second equation, and subtract v_{1i} from both sides:

$$v_{1f} = v_{2i} + v_{2f} - v_{1i}$$

We have now solved one equation for one of the unknown variables (v_{1f}), in terms of the other unknown variable (v_{2f}) and things that we know the values of.

Next, put the equation for v_{1f} in for v_{1f} in the first equation:

$$m_1 \cdot v_{1i} + m_2 \cdot v_{2i} = m_1 \cdot (v_{2i} + v_{2f} - v_{1i}) + m_2 \cdot v_{2f}$$

Multiply out m_1 times the three velocities in parentheses:

$$m_1 \cdot v_{1i} + m_2 \cdot v_{2i} = m_1 \cdot v_{2i} + m_1 \cdot v_{2f} - m_1 \cdot v_{1i} + m_2 \cdot v_{2f}$$

v_{2f} is now the only unknown variable in this equation. We want to get it by itself. Start by moving all terms on the right that do not have v_{2f} in it.

Add $m_1 \cdot v_{1i}$ to both sides, and subtract $m_1 \cdot v_{2i}$ from both sides:

$$m_1 \cdot v_{1i} + m_2 \cdot v_{2i} + m_1 \cdot v_{1i} - m_1 \cdot v_{2i} = m_1 \cdot v_{2f} + m_2 \cdot v_{2f}$$

We can rewrite the right side as $(m_1 + m_2) \cdot v_{2f}$, since both sides have v_{2f} multiplied in it

$$m_1 \cdot v_{1i} + m_2 \cdot v_{2i} + m_1 \cdot v_{1i} - m_1 \cdot v_{2i} = (m_1 + m_2) \cdot v_{2f}$$

Divide both sides by $(m_1 + m_2)$:

$$(m_1 \cdot v_{1i} + m_2 \cdot v_{2i} + m_1 \cdot v_{1i} - m_1 \cdot v_{2i})/(m_1 + m_2) = v_{2f}$$

Put in numbers:

$v_{2f} = (5.00 \text{ kg} \cdot 3.00 \text{ m/s} + 6.50 \text{ kg} \cdot -3.75 \text{ m/s} + 5.00 \text{ kg} \cdot 3.00 \text{ m/s} - 5.00 \text{ kg} \cdot -3.75 \text{ m/s})/(5.00 \text{ kg} + 6.50 \text{ kg}) = \underline{2.12 \text{ m/s}}$

We now have v_{2f}. To find v_{1f}, use the equation for v_{1f} that we solved for earlier in this problem:

$v_{1f} = v_{2i} + v_{2f} - v_{1i} = -3.75 \text{ m/s} + 2.12 \text{ m/s} - 3.00 \text{ m/s} = \underline{-4.63 \text{ m/s}}$

Note: negative means to the left, positive means to the right.

Section 7.7 (Perfectly Inelastic Collisions with Two Dimensions)

13. Two people are running after a ball, and they run into each other. No one is hurt, but the people stick together for a short time after the collision. The first person has mass 100.0 kg, and is traveling at 3.50

m/s at an angle of 30.0° north of east. The second person has mass 65.0 kg, and is traveling at 4.50 m/s at an angle of 57.0° south of west.

A. What are the x and y components of the final velocity of both carts? Hint: since the carts stick together, the final velocity is the same for both carts.

B. What is the magnitude and angle of the final velocity?

Solution:

Part A:

We need to solve this problem twice: once for the x components and once for the y components. The x component solution will give us the x component of final velocity; the y component solution will give the y component. We'll then combine those components to get the magnitude and angle (Part B)

x components: to do the x component, we'll need the x components of the initial velocities:

(magnitude of v_{1ix}) = v_{1i}·cos(angle) = 3.50 m/s·cos(30.0°) = 3.03 m/s

Since v_1 is towards the north (+y) and east (+x), the x component is positive: v_{1ix} = 3.03 m/s

Similarly, for the x component of v_{2i}, (magnitude of v_{2ix}) = 4.50 m/s·cos(5.0°) = 2.45 m/s

Since v_2 is towards the south (–y) and west (–x), the x component is negative: v_{2ix} = –2.45 m/s

We now combine $m_1 \cdot v_{1i} + m_2 \cdot v_{2i} = m_1 \cdot v_{1f} + m_2 \cdot v_{2f}$ and $v_{1f} = v_{2f}$ for the x components

Since the two final velocities are equal, let's substitute in v_{1f} for v_{2f} in the first equation:

$$m_1 \cdot v_{1i} + m_2 \cdot v_{2i} = m_1 \cdot v_{1f} + m_2 \cdot v_{1f}$$

We can rewrite the right side: $m_1 \cdot v_{1i} + m_2 \cdot v_{2i} = (m_1 + m_2) \cdot v_{1f}$

Divide both sides by $(m_1 + m_2)$: $v_{1f} = (m_1 \cdot v_{1i} + m_2 \cdot v_{2i})/(m_1 + m_2)$

Put in numbers: v_{1f} = (100 kg·3.03 m/s + 65.0 kg·–2.45 m/s)/(100 kg + 65.0 kg) = 0.872 m/s (x component)

Since the people stick together, this is the x component of final velocity for both people.

The y component is the same process:

(magnitude of v_{1iy}) = v_{1i}·sin(angle) = 3.50 m/s·sin(30.0°) = 1.75 m/s

Since v_1 is towards the north (+y) and east (+x), the y component is positive: v_{1iy} = 1.75 m/s

(magnitude of v_{2iy}) = 4.50 m/s·sin(57.0°) = 3.77 m/s

Since v_2 is towards the south (–y) and west (–x), the y component is negative: v_{2iy} = –3.77 m/s

We now combine $m_1 \cdot v_{1i} + m_2 \cdot v_{2i} = m_1 \cdot v_{1f} + m_2 \cdot v_{2f}$ and $v_{1f} = v_{2f}$ for the y components.

Since the two final velocities are equal, let's substitute in v_{1f} for v_{2f} in the first equation:

$$m_1 \cdot v_{1i} + m_2 \cdot v_{2i} = m_1 \cdot v_{1f} + m_2 \cdot v_{1f}$$

We can rewrite the right side: $m_1 \cdot v_{1i} + m_2 \cdot v_{2i} = (m_1 + m_2) \cdot v_{1f}$

Divide both sides by $(m_1 + m_2)$: $v_{1f} = (m_1 \cdot v_{1i} + m_2 \cdot v_{2i})/(m_1 + m_2)$

Put in numbers: v_{1f} = (100 kg·1.75 m/s + 65.0 kg·–3.77 m/s)/(100 kg + 65.0 kg) = <u>–0.426 m/s</u> (y component)

Since the people stick together, this is the y component of final velocity for both people.

Part B:

We now have the x and y components: x component 0.872 m/s, y component –0.426 m/s.

$$(\text{magnitude}) = \sqrt{(\text{x component})^2 + (\text{y component})^2} \qquad (3.3.2)$$

Put in numbers: $(\text{magnitude}) = \sqrt{(0.872 \, m/s)^2 + (-0.426 \, m/s)^2}$
= <u>0.970 m/s</u>

$$\theta = \tan^{-1}(|\text{y component}|/|\text{x component}|) \qquad (3.3.3)$$

Put in numbers: $\theta = \tan^{-1}(|-0.426 \text{ m/s}|/|0.872 \text{ m/s}|) = 26.0°$

Since the x component is positive and the y component is negative, the velocity is towards the south (–y) and east (+x). We can say that the angle is <u>26.0° south of east</u>

Final solution: the velocity of the people stuck together is 0.970 m/s, 23.7° south of east

14. Two people walk into each other, in a busy shopping area. No one is hurt, but the people stick together for a short time after the collision. The first person has mass 90.0 kg, and is traveling at 1.50 m/s at an angle of 30.0° north of east. The second person has mass 75.0 kg, and is traveling at 1.25 m/s at an angle of 55.0° south of east.

A. What are the x and y components of the final velocity of both carts? Hint: since the carts stick together, the final velocity is the same for both carts.

B. What is the magnitude and angle of the final velocity?

Solution:

Part A:

We need to solve this problem twice: once for the x components and once for the y components. The x component solution will give us the x component of final velocity; the y component solution will give the y component. We'll then combine those components to get the magnitude and angle (Part B)

x components: to do the x component, we'll need the x components of the initial velocities:

(magnitude of v_{1ix}) = v_{1i}·cos(angle) = 1.50 m/s·cos(30.0°) = 1.30 m/s

Since v_1 is towards the north (+y) and east (+x), the x component is positive: v_{1ix} = 1.30 m/s

Similarly, for the x component of v_{2i}, (magnitude of v_{2ix}) = 1.25 m/s·cos(55.0°) = 0.717 m/s

Since v_2 is towards the south (–y) and east (+x), the x component is positive: v_{2ix} = 0.717 m/s

We now combine $m_1 \cdot v_{1i} + m_2 \cdot v_{2i} = m_1 \cdot v_{1f} + m_2 \cdot v_{2f}$ and $v_{1f} = v_{2f}$ for the x components

Since the two final velocities are equal, let's substitute in v_{1f} for v_{2f} in the first equation:

$$m_1 \cdot v_{1i} + m_2 \cdot v_{2i} = m_1 \cdot v_{1f} + m_2 \cdot v_{1f}$$

We can rewrite the right side: $m_1 \cdot v_{1i} + m_2 \cdot v_{2i} = (m_1 + m_2) \cdot v_{1f}$

Divide both sides by $(m_1 + m_2)$: $v_{1f} = (m_1 \cdot v_{1i} + m_2 \cdot v_{2i})/(m_1 + m_2)$

Put in numbers: $v_{1f} = (90.0 \text{ kg} \cdot 1.30 \text{ m/s} + 75.0 \text{ kg} \cdot 0.717 \text{ m/s})/(90.0 \text{ kg} + 75.0 \text{ kg}) = \underline{1.03 \text{ m/s}}$ (x component)

Since the people stick together, this is the x component of final velocity for both people.

The y component is the same process:

(magnitude of v_{1iy}) $= v_{1i} \cdot \sin(\text{angle}) = 1.50 \text{ m/s} \cdot \sin(30.0°) = 0.750 \text{ m/s}$

Since v_1 is towards the north (+y) and east (+x), the y component is positive: $v_{1iy} = 0.750 \text{ m/s}$

(magnitude of v_{2iy}) $= 1.25 \text{ m/s} \cdot \sin(55.0°) = 1.02 \text{ m/s}$

Since v_2 is towards the south (–y) and east (+x), the y component is negative: $v_{2iy} = -1.02 \text{ m/s}$

We now combine $m_1 \cdot v_{1i} + m_2 \cdot v_{2i} = m_1 \cdot v_{1f} + m_2 \cdot v_{2f}$ and $v_{1f} = v_{2f}$ for the y components

Since the two final velocities are equal, let's substitute in v_{1f} for v_{2f} in the first equation:

$$m_1 \cdot v_{1i} + m_2 \cdot v_{2i} = m_1 \cdot v_{1f} + m_2 \cdot v_{1f}$$

We can rewrite the right side: $m_1 \cdot v_{1i} + m_2 \cdot v_{2i} = (m_1 + m_2) \cdot v_{1f}$

Divide both sides by $(m_1 + m_2)$: $v_{1f} = (m_1 \cdot v_{1i} + m_2 \cdot v_{2i})/(m_1 + m_2)$

Put in numbers: $v_{1f} = (90.0 \text{ kg} \cdot 0.750 \text{ m/s} + 75.0 \text{ kg} \cdot -1.02 \text{ m/s})/(90.0 \text{ kg} + 75.0 \text{ kg}) = \underline{-0.0563 \text{ m/s}}$ (y component)

Since the people stick together, this is the y component of final velocity for both people.

Part B:

We now have the x and y components: x component 1.03 m/s, y component –0.0563 m/s.

$$(\text{magnitude}) = \sqrt{(\text{x component})^2 + (\text{y component})^2} \qquad (3.3.2)$$

Put in numbers: $(\text{magnitude}) = \sqrt{(1.03 \text{ m/s})^2 + (-0.0563 \text{ m/s})^2}$

$= \underline{1.04 \text{ m/s}}$

$$\theta = \tan^{-1}(|y \text{ component}|/|x \text{ component}|) \qquad (3.3.3)$$

Put in numbers: $\theta = \tan^{-1}(|-0.0563 \text{ m/s}|/|1.03 \text{ m/s}|) = 3.12°$

Since the x component is positive and the y component is negative, the velocity is towards the south (−y) and east (+x). We can say that the angle is <u>3.12° south of east</u>

Final solution: the velocity of the people stuck together is <u>1.04 m/s, 3.12° south of east</u>

15. During the shootaround before a basketball game, two basketballs collide with each other just over the rim. The first basketball is traveling 5.00 m/s, at an angle of 25.0° below "to the right". The second basketball is traveling at 3.00 m/s, at an angle of 45.0° below "to the left". The basketballs have the same mass, which means that it will cancel out in all of the collision equations. The two balls stick together for a short time after the collision. What are the x and y components of the final velocity?

Solution:

We need to solve this problem twice: once for the x components and once for the y components. The x component solution will give us the x component of final velocity; the y component solution will give the y component. We'll make "up" the +y direction and "to the right" the +x direction.

x components: to do the x component, we'll need the x components of the initial velocities:

(magnitude of v_{1ix}) = v_{1i}·cos(angle) = 5.00 m/s·cos(25.0°) = 4.53 m/s

Since v_1 is towards down (−y) and right (+x), the x component is positive: $v_{1ix} = 4.53$ m/s

Similarly, for the x component of v_{2i}, (magnitude of v_{2ix}) = 3.00 m/s·cos(45.0°) = 2.12 m/s

Since v_2 is towards down (−y) and left (−x), the x component is negative: $v_{2ix} = -2.12$ m/s

We now combine $m_1 \cdot v_{1i} + m_2 \cdot v_{2i} = m_1 \cdot v_{1f} + m_2 \cdot v_{2f}$ and $v_{1f} = v_{2f}$ for the x components

Since the two final velocities are equal, let's substitute in v_{1f} for v_{2f} in the first equation:

$$m_1 \cdot v_{1i} + m_2 \cdot v_{2i} = m_1 \cdot v_{1f} + m_2 \cdot v_{1f}$$

Since m_1 and m_2 are equal here, we can divide by the mass (m_1 and m_2) and cancel all of the mass terms: $v_{1i} + v_{2i} = v_{1f} + v_{1f}$

We can rewrite the right side: $v_{1i} + v_{2i} = 2 \cdot v_{1f}$

Divide both sides by 2: $v_{1f} = (v_{1i} + v_{2i})/2$

Put in numbers: $v_{1f} = (4.53 \text{ m/s} - 2.12 \text{ m/s})/2 = \underline{1.21 \text{ m/s}}$ (x component)

Since the basketballs stick together, this is the x component of final velocity for both basketballs.

The y component is the same process:

(magnitude of v_{1iy}) = $v_{1i} \cdot \sin(\text{angle}) = 5.00 \text{ m/s} \cdot \sin(25.0°) = 2.11 \text{ m/s}$

Since v_1 is towards down (–y) and right (+x), the y component is negative: $v_{1iy} = -2.11 \text{ m/s}$

(magnitude of v_{2iy}) = $3.00 \text{ m/s} \cdot \sin(45.0°) = 2.12 \text{ m/s}$

Since v_2 is towards down (–y) and left (–x), the y component is negative: $v_{2iy} = -2.12 \text{ m/s}$

We now combine $m_1 \cdot v_{1i} + m_2 \cdot v_{2i} = m_1 \cdot v_{1f} + m_2 \cdot v_{2f}$ and $v_{1f} = v_{2f}$ for the y components

The math would look exactly the same as the x direction, and comes to exactly the same answer: $v_{1f} = (v_{1i} + v_{2i})/2$

Put in numbers: $v_{1f} = (-2.11 \text{ m/s} + -2.12 \text{ m/s})/2 = \underline{-2.12 \text{ m/s}}$

Section 7.8 (How Much Force Happened)

16. An acorn with mass $1.00 \cdot 10^{-2}$ kg approaches the ground at a velocity of 10.0 m/s. As it hits the ground, the velocity goes to zero over $5.00 \cdot 10^{-2}$ s. What is the magnitude of the average force?

Solution:

$$F_{average} = \Delta p / \Delta t \qquad (7.8.1)$$

$$\Delta p = m \cdot v_f - m \cdot v_i \qquad (7.3.1)$$

Put the equation for Δp into the equation for force: $F_{average} = (m \cdot v_f - m \cdot v_i)/\Delta t$

Put in numbers: $F_{average} = (1.00 \cdot 10^{-2} \text{ kg} \cdot 0 \text{ m/s} - 1.00 \cdot 10^{-2} \text{ kg} \cdot 10.0 \text{ m/s})/(5.00 \cdot 10^{-2} \text{ s}) = \underline{-2.00 \text{ N}}$

Note on units: remember that 1 Newton is the same as $1 \text{ kg} \cdot \text{m/s}^2$

17. One person collides with another person. One of the people in the collision has mass 75.0 kg, and their velocity changes from 1.55 m/s to the north (before the collision) to 0.150 m/s to the north after the collision. If the collision happens over 0.500 s, what is the average force on this person?

Solution:

$$F_{average} = \Delta p / \Delta t \qquad (7.8.1)$$

$$\Delta p = m \cdot v_f - m \cdot v_i \qquad (7.3.1)$$

Put the equation for Δp into the equation for force: $F_{average} = (m \cdot v_f - m \cdot v_i)/\Delta t$

Put in numbers: $F_{average} = (75.0 \text{ kg} \cdot 0.150 \text{ m/s} - 75.0 \text{ kg} \cdot 1.55 \text{ m/s})/(0.500 \text{ s}) = \underline{-182 \text{ N}}$

18. One problem in Section 7.5 involves a person who collides with another person. One of the people in the collision has mass 79.0 kg, and their velocity changes from 1.45 m/s to the south (before the collision) to 0.335 m/s to the north after the collision. If the collision happens over 0.500 s, what is the average force on this person?

Solution:

$$F_{average} = \Delta p / \Delta t \qquad (7.8.1)$$

$$\Delta p = m \cdot v_f - m \cdot v_i \qquad (7.3.1)$$

Put the equation for Δp into the equation for force: $F_{average} = (m \cdot v_f - m \cdot v_i)/\Delta t$

Put in numbers: $F_{average} = (79.0 \text{ kg} \cdot 0.335 \text{ m/s} - 79.0 \text{ kg} \cdot -1.45 \text{ m/s})/(0.500 \text{ s}) = \underline{253 \text{ N}}$

Section 7.9 (Center of Mass)

19. People stand in a line, waiting to get into an event. Person 1 is 1.00 m from the door, with mass 50.0 kg. Person 2 is 3.00 m from the door, with mass 95.0 kg. Person 3 is 5.00 m from the door, with mass 75.0 kg. Person 4 is 8.00 m from the door, with mass 112 kg. Where is the center of mass of the four people, in the direction along the line?

Solution:

$$x_{com} = \frac{\sum_{i=1}^{N} m_i \cdot x_i}{m_{total}}$$ (7.9.1)

Put in numbers: $x_{com} = $ (50.0 kg·1.00 m + 95.0 kg·3.00 m + 75.0 kg ·5.00 m + 112 kg·8.00 m)/(50.0 kg + 95.0 kg + 75.0 kg + 112 kg) = <u>4.84 m</u> (from the door)

20. The same people as Problem 19 are now moving. Person 1 (with mass 50.0 kg) is moving at 1.75 m/s, towards the door. Person 2 (with mass 95.0 kg) is moving at 1.50 m/s, towards the door. Person 3 (with mass 75.0 kg) is moving at 1.25 m/s, towards the door. Person 4 (with mass 112 kg) is moving at 1.00 m/s, towards the door. What is the velocity of the center of mass?

Solution:

$$v_{com} = \frac{\sum_{i=1}^{N} m_i \cdot v_i}{m_{total}}$$ (7.9.2)

Put in numbers: $v_{com} = $ (50.0 kg·1.75 m/s + 95.0 kg·1.50 m/s + 75.0 kg·1.25 m/s + 112 kg·1.00 m/s)/(50.0 kg + 95.0 kg + 75.0 kg + 112 kg) = <u>1.31 m/s</u>

Note: in this solution, the positive direction means towards the door.

21. People stand in a line, getting out of an event. Person 1 is 7.00 m outside the door, with mass 70.0 kg. Person 2 is 5.00 m outside the door, with mass 115 kg. Person 3 is 3.00 m outside the door, with mass 75.0 kg. Person 4 is 1.00 m outside the door, with mass 55.0 kg. Where is the center of mass of the four people, in the direction along the line?

Solution:

$$x_{com} = \frac{\sum_{i=1}^{N} m_i \cdot x_i}{m_{total}}$$ (7.9.1)

Put in numbers: $x_{com} = $ (70.0 kg·7.00 m + 115 kg·5.00 m + 75.0 kg·3.00 m + 55.0 kg·1.00 m)/(70.0 kg + 115 kg + 75.0 kg + 55.0 kg) = <u>4.27 m</u> (from the door)

22. The same people as Problem 21 are now moving. They are all moving towards outside. Person 1 (with mass 70.0 kg) is moving at 1.50 m/s. Person 2 (with mass 115 kg) is moving at 1.35 m/s. Person 3 (with mass 75.0 kg) is moving at 1.15 m/s. Person 4 (with mass 55.0 kg) is moving at 1.00 m/s. What is the velocity of the center of mass?

Solution:

$$v_{com} = \frac{\sum_{i=1}^{N} m_i \cdot v_i}{m_{total}} \qquad (7.9.2)$$

Put in numbers: v_{com} = (70.0 kg·1.50 m/s + 115 kg·1.35 m/s + 75.0 kg·1.15 m/s + 55.0 kg·1.00 m/s)/(70.0 kg + 115 kg + 75.0 kg + 55.0 kg) = <u>1.27 m/s</u>

Note: in this solution, the positive direction means "towards outside".

Uniform Circular Motion (Moving in a Circle at Constant Speed)

8.1 INTRODUCTION: SOMETIMES THINGS MOVE IN A CIRCLE

Many motion problems that we've covered so far have involved objects that are moving linearly – meaning, in a straight line. In the world, this is definitely not always the case. We're now going to focus on things that move in a circle.

In this chapter, we're going to look at a very specific type of motion in a circle. **Uniform circular motion** is when an object is moving in a circle at a constant speed (where speed means the magnitude of velocity). The direction changes as an object moves in a circle, but the speed itself stays the same.

8.2 CENTRIPETAL AND TANGENTIAL DIRECTIONS

For our linear motion, we've been using x and y coordinates. For circular motion, we'll use a different set of coordinates called **centripetal** and **tangential**. Picture an object traveling in a circle. The direction of travel at that instant is the tangential direction. The direction towards the center of the circle is the centripetal direction.

DOI: 10.1201/9781003005049-8

8.3 CENTRIPETAL ACCELERATION

For any object traveling in a circle, the acceleration in the centripetal direction is always equal to:

$$a_c = v^2/r \qquad (8.3.1)$$

a_c is the acceleration in the centripetal direction, v is the magnitude of velocity, and r is the radius of the circle that the object is going in.

Equation 8.3.1 applies to all objects traveling in a circle. For the specific case of uniform circular motion, where the speed is the same, two additional equations also apply.

$$a_t = 0 \qquad (8.3.2)$$

$$v = (\text{circumference})/(\text{time for one circle}) = (2 \cdot \pi \cdot r)/T \qquad (8.3.3)$$

a_t is the tangential direction acceleration, and it is zero for uniform circular motion. (Zero acceleration in that direction is the reason why the speed stays the same.) Since the magnitude of velocity is the same, it is equal to the distance it travels divided by time. One way to calculate it is to divide the circumference of the circle ($2 \cdot \pi \cdot r$) by the time it takes to go around the circle once (called T).

Example 8.3.1

A car turns off the highway onto a circular off ramp. It goes all the way around in a circle over 20.0 s. The radius of the circle is 30.0 m. What is the centripetal acceleration?

Solution:

We'll use $a_c = v^2/r$ and $v = (2 \cdot \pi \cdot r)/T$

We'll find v first: $v = (2 \cdot \pi \cdot r)/T = (2 \cdot \pi \cdot 30.0 \text{ m})/(20.0 \text{ s}) = 9.42 \text{ m/s}$

Put this into the equation for centripetal acceleration: $a_c = v^2/r = (9.42 \text{ m/s})^2/(30.0 \text{ m}) = \underline{2.96 \text{ m/s}^2}$

Problem to Try Yourself

A person is walking down a hallway, and their path ends up being ¼ of a circle as they turn to go into a room. The circle has radius 0.600 m, and it takes the person 3.50 s to make the ¼ of a circle. What is the centripetal acceleration? Hint: if it takes 2.50 seconds to make ¼ of a circle, it will take 10.0 s to do the entire circle.

Solution:
We'll use $a_c = v^2/r$ and $v = (2 \cdot \pi \cdot r)/T$

We'll find v first: $v = (2 \cdot \pi \cdot r)/T = (2 \cdot \pi \cdot 0.60\ 0\ m)/ (10.0\ s) = 0.377\ m/s$

Put this into the equation for centripetal acceleration: $a_c = v^2/r = (0.377\ m/s)^2/(0.600\ m) = \underline{0.237\ m/s^2}$

(Note: your answer may vary slightly, depending on rounding.)

8.4 NET FORCE IN THE CENTRIPETAL DIRECTION (ALSO CALLED CENTRIPETAL FORCE)

After reading this, you might be wondering what centripetal acceleration means. For an object to stay going in a circle, it must be pulled towards the center by a force. That force can be a lot of things – friction against the road for a car going in a circle, gravity for the moon going around Earth, and so on. It can be more than one force at a time.

No matter what is causing the force, we can find the total amount of force using the acceleration given by Equation 8.3.1:

$$a_c = v^2/r \qquad (8.3.1)$$

Newton's second law is that the net force is equal to:

$$F_{net} = m \cdot a$$

This is true for any direction, including the centripetal direction.

In this case, for the centripetal direction:

$$F_{c,\,net} = m \cdot v^2/r \qquad (8.4.1)$$

$F_{c,\,net}$ is the net force in the centripetal direction, m is mass, v is the magnitude of velocity (speed), and r is the radius of the circle that is being made by the moving object. $m \cdot v^2/r$ is the same thing as $m \cdot a_c$, since $a_c = v^2/r$.

When using this equation, it is very important to be careful about the sign of $m \cdot v^2/r$. The direction of the net force will always be "towards the center". That means that the net force can be in any direction, depending on the situation. It can sometimes be negative, for example if it is down or to the left.

Side note 1: the net force in the centripetal direction is often called the "centripetal force". However, it might be clearer to think of it as "net force in the centripetal direction", particularly since several of the upcoming problems will have more than one force.

Side note 2: in the examples below, the speed is always constant. If we look at one specific time, the strategies in this section would work even without constant speed.

Example 8.4.1: Centripetal acceleration from one force

The "problem to try yourself" problem from Section 8.3 had a person walking in a circle at 0.377 m/s, in a circle with radius 0.600 m. If the person has a mass of 75.0 kg, what is the net force in the centripetal direction?

Solution:

$F_{c, net} = m \cdot a_c = m \cdot v^2 / r$

Put in numbers: $F_{c,net} = 75.0 \, kg \cdot (0.377 \, m/s)^2 / (0.600 \, m) = \underline{17.8 \, Newtons}$

This force is about 2.4% of the person's gravity force (weight).

Note that this problem could have been asked in an easier way (if we'd started from the centripetal acceleration that was found) or asked in a harder way (if the velocity needed to be calculated from information in the original problem).

Example 8.4.2 Centripetal acceleration from two forces:

A 51.0 kg person is on a Ferris wheel, going in a vertical circle at a velocity of 5.00 m/s. The radius of the circle is 4.50 m. At the very bottom of the circle, there are two forces on the person: normal force upward and gravity downward. What is the magnitude of the normal force?

Solution:

Because the motion is in a circle, the net force in the centripetal direction must equal $m \cdot v^2 / r$, towards the center of the motion. At this point in time, when the person is at the bottom of the circle, the centripetal direction is up and down. Towards the center would be up. This means that the net force must equal $m \cdot v^2 / r$, upward.

The two forces are normal force (up) and gravity (m·g, down). If we set upward as positive, the net force would be $F_{normal} - m·g$.

We now have two equations for the net force: $F_{normal} - m·g$ and $m·v^2/r$, both with "upward" having a positive value. We can set them equal to each other:

$$F_{normal} - m·g = m·v^2/r$$

Add m·g to both sides: $F_{normal} = m·g + m·v^2/r$

Put in numbers: $F_{normal} = 51.0$ kg·9.80 m/s^2 + (51.0 kg)·(5.00 m/s)^2/4.50 m = 783 Newtons

Note 1: the force of gravity is m·g = 51.0 kg·9.80 m/s^2 = 500 Newtons. The normal force is larger in order to have the net force be upward (and equal to m·v^2/r).

Note 2: remember that 1 Newton = 1 kg·m/s^2

Problem to Try Yourself

A 56.0 kg person is on a Ferris wheel, going in a vertical circle at a velocity of 6.00 m/s. The radius of the circle is 4.50 m. At the very top of the circle, there are two forces on the person: normal force upward and gravity downward. What is the magnitude of the normal force?

Hint: now the direction of "towards the center" is downward.

Solution:

Because the motion is in a circle, the net force in the centripetal direction must equal m·v^2/r, towards the center of the motion. At this point in time, when the person is at the bottom of the circle, the centripetal direction is up and down. Towards the center is downward. This means that the net force must equal −m·v^2/r, if we define upward as positive.

The two forces are normal force (up) and gravity (m·g, down). If we set upward as positive, the net force would be $F_{normal} - m·g$.

We now have two equations for the net force: $F_{normal} - m·g$ and $-m·v^2/r$, both with "upward" having a positive value. We can set them equal to each other:

$$F_{normal} - m·g = -m·v^2/r$$

Add m·g to both sides: $F_{normal} = m·g - m·v^2/r$
Put in numbers: $F_{normal} = 56.0$ kg·9.80 m/s^2 − (56.0 kg)·(6.00
m/s)^2/4.50 m = <u>101 Newtons</u>

8.5 CHAPTER 8 SUMMARY

Some objects move in a circle. This chapter looks at objects that move in a circle at a constant speed. This is called **uniform circular motion**.

A new coordinate system was defined. Picture an object traveling in a circle. The direction of travel is called the **tangential direction**. The direction towards the center of the circle is the **centripetal direction**.

For any object going in a circle, we can write the acceleration and net force in the centripetal direction:

$$a_c = v^2/r \tag{8.3.1}$$

$$F_{c, net} = m·v^2/r \tag{8.4.1}$$

For the specific case of uniform circular motion, the tangential direction acceleration is zero and we can calculate the speed (v) using the circumference of the circle being made:

$$a_t = 0 \tag{8.3.2}$$

$$v = (circumference)/(time for one circle) = (2·\pi·r)/T \tag{8.3.3}$$

Chapter 8 problem types:

For the situation where an object is traveling in a circle at constant magnitude of velocity:

Calculate using the acceleration in the centripetal acceleration (Section 8.3)

Calculate using the net force in the centripetal direction (Section 8.4)

Practice Problems

Section 8.3 (Centripetal Acceleration)

1. An airplane flies in a vertical circle, with the pilot going upside down at the top. The circle is radius 200.0 m, and the velocity is constant and 40.0 m/s.

A. What is the magnitude of the centripetal acceleration of the pilot?

B. How much time does the plane take to go in a complete circle?

Solution:

Part A:

$$a_c = v^2/r \qquad (8.3.1)$$

Put in numbers: $a_c = (40.0 \text{ m/s})^2/(200.0 \text{ m}) = \underline{8.00 \text{ m/s}^2}$

Part B:

$$v = (\text{circumference})/(\text{time for one circle}) = (2 \cdot \pi \cdot r)/T \qquad (8.3.3)$$

T is the time for one circle. To solve for T in $v = (2 \cdot \pi \cdot r)/T$, multiply both sides by T and divide both sides by v: $T = (2 \cdot \pi \cdot r)/v$

Put in numbers: $T = (2 \cdot \pi \cdot 200.0 \text{ m})/(40.0 \text{ m/s}) = \underline{31.4 \text{ s}}$

2. An airplane flies in a vertical circle, with the pilot going upside down at the top. The circle has radius 200.0 m. If the pilot has a centripetal acceleration of 6.50 m/s², and is going at constant speed, what is the magnitude of that speed?

Solution:

$$a_c = v^2/r \qquad (8.3.1)$$

Solve for v. Multiply both sides by r: $v^2 = a_c \cdot r$

Take the square root: $v = \pm\sqrt{a_c \cdot r}$

Take the positive root, since we are calculating a magnitude and magnitudes are always positive.

Put in numbers: $v = +\sqrt{\left(6.50 \text{ m}/s^2\right) \cdot \left(200.0 \text{ m}\right)} = \underline{36.1 \text{ m/s}}$

3. A toy car goes over a loop-the-loop – a vertical circle where it is upside down at the top. The car is traveling at a constant speed of 1.75 m/s, and it takes 0.900 s to go in a full circle.

A. What is the radius?

B. What is the centripetal acceleration?

Solution:

Part A:

We have velocity and time to go in a full circle. To get to radius, we will use Equation 8.3.3:

$$v = (\text{circumference})/(\text{time for one circle}) = (2 \cdot \pi \cdot r)/T \qquad (8.3.3)$$

Multiply both sides by T: $v \cdot T = 2 \cdot \pi \cdot r$

Divide both sides by $(2 \cdot \pi)$: $r = v \cdot T/(2 \cdot \pi)$

Put in numbers: $r = (1.75 \text{ m/s}) \cdot (0.900 \text{ s})/(2 \cdot \pi) = \underline{0.251 \text{ m}}$

Part B:

$$a_c = v^2/r \qquad (8.3.1)$$

Put in numbers: $a_c = (1.75 \text{ m/s})^2/(0.251 \text{ m}) = \underline{12.2 \text{ m/s}^2}$

Section 8.4 (Net Force in the Centripetal Direction; Also Called Centripetal Force)

4. An airplane flies in a vertical circle, with the pilot going upside down at the top. The circle is radius 200.0 m, and the velocity is constant and 40.0 m/s. At the top of the circle, the pilot has two forces in the centripetal direction: force from a seatbelt (upward) and force from gravity (downward; pilot's mass is 75.0 kg).

   A. What is the net force in the centripetal direction?

   B. What is the force from the seatbelt?

Solution:

Part A:

We will use:

$$F_{c,\text{net}} = m \cdot v^2/r \qquad (8.4.1)$$

But we need to be very careful about directions. In this situation, the centripetal direction is up and down. The direction of the net force is towards the center, which is down. So the term $m \cdot v^2/r$ will be negative (downward).

Put in numbers: $F_{c,\ net} = (75.0\ kg) \cdot (40.0\ m/s)^2/(200.0\ m) = \underline{600\ N,}$ downward

Part B:

Writing out the individual forces:

(harness force) $- m \cdot g = -600\ N$

The negative sign on 600 N is because it is downward. $m \cdot g$ is the magnitude of the gravity force, from Equation 5.4.1

Add $m \cdot g$ to both sides: (harness force) $= m \cdot g - 600\ N$

Put in numbers: (harness force) $= (75.0\ kg) \cdot (9.80\ m/s^2) - 600$ $N = \underline{135\ N}$

5. An airplane flies in a vertical circle, with the pilot going upside down at the top. The circle is radius 200.0 m, and the velocity is constant. At the bottom of the circle, the pilot has two forces in the centripetal direction: force from the seat (upward) and force from gravity (downward; pilot's mass is 75.0 kg).

   A. If the force from the seat is $1.07 \cdot 10^3$ Newtons, what is the magnitude of the velocity of the plane?

   B. What is the net force in the centripetal direciton on the pilot?

   Solution:

   Part A:

   $$F_{c,\ net} = m \cdot v^2/r \qquad (8.4.1)$$

   Here, the net force in terms of forces is (seat force) $- m \cdot g$. $m \cdot v^2/r$ is always towards the center, which in this case is upward. So it will have a positive sign.

   Putting this into Equation 8.4.1:

   $$\text{(seat force)} - m \cdot g = m \cdot v^2/r$$

   We know the seat force, m, and r (and g). That makes v the only variable we don't know. Multiply both sides by r, and divide both sides by m:

   $$(r/m) \cdot (\text{(seat force)} - m \cdot g) = v^2$$

Take the square root: $v = \pm\sqrt{(r/m)\cdot\left((\text{seat force})-m\cdot g\right)}$

Take the positive number, since we are calculating the magnitude and the magnitude is positive.

Put in numbers: $v = +\sqrt{(200.0\,\text{m}/75.0\,\text{kg})\cdot\left(\dfrac{(1.07\cdot10^{3}\,\text{N})}{-75.0\,\text{kg}\cdot9.80\,\text{m}/\text{s}^{2}}\right)}$

$v = \underline{29.9\ \text{m/s}}$

Part B:

$$F_{c,\,net} = m\cdot v^{2}/r \tag{8.4.1}$$

Here, the net force in terms of forces is (seat force) – m·g. m·v²/r is always towards the center, which in this case is upward. So we'll give it a positive sign.

Put in numbers for m·v²/r: (75.0 kg)·(29.9 m/s)²/(200.0 m) = $\underline{335\ N}$ $\underline{\text{(upward)}}$

6. A toy car goes over a loop-the-loop – a vertical circle where it is upside down at the top. The track makes a circle with radius 0.200 m. The mass of the car is 5.00·10⁻² kg. At the bottom of the loop, there are two forces on the car in the centripetal direction: gravity (downward) and a force upward from the track. The force upward from the track is 1.00 N. Assume that the car has a constant velocity.

A. What is the net force in the centripetal direction?

B. What is the speed of the car?

Solution:

Part A:

$$F_{c,\,net} = m\cdot v^{2}/r \tag{8.4.1}$$

In this case, we don't know v, so we can't use m·v²/r. However, we do know both of the forces that go into the net force, so we can use the forces to calculate it.

$F_{c,\,net}$ = (force from track) – m·g = (1.00 N) – (5.00·10⁻² kg)·(9.80 m/ s²) = $\underline{0.51\ N\ \text{(upward)}}$

Note: getting a positive number (meaning upward) makes sense because it is towards the center.

Part B:

$$F_{c, net} = m \cdot v^2 / r \tag{8.4.1}$$

We know $F_{c, net}$ from Part A (0.51 N), so v is the only variable that we don't know.

Multiply both sides by r, and divide both sides by m:

$$(r/m) \cdot F_{c, net} = v^2$$

Take the square root: $v = \pm \sqrt{(r/m) \cdot F_{c,net}}$

Take the positive number, since we are calculating the magnitude and the magnitude is positive.

Put in numbers: $v = +\sqrt{(0.200\,m / 5.00 \cdot 10^{-2}\,kg) \cdot (0.51\,N)} = \underline{1.4\ m/s}$

7. Note: this problem is a bit harder than the others. A toy car goes over a loop-the-loop – a vertical circle where it is upside down at the top. The car is traveling at a constant speed of 1.75 m/s, and the track makes a circle with radius 0.200 m. At the top of the loop, there are two forces on the car in the centripetal direction: gravity (downward) and a force downward from the track (0.230 N, downward). What is the mass of the car?

Solution:

$$F_{c, net} = m \cdot v^2 / r \tag{8.4.1}$$

We can rewrite the sum of forces as –(track force) – m·g. Both are negative because both are downward. m·v²/r will also be downward, since "down" is towards the center. We know v, r, g, and the track force, so m is the only variable that we don't know.

Putting the forces and signs into Equation 8.4.1:

–(track force) – m·g = – m·v²/r

Add m·g to both sides: −(track force) = m·g − m·v²/r

Since m is in both terms on the right, we can rewrite it as:

−(track force) = m·(g − v²/r)

Divide both sides by (g − v²/r): m = −(track force)/(g − v²/r)

Put in numbers: m = −(0.230 N)/(9.80 m/s² − (1.75 m/s)²/0.200 m)) = <u>4.17·10⁻² kg</u>

Rotation Motion and Forces

9.1 INTRODUCTION: ROTATIONAL MOTION IS LIKE LINEAR MOTION

Chapters 9 and 10 are about rotational motion, meaning things that rotate in a circle. Even though this motion looks different, you will see that the math is very similar to things we've done before.

One example of this: instead of position being something that changes with motion, it is replaced with angle. Then, instead of velocity (how much the distance is changing with time), you will have an angular velocity, which is how much the angle is changing with time.

We will end up repeating many of the things we have done before, including the equations from Chapter 2 that describe motion, and Newton's second law from Chapter 4. In a way, it will be like reviewing almost everything that we have already done.

9.2 UNITS FOR ANGLE: RADIANS AND REVOLUTIONS

The physics equations for rotation done in Chapters 9 and 10 will assume that the angles are in the unit of radians (rad). The unit conversion is:

$$180 \text{ degrees} = \pi \text{ radians} \tag{9.2.1}$$

π is approximately equal to 3.14.

DOI: 10.1201/9781003005049-9

Another unit for angle that is used often is revolutions (rev). The conversion is:

$$360 \text{ degrees} = 1 \text{ revolution} \qquad (9.2.2)$$

This conversion is because 1 revolution is the number of degrees in a full circle.

You can combine Equations 9.2.1 and 9.2.2 to get a conversion between radians and revolutions:

$$1 \text{ revolution} = 2 \cdot \pi \text{ radians} \qquad (9.2.3)$$

Example 9.2.1

Convert 279 degrees to (A) radians and to (B) revolutions.

A. $279 \text{ degrees} \cdot \dfrac{\pi \text{ radians}}{180 \text{ degrees}} = 4.87 \text{ rad}$

B. $279 \text{ degrees} \cdot \dfrac{1 \text{ revolution}}{360 \text{ degrees}} = 0.775 \text{ rev}$

Problem to Try Yourself

A snowboarder does a "1080", covering 1080 degrees, which is three full turns. What is this angle in radians and in revolutions?"

A. $1080 \text{ degrees} \cdot \dfrac{\pi \text{ radians}}{180 \text{ degrees}} = 18.8 \text{ rad}$ (note: this is 6·π)

B. $1080 \text{ degrees} \cdot \dfrac{1 \text{ revolution}}{360 \text{ degrees}} = 3.00 \text{ rev}$

9.3 ROTATION EQUIVALENTS OF POSITION, VELOCITY, AND ACCELERATION

When things rotate, we will look at the change in angle instead of position. This means that the rotation equivalent of position is **angle**. The variable for angle is θ (Greek letter theta). Similarly, the **angular displacement** is the equivalent of displacement (Δx) in linear motion, and is called Δθ.

The units of angle are radians for Chapters 9 and 10.

In previous chapters, we had "velocity" (v), which more specifically is **linear velocity**. The rotation equivalent of linear velocity is **angular velocity**, which we call ω (Greek letter omega). Conceptually, ω is the amount that the angle is changing with time, in the same way that linear velocity is how much displacement is happening with time. The units of angular velocity are radians per second (rad/s). Remember that linear velocity has units of m/s.

The rotation equivalent of linear acceleration (a) is **angular acceleration**, which we call α (Greek letter alpha). The angular acceleration is how much the angular velocity is changing with time, just like how (linear) acceleration a is how much the (linear) velocity v is changing with time. The units of angular acceleration are radians per second squared (rad/s^2), similar to how linear acceleration has units of m/s^2.

Side note: some textbooks calculate average angular velocity and average angular acceleration. The equations for these are:

$$\omega_{average} = \Delta\theta / \Delta t \qquad (9.3.1)$$

$$\alpha_{average} = \Delta\omega / \Delta t \qquad (9.3.2)$$

Δθ (angular displacement), ω (angular velocity), and α (angular velocity) are all as described above; Δt is the amount of time the average is over and Δω would be the amount of change in the angular velocity over the period of time. This book will skip any examples with these equations, and instead focus on the problems in the next section that use values that are specific to a given time point.

9.4 MOTION WITH CONSTANT ANGULAR ACCELERATION

In Chapter 2, we covered problems with constant linear acceleration. There were four equations:

$$v_f = v_i + a \cdot \Delta t \qquad (2.6.1)$$

$$\Delta x = v_i \cdot \Delta t + \frac{1}{2} \cdot a \cdot (\Delta t)^2 \qquad (2.6.2)$$

$$\Delta x = \frac{1}{2} \cdot \left(v_i + v_f \right) \cdot \Delta t \qquad (2.6.3)$$

$$v_f^{\,2} = v_i^{\,2} + 2 \cdot a \cdot \Delta x \qquad (2.6.4)$$

There are five variables (a, v_i, v_f, Δt and Δx), and these equations give you the ability to solve for two variables so you expect the problem statement to give you the other three.

For angular motion with constant angular acceleration, we can use essentially the same equations. The only difference is that we replace displacement with angular displacement, replace linear velocity with angular velocity, and replace linear acceleration with angular acceleration. The equations now look like this:

$$\omega_f = \omega_i + \alpha \cdot \Delta t \qquad (9.4.1)$$

$$\Delta\theta = \omega_i \cdot \Delta t + \tfrac{1}{2} \cdot \alpha \cdot (\Delta t)^2 \qquad (9.4.2)$$

$$\Delta\theta = \tfrac{1}{2} \cdot (\omega_i + \omega_f) \cdot \Delta t \qquad (9.4.3)$$

$$\omega_f^{\,2} = \omega_i^{\,2} + 2 \cdot \alpha \cdot \Delta\theta \qquad (9.4.4)$$

Now, the five variables are ω_i (initial time point angular velocity), ω_f (final time point angular velocity), α (angular acceleration, which has only one value throughout the problem), Δt (time between initial and final time points), and $\Delta\theta$ (angular displacement between initial and final time points). You'll expect to find at least three of the variables given in the problem, and two left to solve for with these equations.

In Chapter 2, we had to define a positive direction and a negative direction. We usually chose "to the right" or "up" as positive, and "to the left" or "down" as negative. For rotations, you can have "counterclockwise" as positive, and "clockwise" as negative. Clockwise is the direction that the hands move on a clock; counterclockwise is the opposite direction.

Example 9.4.1

At a summer job, you turn on a string trimmer to chop some weeds. The string goes around in a circle as the string trimmer operates. The string speeds up from zero over a period of 0.870 s, using a constant angular acceleration. If it goes through an angle of 48.0 radians during this time, what was the value of the angular acceleration?

Solution:
We are given $\omega_i = 0$, $\Delta t = 0.870$ s, and $\Delta \theta = 48.0$ radians. We want to find α. The equation with only α and things that we know is Equation 9.4.2:

$$\Delta \theta = \omega_i \cdot \Delta t + \tfrac{1}{2} \cdot \alpha \cdot (\Delta t)^2$$

We want to solve for α. First, to make the math easier, put 0 in for ω_i.

$$\Delta \theta = 0 \cdot \Delta t + \tfrac{1}{2} \cdot \alpha \cdot (\Delta t)^2$$

This becomes $\Delta \theta = \tfrac{1}{2} \cdot \alpha \cdot (\Delta t)^2$
  Next, divide both sides by $(\tfrac{1}{2} \cdot (\Delta t)^2)$, which is everything next to α:

$$\alpha = \Delta \theta / (\tfrac{1}{2} \cdot (\Delta t)^2)$$

Put in numbers: $\alpha = 48.0$ rad$/(\tfrac{1}{2} \cdot (0.870$ s$)^2) = \underline{127 \text{ rad/s}^2}$
  Note 1: we assumed that the direction that the string was spinning in was the positive direction.
  Note 2: we can also use the information in this problem to find ω_f, and it comes to 110 rad/s.

Problem to Try Yourself

At a similar summer job, you are using a brush cutter, which involves a small cutter spinning around in a circle. The cutter is spinning at 24.0 rad/s, and then you turn it off. The angular velocity slows down with an acceleration of -10.0 rad/s^2. How much time does the cutter take the stop rotating?

Solution:
We are given $\omega_i = 24.0$ rad/s, $\alpha = -10.0$ rad/s^2, and $\omega_f = 0$. We are asked to find Δt. The equation with Δt and variables that we know is Equation 9.4.1:

$$\omega_f = \omega_i + \alpha \cdot \Delta t$$

To solve for Δt, first subtract ω_i from both sides: $\omega_f - \omega_i = \alpha \cdot \Delta t$
Next, divide both sides by α: $\Delta t = (\omega_f - \omega_i)/\alpha$
Put in numbers: $\Delta t = (0 - 24.0 \text{ rad/s})/(-10.0 \text{ rad/s}^2) = \underline{2.40 \text{ s}}$
Note: we could also use the information in this problem to find $\Delta \theta$; it turns out to be 28.8 rad.

9.5 MOMENT OF INERTIA (ROTATION EQUIVALENT OF MASS)

In order to talk about forces in Chapter 4, we needed mass. The rotation equivalent of mass is called the **moment of inertia**. For a "point mass", meaning something that is very small, the moment of inertia is equal to:

$$I = m \cdot r^2 \tag{9.5.1}$$

I is moment of inertia, m is mass, and r is the distance (radius) that the mass is from the center of rotation. The units of moment of inertia are kg·m² (kg from the mass and m² from the r²).

For an object that is larger than a point mass, we have equations for different shapes. These equations come from adding every single point mass inside of the larger objects. There are many equations, but we will keep our examples to using these shapes:

Solid sphere, spinning around a line through the middle:

$$I = (2/5) \cdot m \cdot R^2 \tag{9.5.2}$$

Solid disk, spinning around a perpendicular line

$$\text{through the middle: } I = (1/2) \cdot m \cdot R^2 \tag{9.5.3}$$

$$\text{Rod, spun around one of its ends: } I = (1/3) \cdot m \cdot L^2 \tag{9.5.4}$$

Example 9.5.1 – Point Mass Moment of Inertia

A point mass has moment of inertia 1.50 kg·m² and mass 0.125 kg. How far away is it from the center of rotation?

Solution:

$$I = m \cdot r^2$$

We have I and m, and are solving for r.

Divide both sides by m: $r^2 = I/m$

Take the square root: $r = \sqrt{I/m}$

Note: The ± sign that normally comes with a square root can be left out here, because the radius (distance) is always positive.

Put in numbers: $r = \sqrt{\left(1.50\,\text{kg} \cdot \text{m}^2\right)/\left(0.125\,\text{kg}\right)} = 3.46\,\text{m}$

Example 9.5.2 – Larger Shape moment of Inertia

Picture a string trimmer as two rods spinning in a circle. Each rod is 0.120 m long, and has a mass of $1.50 \cdot 10^{-2}$ kg. What is the total moment of inertia?

Solution:

For each rod, $I = (1/3) \cdot m \cdot L^2$. For the two rods that are the same, it will be twice as much

$I = 2 \cdot (1/3) \cdot m \cdot L^2 = 2 \cdot (1/3) \cdot (1.50 \cdot 10^{-2} \text{ kg}) \cdot (0.120 \text{ m})^2 = \underline{1.44 \cdot 10^{-4} \text{ kg} \cdot m^2}$

Note: if the two rods were different in some way, you could calculate the moment of inertia separate for each one and then add them.

Example 9.5.3 – Combining More than One Shape

Picture the grass-cutting blade of a lawnmower as being made of three parts: two rods of length 0.350 m and mass 1.50 kg for each, and one disk of radius 0.100 m and mass 3.00 kg. What is the total moment of inertia?

Solution:

Calculate separately, and then add.

The rods are each $I = (1/3) \cdot m \cdot L^2$; the disk is $I = (1/2) \cdot m \cdot R^2$

Total moment of inertia $= 2 \cdot (1/3) \cdot m \cdot L^2 + (1/2) \cdot m \cdot R^2$, where the first m is rod mass and the second m is disk mass

Put in numbers: $I = 2 \cdot (1/3) \cdot (1.50 \text{ kg}) \cdot (0.350 \text{ m})^2 + (1/2) \cdot (3.00 \text{ kg}) \cdot (0.100 \text{ m})^2 = \underline{0.138 \text{ kg} \cdot m^2}$

Problem to Try Yourself

A blender's cutting blade is made of four rods that spin around their ends, and a disk on top of them. The rods are each $8.50 \cdot 10^{-2}$ m long and each has mass 0.100 kg. The disk has radius $2.75 \cdot 10^{-2}$ m, and mass 0.170 kg. What is the total moment of inertia?

Solution:

Calculate separately, and then add.

The rods are each $I = (1/3) \cdot m \cdot L^2$; the disk is $I = (1/2) \cdot m \cdot R^2$

Total moment of inertia $= 4 \cdot (1/3) \cdot m \cdot L^2 + (1/2) \cdot m \cdot R^2$, where the first m is rod mass and the second m is disk mass

Put in numbers: $I = 4 \cdot (1/3) \cdot (0.100 \text{ kg}) \cdot (8.50 \cdot 10^{-2} \text{ m})^2 + (1/2) \cdot (0.170 \text{ kg}) \cdot (2.75 \cdot 10^{-2} \text{ m})^2 = \underline{1.03 \cdot 10^{-3} \text{ kg} \cdot m^2}$

9.6 TORQUE (ROTATION EQUIVALENT OF FORCE)

Torque is the rotation equivalent of force. Each force creates an amount torque given by:

$$\tau = r \cdot F \cdot \sin(\varphi) \qquad (9.6.1)$$

τ (Greek letter tau) is the amount of torque, r is the distance (radius) from the center of rotation to where the force happens, F is the amount (magnitude) of force, and φ (Greek letter phi) is the angle between the direction of F and the direction of a line from the center of rotation to where the force happens.

The torques can also have a sign, just like forces were positive if upward and negative if downward in Chapter 4. If the torque is in the direction of a counterclockwise rotation, it is positive. If the torque is in the direction of a clockwise rotation, it is negative.

Example 9.6.1

Inside of a motor, a gear is spun by touching another gear. Approximate that the gear is a disk, and imagine that you are looking at the gear from above. The force is almost on the edge of the gear, $7.00 \cdot 10^{-2}$ m from the center. The force is in the tangential direction, would cause a counterclockwise torque, and is equal to 1.25 N. What is the torque? Be certain to get the sign correct.

Solution:

$$\tau = r \cdot F \cdot \sin(\varphi) \qquad (9.6.1)$$

The distance from center to where the force is applied is $r = 7.00 \cdot 10^{-2}$ m. The force is given as 1.25 Newtons. The angle φ is 90°, because the direction of the line from center to where the force happens is along a radius (the centripetal direction), and the force's direction (tangential) is 90° away from this. The torque's sign is positive since it would cause a counterclockwise rotation.

Put in numbers: $\tau = (7.00 \cdot 10^{-2}$ m$) \cdot (1.25$ N$) \cdot \sin(90°) = \underline{8.75 \cdot 10^{-2}$ N·m}

Note: torque calculations with tangential forces happen often, and the angle in the torque equation is always 90° for those forces.

Example 9.6.2

Inside of a motor, a gear is spun by touching another gear. Approximate that the gear is a disk, and imagine that you are looking at the gear from above. The force is almost on the edge of the gear, $7.00 \cdot 10^{-2}$ m from the center. The force is equal to 1.25 N, but is now 50.0° from the centripetal direction and would cause a clockwise torque. What is the torque? Be certain to get the sign correct.

Solution:

$$\tau = r \cdot F \cdot \sin(\varphi)$$

The distance from center to where the force is applied is $r = 7.00 \cdot 10^{-2}$ m. The force is given as 1.25 Newtons. The angle φ is 50°, because the direction of the line from center to where the force happens is the centripetal direction, and the force's direction is 50.0° away from this. The torque's sign is negative since it would cause a counterclockwise rotation.

Put in numbers: $\tau = -(7.00 \cdot 10^{-2}$ m$) \cdot (1.25$ N$) \cdot \sin(50.0°) = \underline{-6.70 \cdot 10^{-2}}$ N·m

Example 9.6.3

Inside of a motor, one gear is unexpectedly putting a force on a second gear. The 4.50 N force happens at the edge of the gear, $7.35 \cdot 10^{-2}$ m from the center. This force is along the centripetal (radial) direction. What is the torque?

Solution:

$$\tau = r \cdot F \cdot \sin(\varphi)$$

Here, we are given that $r = 7.35 \cdot 10^{-2}$ m and $F = 4.50$ N. However, φ is either 0° or 180°, since both the force and the line from center to where the force happens are both in the centripetal direction. (You could picture one gear putting a force on the left side of one gear, which would result in a line from center being to the left, and the force being to the right, for an angle of 180°.) Either way, $\sin(0°)$ and $\sin(180°)$ are both equal to zero.

$$\underline{\tau = 0}$$

Note: any time a force is in the centripetal direction, there is no torque.

Problem to Try Yourself

Picture a spinning wheel on top of a table, for a board game. There is a spinner (like a rod), that spins around and lands on different numbers or colors. You flick the spinner with a force of 3.25 N, in the tangential direction, just barely inside the edge of the spinner ($2.75 \cdot 10^{-2}$ m from the center). A clockwise spin comes from this. What was the torque?

Solution:

$$\tau = r \cdot F \cdot \sin(\varphi)$$

The force (F = 3.25 N) and distance ($r = 2.75 \cdot 10^{-2}$ m) are given. The angle is 90° since the force is tangential. The torque is negative because it must have been in the clockwise direction in order to cause a clockwise spin.

Put in numbers: $\tau = -(2.75 \cdot 10^{-2}$ m$) \cdot (3.25$ N$) \cdot \sin(90°) = \underline{-8.94 \cdot 10^{-2}$ N·m}

9.7 NEWTON'S SECOND LAW (ROTATION EQUIVALENT)

In Chapter 4, we covered Newton's second law:

$$F_{net} = m \cdot a \tag{4.2.1}$$

In words, adding all of the forces was equal to mass times acceleration.

For rotation, Newton's second law is the same type of equation, but now with rotation equivalent quantities instead of linear quantities.

$$\tau_{net} = I \cdot \alpha \tag{9.7.1}$$

τ_{net} is the **net torque** on an object (equivalent of force), I is moment of inertia of the object (equivalent of mass), and α is the angular acceleration of the object (equivalent of a). To calculate net torque, just add up all of the torques on an object, making sure the torques toward the counterclockwise direction are positive and torques toward the clockwise direction are negative.

Some problems use this equation along with Equations 9.4.1 to 9.4.4. Assuming that the net torque and moment of inertia stay the same during the problem, the angular acceleration (α) is constant and all four of those equations can be used with Newton's second law (Equation 9.7.1). You could also add in a calculation to find the moment of inertia, I.

Example 9.7.1

A circular saw blade has three torques on it as it cuts through wood during the construction of a house. The first torque comes from the motor, and is a counterclockwise torque equal to 0.100 N·m. The second torque is from friction as the saw blade turns, and is a clockwise torque equal to $2.00 \cdot 10^{-2}$ N·m. The third torque is from where the saw is touching the wood – it's a clockwise torque from a 1.00 N force that is 0.110 m from the center of the saw blade and the force is in the tangential direction. You can picture the saw as a disk with radius 0.110 m and mass 0.125 kg. What is the acceleration of the saw blade?

Solution:

$$\tau_{net} = I \cdot \alpha \qquad (9.7.1)$$

Divide both sides by I: $\alpha = \tau_{net}/I$

We need to calculate τ_{net} and I, and then calculate α after. We can leave things in equations until the last step.

For a disk, $I = \frac{1}{2} \cdot m \cdot r^2$

The net torque is equal to: $\tau_{net} = \tau_{motor} - \tau_{friction} - r \cdot F_{wood} \cdot \sin(\varphi)$

Note 1: the signs for torques are positive for counterclockwise and negative for clockwise.

Note 2: Equation 9.6.1 was used for the wood force's torque

Put these equations in for τ_{net} and I:

$$\alpha = \tau_{net}/I = (\tau_{motor} - \tau_{friction} - r \cdot F_{wood} \cdot \sin(\varphi))/(\frac{1}{2} \cdot m \cdot r^2)$$

Put in numbers:

$\alpha = (0.100$ N·m $- 2.00 \cdot 10^{-2}$ N·m $- 0.110$ m·1.00 N·sin(90°))/($\frac{1}{2}$·0.125 kg·$(0.110$ m$)^2) = \underline{-39.7 \text{ rad/s}^2}$

Note: you could now use this value of α in a kinematics problem (like Section 9.4).

Problem to Try Yourself

A game show contestant spins a large disk. The wheel has radius 1.75 m, and mass 26.0 kg. The contestant puts a force of 50.0 N on the wheel, 1.70 m from the edge, in the tangential direction and causing a counterclockwise spin. In addition to the torque from this force, there is also a friction force that puts a 15.0 N·m torque in the clockwise direction.

A. What is the angular acceleration of the disk?

B. If the disk starts at rest, what is the angular velocity after 0.750 seconds? Hint: this is a kinematics problem that will make use of your answer to Part A.

Solution:

A.

$$\tau_{net} = I \cdot \alpha$$

Divide both sides by I: $\alpha = \tau_{net}/I$

We need to calculate τ_{net} and I, and then calculate α after. We can leave things in equations until the last step.

For a disk, $I = \frac{1}{2} \cdot m \cdot r^2$

The net torque is equal to $r \cdot F \cdot \sin(\varphi) - \tau_{friction}$

Note: the r in the torque and the r in the moment of inertia are slightly different, based on the problem statement. It's 1.70 m for the torque (we'll call it r_{force}) and 1.75 m for the moment of inertia (we'll call it r_{disk}).

$\varphi = 90°$, since the force is in the tangential direction.

Put the equations for moment of inertia and torque in:

$$\alpha = (r_{force} \cdot F \cdot \sin(\varphi) - \tau_{friction})/(\frac{1}{2} \cdot m \cdot r_{disk}^2)$$

Put in numbers:

$\alpha = (1.70 \text{ m} \cdot 50.0 \text{ N} \cdot \sin(90°) - 15.0 \text{ N·m})/(\frac{1}{2} \cdot 26.0 \text{ kg} \cdot (1.75 \text{ m})^2) = \underline{1.76 \text{ rad/s}^2}$

B.

In the setup of a kinematics problem, we now have $\alpha = 1.76 \text{ rad/s}^2$, $\Delta t = 0.750 \text{ s}$, and $\omega_i = 0$ (starts at rest). We want to find ω_f.

The equation that has ω_f and the variables we know is Equation 9.4.1:

$$\omega_f = \omega_i + \alpha \cdot \Delta t$$

Put in numbers: $\omega_f = 0 + (1.76 \text{ rad/s}^2) \cdot (0.750 \text{ s}) = \underline{1.32 \text{ rad/s}}$

9.8 RELATING ANGULAR DISPLACEMENT, ANGULAR VELOCITY, AND ANGULAR ACCELERATION TO LINEAR DISPLACEMENT, LINEAR VELOCITY, AND LINEAR ACCELERATION

When an object is going in a circle, you can describe its motion using either linear quantities (for example v or a), or rotational quantities ($\Delta\theta$, ω, α). The relationships depend on the radius of the circle (r) as well.

Relationship of displacements:

$$\Delta s = r \cdot \Delta\theta \qquad (9.8.1)$$

Δs is called "arc length", and is a special case of linear displacement (Δx). Going all the way around a circle would have an arc length equal to the circumference of the circle.

Relationship of velocities:

$$v = \omega \cdot r \qquad (9.8.2)$$

The relationship of accelerations is more complicated, because there can be an acceleration in two directions. Remember that in Chapter 8 we discussed two coordinates for moving in a circle. The direction of travel is called the tangential direction. The direction towards the center of the circle is the centripetal direction. It turns out that there can be a linear acceleration in each of the tangential and centripetal directions:

$$a_t = \alpha \cdot r \qquad (9.8.3)$$

$$a_c = v^2/r \text{ (just as before, always true for circular motion)} \qquad (9.8.4)$$

In Chapter 3, when there were two components of acceleration (a_x and a_y), we used Equation 3.3.2 to find the total magnitude using x and y components. Here, we can use the idea, with tangential and centripetal components:

$$a = \sqrt{a_c^2 + a_t^2} \qquad (9.8.5)$$

Side note: Δs and v are in the tangential direction; you can think of them as having a component in the centripetal direction equal to 0. If they were moving towards or away or from the center of the circle, it wouldn't be making a circle!

Example 9.8.1

A bird flies in a circular path, on the way back to a nest. The bird's path is a half circle, of radius 10.0 m. At the start of the path, the bird has a linear velocity of 5.00 m/s. At the end of the path, the bird has a linear velocity of zero (so that it can land in the nest). Assume that the change in angular velocity happens at a constant rate.

A. What is the arc length of the path? Hint: it's half of a circle.
B. What is the angular displacement?
C. What is the angular velocity at the start?
D. What is the angular velocity at the end?
E. What is the angular acceleration? Hint: it's one value throughout the problem, and you can find it using kinematics.
F. What is the tangential acceleration? Hint: it's one value throughout the problem.
G. What is the centripetal acceleration at the start?
H. What is the linear acceleration, at the start?

Solutions:

A. Half of a circle's circumference: $\frac{1}{2} \cdot (2 \cdot \pi \cdot r) = \frac{1}{2} \cdot (2 \cdot \pi \cdot 10.0 \text{ m}) = \underline{31.4}$ $\underline{m}$

B. Take $\Delta s = \Delta\theta \cdot r$ and divide both sides by r: $\Delta\theta = \Delta s/r = (31.4 \text{ m})/(10.0 \text{ m}) = 3.14$ rad

C. $v = \omega \cdot r$ Divide both sides by r: $\omega = v/r = (5.00 \text{ m/s})/(10.0 \text{ m}) = \underline{0.500}$ $\underline{\text{rad/s}}$

D. $\omega = v/r = (0 \text{ m/s})/(10.0 \text{ m}) = \underline{0 \text{ rad/s}}$

E. This is a kinematics problem. Use Equation 9.4.4: $\omega_f^2 = \omega_i^2 + 2 \cdot \alpha \cdot \Delta\theta$. Subtract ω_i from both sides, and then divide both sides by $(2 \cdot \Delta\theta)$ to find $\alpha = (\omega_f^2 - \omega_i^2)/(2 \cdot \Delta\theta)$. Put in numbers: $(0^2 - (0.500 \text{ rad/s})^2)/(2 \cdot 3.14 \text{ rad}) = \underline{-3.98 \cdot 10^{-2} \text{ rad/s}^2}$

F. $a_t = \alpha \cdot r = (-3.98 \cdot 10^{-2} \text{ rad/s}^2) \cdot (10.0 \text{ m}) = \underline{-3.98 \cdot 10^{-1} \text{ m/s}^2}$

G. $a_c = v^2/r = (5.00 \text{ m/s})^2/(10.0 \text{ m}) = \underline{2.50 \text{ m/s}^2}$

H. $a = \sqrt{a_c^2 + a_t^2} = \sqrt{\left(2.50 \text{m}/\text{s}^2\right)^2 + \left(-3.98 \cdot 10^{-1} \text{m}/\text{s}^2\right)^2} = 2.53 \text{m/s}^2$

Note: with units, radians are included for angular quantities and not included for linear quantities.

Problem to Try Yourself

A person walks in a circle, thinking about something important. The circle has radius 5.00 m. At the beginning, the person is walking at 0.500 m/s, and the end of one complete circle the person is walking at 1.50 m/s after changing angular velocities at a constant angular acceleration.

A. What is the angular velocity, both at the start and at the end?
B. What is the angular acceleration? Hint: what is $\Delta\theta$?
C. What is the magnitude of the linear acceleration at the end? Don't forget to calculate both components before calculating a magnitude.

Solution:

A. $v = \omega \cdot r$ Divide both sides by r: $\omega = v/r$. Start: (0.500 m/s)/(5.00 m) = 0.100 rad/s. End: (1.50 m/s)/(5.00 m) = 0.300 rad/s.

B. $\Delta\theta$ is $2\cdot\pi$ radians for one complete circle. We now do a kinematics problem. We know $\Delta\theta$, ω_i, and ω_f and we want to know α. The equation with α and variables that we know is Equation 9.4.4: $\omega_f^2 = \omega_i^2 + 2\cdot\alpha\cdot\Delta\theta$. Subtract ω_i from both sides, and then divide both sides by $(2\cdot\Delta\theta)$ to find $\alpha = (\omega_f^2 - \omega_i^2)/(2\cdot\Delta\theta)$. Put in numbers: $((0.300 \text{ rad/s})^2 - (0.100 \text{ rad/s})^2)/(2\cdot2\cdot\pi \text{ rad}) = 6.37\cdot10^{-3}$ rad/s²

C. We need the tangential and centripetal components. Tangential: $a_t = \alpha \cdot r = (6.37\cdot10^{-3} \text{ rad/s}^2)\cdot(5.00 \text{ m}) = 3.19\cdot10^{-2}$ m/s². Centripetal: $a_c = v^2/r = (1.50 \text{ m/s})^2/(5.00 \text{ m}) = 0.450 \text{ m/s}^2$.

$$a = \sqrt{a_c^2 + a_t^2} = \sqrt{\left(0.450\,\text{m}/\text{s}^2\right)^2 + \left(3.19\cdot10^{-2}\,\text{m}/\text{s}^2\right)^2} = 0.451\,\text{m/s}^2$$

9.9 EXTRA TOPIC: TORQUE IS A VECTOR; CALCULATING ITS COMPONENTS

In Section 9.6, Equation 9.6.1 was given to find the magnitude (amount) of torque from a force, given the magnitude of the distance from the center of rotation to where the force happens (r), magnitude of the force (F), and angle between the direction of F and the direction from the center of rotation to where the force happens:

$$\tau = r \cdot F \cdot \sin(\varphi) \tag{9.6.1}$$

It turns out that torque is a vector. Equation 9.6.1 is for the magnitude of the vector. We can also calculate the components. Imagine that the vector τ has x, y and z components τ_x, τ_y, τ_z, that the vector **r** has components r_x, r_y, r_z, and that the vector **F** has components F_x, F_y, F_z. The x, y and z components of torque (τ) are:

$$\tau_x = r_y \cdot F_z - r_z \cdot F_y \tag{9.9.1}$$

$$\tau_y = r_z \cdot F_x - r_x \cdot F_z \tag{9.9.2}$$

$$\tau_z = r_x \cdot F_y - r_y \cdot F_x \tag{9.9.3}$$

Side note 1: the direction of the torque vector is always perpendicular to the circular motion. If the motion is all on flat ground, the torque vector's direction is either into the ground or out of the ground.

Example 9.9.1

You spin a chair around, so that you can sit in it. Making the center of the chair's seat have position 0 in every direction, the force happens at $r = 0.250$ m $i + 0.250$ m $j + 0$ k. The force is $F = 5.00$ N $i - 5.00$ N $j + 0$ k. What are the components of the torque?

Solution:

The x, y, and z components are given in the vectors. $r_x = 0.250$ m, $r_y = 0.250$ m, $r_z = 0$, $F_x = 5.00$ N, $F_y = -5.00$ N, and $F_z = 0$ N. Using Equations 9.9.1 to 9.9.3,

$$\tau_x = r_y \cdot F_z - r_z \cdot F_y = (0.250 \text{ m}) \cdot (0) - (0) \cdot (-5.00 \text{ N}) = \underline{0}$$

$$\tau_y = r_z \cdot F_x - r_x \cdot F_z = (0) \cdot (5.00 \text{ N}) - (0.250 \text{ m}) \cdot (0) = \underline{0}$$

$$\tau_z = r_x \cdot F_y - r_y \cdot F_x = (0.250 \text{ m}) \cdot (-5.00 \text{ N}) - (0.250 \text{ m}) \cdot (5.00 \text{ N}) = \underline{-2.50 \text{ N·m}}$$

Note: if you picture the described situation, you can see that the torque is clockwise. Using "north" as $+y$ and "east" as $+x$, the force happens in the northeast corner and the force itself is in the southeast direction.

Example 9.9.2

You spin an object by putting a force equal to $0\,i + 0\,j + 5.00\,N\,k$ at location $0\,i + 0.100\,m\,j + 0\,k$. This would be like rolling something from at the top (+y direction) with a force forward (+z direction). What are the x, y, and z components of the torque?

Solution:

$$\tau_x = r_y \cdot F_z - r_z \cdot F_y = (0.100\,m) \cdot (5.00\,N) - (0) \cdot (0) = \underline{0.500\ N \cdot m}$$
$$\tau_y = r_z \cdot F_x - r_x \cdot F_z = (0) \cdot (0) - (0) \cdot (5.00\,N) = \underline{0\ N \cdot m}$$
$$\tau_z = r_x \cdot F_y - r_y \cdot F_x = (0) \cdot (0) - (0.100\,m) \cdot (0) = \underline{0\ N \cdot m}$$

Problem to Try Yourself

While eating dinner, you turn your plate so that you can more easily get the last peas with your fork. You put a force of $-1.00\,N\,i + 0\,j + 0\,k$ at location $0\,i - 0.170\,m\,j + 0\,k$. (The center of the plate is set as position zero for each dimension.) What is the torque?

Solution:

The x, y, and z components are given in the vectors. $r_x = 0$, $r_y = -0.170\,m$, $r_z = 0$, $F_x = -1.00\,N$, $F_y = 0$, and $F_z = 0\,N$. Using Equations 9.9.1 to 9.9.3,

$$\tau_x = r_y \cdot F_z - r_z \cdot F_y = (-0.170\,m) \cdot (0) - (0) \cdot (0) = \underline{0}$$
$$\tau_y = r_z \cdot F_x - r_x \cdot F_z = (0) \cdot (-1.00\,N) - (0) \cdot (0) = \underline{0}$$
$$\tau_z = r_x \cdot F_y - r_y \cdot F_x = (0) \cdot (0) - (-0.170\,m) \cdot (-1.00\,N) = \underline{-0.170\ N \cdot m}$$

9.10 CHAPTER 9 SUMMARY

When things rotate, the math is similar to linear motion.

Displacement (Δx) becomes angular displacement ($\Delta\theta$), with units of radians

Linear velocity (v) becomes angular velocity (ω), with units of radians/second

Linear acceleration (a) becomes angular acceleration (α), with units of radians/second^2

Mass (m) becomes moment of inertia (I), which is calculated using $I = m \cdot r^2$ for point masses and other formulas for larger shapes.

Force (F) becomes torque (τ), where the amount is calculated using $\tau = r \cdot F \cdot \sin(\varphi)$

Using these substitutions, the equations for motion with a constant (angular) acceleration become:

$$\omega_f = \omega_i + \alpha \cdot \Delta t \qquad (9.4.1)$$

$$\Delta\theta = \omega_i \cdot \Delta t + \tfrac{1}{2} \cdot \alpha \cdot (\Delta t)^2 \qquad (9.4.2)$$

$$\Delta\theta = \tfrac{1}{2} \cdot (\omega_i + \omega_f) \cdot \Delta t \qquad (9.4.3)$$

$$\omega_f^2 = \omega_i^2 + 2 \cdot \alpha \cdot \Delta\theta \qquad (9.4.4)$$

Newton's second law becomes:

$$\tau_{net} = I \cdot \alpha \qquad (9.7.1)$$

τ_{net} is net torque, where all torques are added with torques towards counterclockwise given a positive sign and torques towards clockwise given a negative sign.

For an object moving a circle, you can describe the motion using rotational quantities ($\Delta\theta$, ω, α), or linear quantities:

$$\Delta s = r \cdot \Delta\theta \qquad (9.8.1)$$

Δs is called "arc length", and is the distance moved for an object moving in a circle.

$$v = \omega \cdot r \qquad (9.8.2)$$

$$a = \sqrt{a_c^2 + a_t^2} \qquad (9.8.5)$$

a_c is acceleration in the centripetal direction, equal to v^2/r just like Chapter 8
a_t is acceleration in the tangential direction, equal to $\alpha \cdot r$
Torque is a vector, and the components of it can be calculated using

$$\tau_x = r_y \cdot F_z - r_z \cdot F_y \qquad (9.9.1)$$

$$\tau_y = r_z \cdot F_x - r_x \cdot F_z \qquad (9.9.2)$$

$$\tau_z = r_x \cdot F_y - r_y \cdot F_x \qquad (9.9.3)$$

Chapter 9 problem types:

Unit conversions between degrees, radians, and revolutions (Section 9.2)

Kinematics problems using rotational equivalents of displacement, velocity and acceleration (Section 9.4)

Calculate using the moment of inertia (Section 9.5)

Calculate using torque (Section 9.6)

Problems using Newton's second law, with rotational equivalents (Section 9.7)

Problems that require you to go between angular displacement, angular velocity and/or angular acceleration, and linear displacement, linear velocity, and/or linear acceleration (Section 9.8)

Calculate using the components of the torque vector (Section 9.9)

Practice Problems
Section 9.2 (Units for Angle)

1. Convert 75.0°:

    A. To radians

    B. To revolutions

    Solution:
    Part A:

    $$75.0° \cdot \frac{\pi \, \text{rad}}{180°} = 1.31 \, \text{rad}$$

    Part B:

    $$75.0° \cdot \frac{1 \, \text{revolution}}{360°} = 0.208 \, \text{rev}$$

2. Convert 1.25 rev:

    A. To radians

   B. To degrees

   Solution:

   Part A:

$$1.25\,\text{rev} \cdot \frac{2 \cdot \pi\,\text{rad}}{1\,\text{rev}} = 7.85\,\text{rad}$$

   Part B:

$$1.25\,\text{rev} \cdot \frac{360°}{1\,\text{rev}} = 450°$$

3. Convert 1.95 rad:

   A. To revolutions

   B. To degrees

   Solution:

   Part A:

$$1.95\,\text{rad} \cdot \frac{1\,\text{rev}}{2 \cdot \pi\,\text{rad}} = 0.310\,\text{rev}$$

   Part B:

$$1.95\,\text{rad} \cdot \frac{180°}{\pi\,\text{rad}} = 112°$$

4. Convert 165.0°:

   A. To radians

   B. To revolutions

   Solution:

   Part A:

$$165° \cdot \frac{\pi\,\text{rad}}{180°} = 2.88\,\text{rad}$$

Part B:

$$165° \cdot \frac{1\,\text{revolution}}{360°} = 0.458\,\text{rev}$$

5. Convert 10.7 rev:

A. To radians
B. To degrees

Solution:
Part A:

$$10.7\,\text{rev} \cdot \frac{2 \cdot \pi\,\text{rad}}{1\,\text{rev}} = 67.2\,\text{rad}$$

Part B:

$$10.7\,\text{rev} \cdot \frac{360°}{1\,\text{rev}} = 3.85 \cdot 10^{3\,°}$$

6. Convert 5.00 rad:

A. To revolutions
B. To degrees

Solution:
Part A:

$$5.00\,\text{rad} \cdot \frac{1\,\text{rev}}{2 \cdot \pi\,\text{rad}} = 0.796\,\text{rev}$$

Part B:

$$5.00\,\text{rad} \cdot \frac{180°}{\pi\,\text{rad}} = 286°$$

Section 9.4 (Motion with Constant Angular Acceleration)

7. A kitchen mixer moves one of its parts in a circle as part of the mixing process. It starts at rest, and accelerates to 8.00 rad/s over 0.700 s using a constant acceleration. What is the angular acceleration?

   Solution:

   We know $\omega_i = 0$, $\omega_f = 8.00$ rad/s, and $\Delta t = 0.700$ s. We want to find α. The equation with the variable that we want to find and the variables that we already know is Equation 9.4.1:

   $$\omega_f = \omega_i + \alpha \cdot \Delta t \tag{9.4.1}$$

   Subtract ω_i from both sides: $\omega_f - \omega_i = \alpha \cdot \Delta t$

   Divide both sides by Δt: $\alpha = (\omega_f - \omega_i)/\Delta t$

   Put in numbers: $\alpha = (8.00 \text{ rad/s} - 0 \text{ rad/s})/(0.700 \text{ s}) = \underline{11.4 \text{ rad/s}^2}$

8. A cement mixer rotates the container that the cement is in. At the end of the mixing, it slows down to a stop over 3.00 s using a constant angular acceleration. If it turned 3.75 rad during the turn, how fast was it going before it began to slow down?

   Solution:

   We know $\omega_f = 0$, $\Delta t = 3.00$ s, and $\Delta\theta = 3.75$ rad. We want to know ω_i. The equation with the variable that we want to know and the variables that we already know is Equation 9.4.3:

   $$\Delta\theta = \tfrac{1}{2} \cdot (\omega_i + \omega_f) \cdot \Delta t \tag{9.4.3}$$

   Divide both sides by $(\tfrac{1}{2} \cdot \Delta t)$: $\Delta\theta/(\tfrac{1}{2} \cdot \Delta t) = \omega_i + \omega_f$

   Subtract ω_f from both sides: $\omega_i = \Delta\theta/(\tfrac{1}{2} \cdot \Delta t) - \omega_f$

   Put in numbers: $\omega_i = (3.75 \text{ rad})/(\tfrac{1}{2} \cdot 3.00 \text{ s}) - 0 = \underline{2.50 \text{ rad/s}}$

9. A clothes washer has a middle column (and attached bottom) that rotate together. When the spin cycle starts, the angular velocity is −0.500 rad/s (negative meaning clockwise). 3.00 seconds later, the angular velocity is 5.50 rad/s (positive meaning counterclockwise).

What angle does the column rotate in those 3.00 seconds? Assume a constant angular acceleration.

Solution:

We know $\omega_i = -0.500$ rad/s, $\Delta t = 3.00$ s, and $\omega_f = 5.50$ rad/s. We want to know $\Delta\theta$. The equation with the variable that we want to know and the variables that we already know is Equation 9.4.3:

$$\Delta\theta = \tfrac{1}{2}\cdot(\omega_i + \omega_f)\cdot\Delta t \qquad (9.4.3)$$

This is already solved for $\Delta\theta$, so we can put in numbers:

$$\Delta\theta = \tfrac{1}{2}\cdot(-0.500 \text{ rad/s} + 5.50 \text{ rad/s})\cdot(3.00 \text{ s}) = \underline{7.50 \text{ rad}}$$

10. The sprinkler inside of a dishwasher rotates at an angular velocity of 2.00 rad/s. At the end of the wash cycle, the angular velocity slows down to zero using a constant angular acceleration. During the slowdown to zero, the sprinkler rotates 2.50 rad. What is the angular acceleration?

Solution:

We know $\omega_i = 2.00$ rad/s, $\omega_f = 0$ rad/s, and $\Delta\theta = 2.50$ rad. We want to know α. The equation with the variable that we want to know and the variables that we already know is Equation 9.4.4:

$$\omega_f^2 = \omega_i^2 + 2\cdot\alpha\cdot\Delta\theta \qquad (9.4.4)$$

Subtract ω_i^2 from both sides: $\omega_f^2 - \omega_i^2 = 2\cdot\alpha\cdot\Delta\theta$

Divide both sides by $(2\cdot\Delta\theta)$: $\alpha = (\omega_f^2 - \omega_i^2)/(2\cdot\Delta\theta)$

Put in numbers: $\alpha = ((0 \text{ rad/s})^2 - (2.00 \text{ rad/s})^2)/(2\cdot2.50 \text{ rad}) = \underline{-0.800}$ rad/s^2

11. The "spinner" for a board game is initially at rest. After it is spun, it speeds up at constant angular acceleration. After 0.250 s, it has turned 3.50 rad. What is the angular acceleration?

Solution:

We know $\omega_i = 0$ rad/s, $\Delta t = 0.250$ s, and $\Delta\theta = 3.50$ rad. We want to find α. The equation with the variable that we want to find and the variables that we already know is Equation 9.4.2:

$$\Delta\theta = \omega_i\cdot\Delta t + \tfrac{1}{2}\cdot\alpha\cdot(\Delta t)^2 \qquad (9.4.2)$$

Subtract $\omega_i \cdot \Delta t$ from both sides: $\Delta \theta - \omega_i \cdot \Delta t = \frac{1}{2} \cdot \alpha \cdot (\Delta t)^2$

Divide both sides by $(\frac{1}{2} \cdot (\Delta t)^2)$: $\alpha = (\Delta \theta - \omega_i \cdot \Delta t)/(\frac{1}{2} \cdot (\Delta t)^2)$

Put in numbers: $\alpha = (3.50 \text{ rad} - 0 \text{ rad/s} \cdot 0.250 \text{ s})/(\frac{1}{2} \cdot (0.250 \text{ s})^2) = \underline{112}$ $\underline{\text{rad/s}^2}$

12. The roller inside the bottom of a vacuum cleaner goes from rest to 12.0 rad/s while accelerating at a constant rate and rotating 14.0 radians. What is the angular acceleration?

   Solution:

   We know $\omega_i = 0$ rad/s, $\omega_f = 12.0$ rad/s, and $\Delta \theta = 14.0$ rad. We want to find α. The equation with the variable that we want to find and the variables that we already know is Equation 9.4.4:

   $$\omega_f^2 = \omega_i^2 + 2 \cdot \alpha \cdot \Delta \theta \tag{9.4.4}$$

   Subtract ω_i^2 from both sides: $\omega_f^2 - \omega_i^2 = 2 \cdot \alpha \cdot \Delta \theta$

   Divide both sides by $(2 \cdot \Delta \theta)$: $\alpha = (\omega_f^2 - \omega_i^2)/(2 \cdot \Delta \theta)$

   Put in numbers: $\alpha = ((12.0 \text{ rad/s})^2 - (0 \text{ rad/s})^2)/(2 \cdot 14.0 \text{ rad}) = \underline{5.14}$ $\underline{\text{rad/s}^2}$

Section 9.5 (Moment of Inertia)

13. A solid ball is rotating, and has a mass of 7.00 kg, and a radius of $7.50 \cdot 10^{-2}$ m. What is its moment of inertia?

   Solution:

   Solid sphere, spinning around a line through the middle:

   $$I = (2/5) \cdot m \cdot R^2 \tag{9.5.2}$$

   Put in numbers: $I = (2/5) \cdot (7.00 \text{ kg}) \cdot (7.50 \cdot 10^{-2} \text{ m})^2 = \underline{1.58 \cdot 10^{-2} \text{ kg} \cdot \text{m}^2}$

14. A point mass has moment of inertia 0.200 kg·m², and mass $5.00 \cdot 10^{-2}$ kg. How far from the center of rotation is it?

   Solution:

   $$I = m \cdot r^2 \tag{9.5.1}$$

   Solve this for r, since we know I and m.

Divide both sides by m: $r^2 = I/m$

Take the square root: $r = \pm\sqrt{I/m}$

Take the positive root, since the distance (radius) is a positive number. Put in numbers:

$$r = +\sqrt{\left(0.200\,kg \cdot m^2\right)/\left(5.00 \cdot 10^{-2}\,kg\right)} = 2.00\,m$$

15. The bottom of a clothes washer is a disk with radius 0.300 m and mass 5.00 kg. What is the moment of inertia?

    Solution:

    Solid disk, spinning around a perpendicular line through the middle: $I = (1/2) \cdot m \cdot R^2$ (Equation 9.5.3)

    Put in numbers: $I = (1/2) \cdot (5.00\ kg) \cdot (0.300\ m)^2 = \underline{0.225\ kg \cdot m^2}$

16. A person rowing makes a motion that is part of a circle, spinning around its end in this setup. If the oar is similar to a rod in shape, has length 2.00 m, and has mass 4.00 kg, what is the moment of inertia?

    Solution:

    Rod, spun around one of its ends: $I = (1/3) \cdot m \cdot L^2$ \hfill (9.5.4)

    Put in numbers: $I = (1/3) \cdot (4.00\ kg) \cdot (2.00\ m)^2 = \underline{5.33\ kg \cdot m^2}$

17. The blade of a food processor is something like a disk (0.100 kg, $1.50 \cdot 10^{-2}$ m radius) plus two rods (the blades; $2.00 \cdot 10^{-2}$ kg mass each, $3.00 \cdot 10^{-2}$ m length each). What is the total moment of inertia?

    Solution:

    Add the moment of inertia for the disk ($I = (1/2) \cdot m \cdot R^2$ from Equation 9.5.3) to the moment of inertia for each rod ($I = (1/3) \cdot m \cdot L^2$ from Equation 9.5.4)

    $I = (1/2) \cdot (0.100\ kg) \cdot (1.50 \cdot 10^{-2}\ m)^2 + (1/3) \cdot (2.00 \cdot 10^{-2}\ kg) \cdot (3.00 \cdot 10^{-2}$ $m)^2 + (1/3) \cdot (2.00 \cdot 10^{-2}\ kg) \cdot (3.00 \cdot 10^{-2}\ m)^2 = \underline{2.33 \cdot 10^{-5}\ kg \cdot m^2}$

18. A weathervane has four rods (mass 0.500 kg each; length 0.300 m each) and each rod has an arrow that we will treat as a point mass

(mass 0.100 kg each; radius from center 0.300 m each). What is the total moment of inertia?

Solution:

Add the moment of inertia for each rod (which is the same for each rod, using $I = (1/3) \cdot m \cdot L^2$ from Equation 9.5.4) and the moment of inertia for each point mass (same for all four, using $I = m \cdot r^2$ from Equation 9.5.1).

$4 \cdot (1/3) \cdot (0.500 \text{ kg}) \cdot (0.300 \text{ m})^2 + 4 \cdot (0.100 \text{ kg}) \cdot (0.300 \text{ m})^2 = \underline{9.60 \cdot 10^{-2} \text{ kg} \cdot \text{m}^2}$

Section 9.6 (Torque)

19. You put torque on a ball (maybe a bowling ball), using a force of 15.0 N, in the tangential direction, at a distance of $4.00 \cdot 10^{-2}$ m from the center of the ball, causing a counterclockwise rotation. How much torque did you put on the ball?

    Solution:

    $$\tau = r \cdot F \cdot \sin(\varphi) \qquad (9.6.1)$$

    Here, $r = 4.00 \cdot 10^{-2}$ m, $F = 15.0$ N, and $\varphi = 90°$ (the tangential direction is 90° from a line from the center to where the force happens). The torque gets a + sign because it is counterclockwise.

    Put in numbers: $\tau = (4.00 \cdot 10^{-2} \text{ m}) \cdot (15.0 \text{ N}) \cdot \sin(90°) = \underline{0.600 \text{ N} \cdot \text{m}}$

20. A child turns a wheel on a toy. The child's hand puts a force of 22.0 N, at a distance of 0.125 m from the center of rotation, causing a counterclockwise torque. The forces is directed at an angle of 65.0° away from the direction of a line from the center to where the force happens. What is the torque?

    Solution:

    $$\tau = r \cdot F \cdot \sin(\varphi) \qquad (9.6.1)$$

    The sign is positive because the torque is counterclockwise.

    $\tau = (0.125 \text{ m}) \cdot (22.0 \text{ N}) \cdot \sin(65.0°) = \underline{2.49 \text{ N} \cdot \text{m}}$

21. You spin an office chair around, so that the seat faces you and you can sit down. You put a force of 12.0 Newtons on it, in the tangential direction, and cause a counterclockwise torque. If you put a torque of 2.40 N·m on the chair, how far from the center of rotation did you put the force on the chair?

    Solution:

    $$\tau = r \cdot F \cdot \sin(\varphi) \qquad\qquad (9.6.1)$$

    We know that $\tau = 2.40$ N·m, $F = 12.0$ N, and $\varphi = 90°$ (tangential direction). To find r, divide both sides by $F \cdot \sin(\varphi)$:

    $$r = \tau/(F \cdot \sin(\varphi))$$

    Put in numbers: $r = (2.40 \text{ N·m})/(12.0 \text{ N} \cdot \sin(90°)) = \underline{0.200 \text{ m}}$

22. An oar rotates as a person rows. In this case, the oar is spinning around a point where it is being held in place on the boat. That point is 0.500 m to the right of the left end of the oar, and 1.50 m to the left of the right end of the oar. The person rowing puts a torque on the oar with each hand. One hand (0.500 m from the center) puts a torque on the oar that is in a clockwise direction, with a force of 25.0 N, in the tangential direction. The second hand (1.00 m from the center) puts a torque on the oar that is in a clockwise direction, with a force of 25.0 N, but at 75.0° from the direction that the oar points. What is the total torque?

    Solution:

    $$\tau = r \cdot F \cdot \sin(\varphi) \qquad\qquad (9.6.1)$$

    We add the torque from the two hands. Both are negative, since they are clockwise. Tangential direction means an angle of 90° for the first hand.

    $\tau = -(0.500 \text{ m}) \cdot (25.0 \text{ N}) \cdot \sin(90°) + -(1.00 \text{ m}) \cdot (25.0 \text{ N}) \cdot \sin(75.0°)$
    $= \underline{-36.6 \text{ N·m}}$

23. A food processor has two torques on it. The first torque comes from the motor, and is $1.00 \cdot 10^{-4}$ N·m in the clockwise direction.

(Note: you might think about this as coming from something like a force of $1.00 \cdot 10^{-2}$ N, at a distance of $1.00 \cdot 10^{-2}$ m.). The last torque is from a piece of food that is stuck on the edge, which gives a force of $3.00 \cdot 10^{-3}$ N at a distance of $3.00 \cdot 10^{-2}$ m in the tangential direction, causing a counterclockwise torque. What is the total torque?

Solution:

$$\tau = r \cdot F \cdot \sin(\varphi) \qquad (9.6.1)$$

Add the two torques. Tangential direction means $\varphi = 90°$, counter-clockwise is the positive direction.

$\tau = 1.00 \cdot 10^{-4}$ N·m $+ (3.00 \cdot 10^{-2}$ m$) \cdot (3.00 \cdot 10^{-3}$ N$) \cdot \sin(90°)$

$\tau = \underline{-1.00 \cdot 10^{-5}$ N·m}

Section 9.7 (Newton's Second Law)

24. Combine the information in problems 13 and 19 to find the angular acceleration of the ball.

    Solution:

    $$\tau_{net} = I \cdot \alpha \qquad (9.7.1)$$

    In problem 13, we found that $I = 1.58 \cdot 10^{-2}$ kg·m²; in problem 19 we found a torque of 0.600 N·m. We will assume that the 0.600 N·m torque is the net torque, meaning that there aren't any other torques on the object.

    Divide both sides of Equation 9.7.1 by I: $\alpha = \tau_{net}/I$

    Put in numbers: $\alpha = (0.600$ N·m$)/(1.58 \cdot 10^{-2}$ kg·m²$) = \underline{38.0$ rad/s²}

25. Combine the information in problems 17 and 23 to find the angular acceleration of the food processor blade.

    Solution:

    $$\tau_{net} = I \cdot \alpha \qquad (9.7.1)$$

    In problem 17, we found that $I = 2.33 \cdot 10^{-5}$ kg·m²; in problem 23 we found a torque of $-1.00 \cdot 10^{-5}$ N·m, with negative meaning clockwise.

Divide both sides of Equation 9.7.1 by I: $\alpha = \tau_{net}/I$

Put in numbers: $\alpha = (-1.00 \cdot 10^{-5}$ N·m$)/(2.33 \cdot 10^{-5}$ kg·m$^2) = \underline{-0.429}$ rad/s^2

26. Use the information in problem 21 to calculate the angular acceleration of the office chair. Assume that the chair part that spins is (mostly) a disk of mass 10.0 kg and radius 0.150 m.

    Solution:

$$\tau_{net} = I \cdot \alpha \tag{9.7.1}$$

In problem 21, we were given that the torque was 2.40 N·m. We will assume that is the net torque. To calculate moment of inertia, we'll use Equation 9.5.3:

Solid disk, spinning around a perpendicular line through the middle: $I = (1/2) \cdot m \cdot R^2$ (Equation 9.5.3)

Put in numbers for moment of inertia: $I = (1/2) \cdot (10.0$ kg$) \cdot (0.150$ m$)^2 = 0.113$ kg·m^2

Divide both sides of Equation 9.7.1 by I: $\alpha = \tau_{net}/I$

Put in numbers: $\alpha = (2.40$ N·m$)/(0.113$ kg·m$^2) = \underline{21.2 \text{ rad/s}^2}$

27. Use the information in problem 20 to calculate the angular acceleration of the wheel, assuming that the wheel is a disk of mass 3.00 kg and radius 0.125 m.

    Solution:

$$\tau_{net} = I \cdot \alpha \tag{9.7.1}$$

In problem 20, the torque was found to be 2.49 N·m. To calculate moment of inertia, we'll use Equation 9.5.3:

Solid disk, spinning around a perpendicular line through the middle: $I = (1/2) \cdot m \cdot R^2$ (Equation 9.5.3)

Put in numbers for moment of inertia: $I = (1/2) \cdot (3.00$ kg$) \cdot (0.125$ m$)^2 = 0.0234$ kg·m^2

Divide both sides of Equation 9.7.1 by I: $\alpha = \tau_{net}/I$

Put in numbers: $\alpha = (2.49$ N·m$)/(0.0234$ kg·m$^2) = \underline{106 \text{ rad/s}^2}$

Section 9.8 (Relating Angular Displacement, Angular Velocity, and Angular Acceleration to Linear Displacement, Linear Velocity, and Linear Acceleration)

28. A kite flies in a circle of radius 7.00 m. At a certain time, the angular velocity is 0.200 rad/s and the angular acceleration is 0.150 rad/s².

   A. What is the linear velocity of the kite?

   B. What is the centripetal direction component of the linear acceleration of the kite?

   C. What is the tangential direction component of the linear acceleration of the kite?

   D. What is the magnitude of the linear acceleration?

   Solution:

   Part A:

   $$v = \omega \cdot r \tag{9.8.2}$$

   Put in numbers: $v = (0.200 \text{ rad/s}) \cdot (7.00 \text{ m}) = \underline{1.40 \text{ m/s}}$

   Part B:

   $$a_c = v^2/r \tag{9.8.4}$$

   Put in numbers: $a_c = (1.40 \text{ m/s})^2/(7.00 \text{ m}) = \underline{0.280 \text{ m/s}^2}$

   Part C:

   $$a_t = \alpha \cdot r \tag{9.8.3}$$

   Put in numbers: $a_t = (0.150 \text{ rad/s}^2) \cdot (7.00 \text{ m}) = \underline{1.05 \text{ m/s}^2}$

   Part D:

   $$a = \sqrt{a_c^2 + a_t^2} \tag{9.8.5}$$

   Put in numbers: $a = \sqrt{\left(0.280 \text{ m/s}^2\right)^2 + \left(1.05 \text{ m/s}^2\right)^2} = 1.09 \text{ m/s}^2$

29. A bicycle rider is making a turn that has radius 5.00 m. The linear velocity is 5.00 m/s, and the tangential component of linear acceleration is 1.50 m/s².

   A. What is the angular velocity?

   B. What is the angular acceleration?

   C. What is the centripetal acceleration?

   D. What is the magnitude of the total linear acceleration?

   Solution:

   Part A:

$$v = \omega \cdot r \tag{9.8.2}$$

   Divide both sides by r: $\omega = v/r$

   Put in numbers: $\omega = (5.00 \text{ m/s})/(5.00 \text{ m}) = \underline{1.00 \text{ rad/s}}$

   Part B:

$$a_t = \alpha \cdot r \tag{9.8.3}$$

   Divide both sides by r: $\alpha = a_t/r$

   Put in numbers: $\alpha = (1.50 \text{ m/s}^2)/(5.00 \text{ m}) = \underline{0.300 \text{ rad/s}^2}$

   Part C:

$$a_c = v^2/r \tag{9.8.4}$$

   Put in numbers: $a_c = (5.00 \text{ m/s})^2/(5.00 \text{ m}) = \underline{5.00 \text{ m/s}^2}$

   Part D:

$$a = \sqrt{a_c^2 + a_t^2} \tag{9.8.5}$$

   Put in numbers: $a = \sqrt{\left(5.00 \,\text{m}/\text{s}^2\right)^2 + \left(1.50 \,\text{m}/\text{s}^2\right)^2} = 5.22 \,\text{m/s}^2$

30. Think about an arrow on a weathervane as it turns. It's a distance of 0.300 m from the center. During a wind gust, the angular velocity is 0.700 rad/s and the angular acceleration is 0.100 rad/s².

A.   What is the linear velocity?

B.   What is the tangential component of linear acceleration?

C.   What is the centripetal component of linear acceleration?

D.   What is the magnitude of linear acceleration?

Solution:

Part A:

$$v = \omega \cdot r \qquad\qquad (9.8.2)$$

Put in numbers: $v = (0.700 \text{ rad/s}) \cdot (0.300 \text{ m}) = \underline{0.210 \text{ m/s}}$

Part B:

$$a_t = \alpha \cdot r \qquad\qquad (9.8.3)$$

Put in numbers: $a_t = (0.100 \text{ rad/s}^2) \cdot (0.300 \text{ m}) = \underline{3.00 \cdot 10^{-2} \text{ m/s}^2}$

Part C:

$$a_c = v^2/r \qquad\qquad (9.8.4)$$

Put in numbers: $a_c = (0.210 \text{ m/s})^2/(0.300 \text{ m}) = \underline{0.147 \text{ m/s}^2}$

Part D:

$$a = \sqrt{a_c{}^2 + a_t{}^2} \qquad\qquad (9.8.5)$$

Put in numbers: $a = \sqrt{\left(0.147 \text{ m/s}^2\right)^2 + \left(3.00 \cdot 10^{-2} \text{ m/s}^2\right)^2} = \underline{0.150}$ $\underline{\text{m/s}^2}$

31. A kite flies in a circle of radius 10.0 m. At first, the kite has a linear velocity of 1.00 m/s. After turning through 4.00 radians, the linear velocity is up to 3.00 m/s. Assume a constant angular acceleration.

A.   What arc length did the kite cover in this time?

B.   What is the angular velocity at the first time point and at the second time point?

C. What is the angular acceleration? Hint: use kinematics.

D. What is the magnitude of the linear acceleration at the final time point? Don't forget to calculate both components before calculating a magnitude.

Solution:

Part A:

$$\Delta s = r \cdot \Delta \theta \tag{9.8.1}$$

Put in numbers: $\Delta s = (10.0 \text{ m}) \cdot (4.00 \text{ rad}) = \underline{40.0 \text{ m}}$

Part B:

$$v = \omega \cdot r \tag{9.8.2}$$

Divide both sides by r: $\omega = v/r$

Put in numbers for the first time point: $\omega = (1.00 \text{ m/s})/(10.0 \text{ m}) = \underline{0.100}$ $\underline{\text{rad/s}}$

Put in numbers for the second time point: $\omega = (3.00 \text{ m/s})/(10.0 \text{ m}) = \underline{0.300 \text{ rad/s}}$

Part C:

Since the angular acceleration is constant, we can use the (kinetmatics) equations from Section 9.4. We know $\Delta \theta = 4.00$ rad, $\omega_i = 0.100$ rad/s, and $\omega_f = 0.300$ rad/s. We want to find α. The equation with the variable that we want to find and the variables that we already know is Equation 9.4.4:

$$\omega_f^2 = \omega_i^2 + 2 \cdot \alpha \cdot \Delta \theta \tag{9.4.4}$$

Subtract ω_i^2 from both sides: $\omega_f^2 - \omega_i^2 = 2 \cdot \alpha \cdot \Delta \theta$

Divide both sides by $(2 \cdot \Delta \theta)$: $\alpha = (\omega_f^2 - \omega_i^2)/(2 \cdot \Delta \theta)$

Put in numbers: $\alpha = ((0.300 \text{ rad/s})^2 - (0.100 \text{ rad/s})^2)/(2 \cdot 4.00 \text{ rad}) = \underline{1.00 \cdot 10^{-2} \text{ rad/s}^2}$

Part D:

$$a = \sqrt{a_c^2 + a_t^2} \tag{9.8.5}$$

$$a_t = \alpha \cdot r \qquad (9.8.3)$$

$$a_c = v^2/r \qquad (9.8.4)$$

Put in numbers:

$a_t = (1.00 \cdot 10^{-2} \text{ rad/s}^2) \cdot (10.0 \text{ m}) = 0.100 \text{ m/s}^2$

$a_c = (3.00 \text{ m/s})^2/10.0 \text{ m} = 0.900 \text{ m/s}^2$

$$a = \sqrt{\left(0.900\,\text{m/s}^2\right)^2 + \left(0.100\,\text{m/s}^2\right)^2} = 0.906\,\text{m/s}^2$$

Section 9.9 (Extra Topic: Torque Vector Components)

32. You flick the spinner to start your turn on a board game. Taking the center of the spinner to be the origin, the +x direction to be to the right, the +y direction to be up, and the +z direction to be forward, the position where the force happens is $r = 5.00 \cdot 10^{-2}$ m $i + 0$ m $j + 0$ m k and the direction of the force is $F = 0$ N $i - 0.200$ N $j + 1.10$ N k. What are the components of torque?

   Solution:

$$\tau_x = r_y \cdot F_z - r_z \cdot F_y \qquad (9.9.1)$$

$$\tau_y = r_z \cdot F_x - r_x \cdot F_z \qquad (9.9.2)$$

$$\tau_z = r_x \cdot F_y - r_y \cdot F_x \qquad (9.9.3)$$

   Put in numbers:

   $\tau_x = (0 \text{ m}) \cdot (1.10 \text{ N}) - (0 \text{ m}) \cdot (-0.200 \text{ N}) = \underline{0 \text{ N·m (x component)}}$

   $\tau_y = (0 \text{ m}) \cdot (0 \text{ N}) - (5.00 \cdot 10^{-2} \text{ m}) \cdot (1.10 \text{ N}) = \underline{-5.50 \cdot 10^{-2} \text{ N·m (y component)}}$

   $\tau_z = (5.00 \cdot 10^{-2} \text{ m}) \cdot (-0.200 \text{ N}) - (0 \text{ m}) \cdot (0 \text{ N}) = \underline{-1.00 \cdot 10^{-2} \text{ N·m (z component)}}$

33. You put a torque on a ball. Taking the center of the ball to be the origin, the +x direction to be to the right, the +y direction to be up, and the +z direction to be forward, the position where the force happens

is $\mathbf{r} = 0$ m $i + 0$ m $j + -4.00 \cdot 10^{-2}$ m k and the direction of the force is $\mathbf{F} = 0$ N $i + 1.20$ N $j + 12.0$ N k. What are the components of torque?

Solution:

$$\tau_x = r_y \cdot F_z - r_z \cdot F_y \qquad (9.9.1)$$

$$\tau_y = r_z \cdot F_x - r_x \cdot F_z \qquad (9.9.2)$$

$$\tau_z = r_x \cdot F_y - r_y \cdot F_x \qquad (9.9.3)$$

Put in numbers:

$\tau_x = (0\text{m}) \cdot (12.0$ N$) - (-4.00 \cdot 10^{-2}$ m$) \cdot (1.20$ N$) = \underline{4.80 \cdot 10^{-2}}$ N m (x component)

$\tau_y = (-4.00 \cdot 10^{-2}$ m$) \cdot (0$ N$) - (0$ m$) \cdot (12.0$ N$) = \underline{0}$ N (y component)

$\tau_z = (0$ m$) \cdot (1.20$ N$) - (0$ m$) \cdot (0$ N$) = \underline{0}$ N (z component)

Rotation: Energy, Momentum, and Rolling

10.1 INTRODUCTION: ENERGY AND MOMENTUM FOR ROTATION

In Chapter 9, we looked at the rotational equivalents for motion and forces. We will now look at the rotation equivalents for energy and momentum. For example: how much rotational energy is contained in a spinning tire? How much rotational momentum does a spinning ball have? We will also look at things that roll.

10.2 ROTATION AND KINETIC ENERGY

In Chapter 6, we introduced kinetic energy for something that was moving linearly:

$$KE = \tfrac{1}{2} \cdot m \cdot v^2 \qquad (6.5.1)$$

m is mass, v is linear velocity.

In Chapter 9, we introduced moment of inertia (I) as the rotation equivalent of mass, and angular velocity (ω) as the rotation equivalent of velocity. We can make those substitutions, and find the equation for kinetic energy and rotation:

$$KE = \tfrac{1}{2} \cdot I \cdot \omega^2 \qquad (10.2.1)$$

Note: all energies always have units of Joules.

DOI: 10.1201/9781003005049-10

We now have an equation for kinetic energy from moving linearly, and an equation for kinetic energy of things that move in rotation. What about something that is rotating and moving linearly at the same time, for example something that is rolling? We just add the two equations:

$$KE = \tfrac{1}{2} \cdot m \cdot v^2 + \tfrac{1}{2} \cdot I \cdot \omega^2 \tag{10.2.2}$$

Example 10.2.1

A string trimmer has a rod spinning around its end. The rod is 0.100 m long and has mass 0.0100 kg. If it spins at 60.0 rad/s, what is the kinetic energy?

Solution:

Here there is no linear motion, just rotation, so $KE = \tfrac{1}{2} \cdot I \cdot \omega^2$

For a rod spinning around its end, $I = (1/3) \cdot m \cdot r^2$. Put this into the kinetic energy equation:

$$KE = \tfrac{1}{2} \cdot (1/3) \cdot m \cdot r^2 \cdot \omega^2$$

Put in numbers: $KE = \tfrac{1}{2} \cdot (1/3) \cdot (0.0100 \ \text{kg}) \cdot (0.100 \ \text{m})^2 \cdot (60.0 \ \text{rad/s})^2 = \underline{0.0600 \ \text{Joules}}$

Notes on units: 1 Joule is equal to 1 $kg \cdot m^2/s^2$. Radians are not written for units of energy, just like for linear quantities as mentioned in Chapter 9.

Example 10.2.2

A ball with mass 0.125 kg and radius 0.100 m is moving linearly at 6.00 m/s, and spinning at 6.00 rad/s (about one turn per second). What is the kinetic energy?

Solution:

This situation has linear motion and rotational motion.

$$KE = \tfrac{1}{2} \cdot m \cdot v^2 + \tfrac{1}{2} \cdot I \cdot \omega^2 \tag{10.2.2}$$

A ball is a sphere, so $I = (2/5) \cdot m \cdot r^2$ Put this into the equation:

$$KE = \tfrac{1}{2} \cdot m \cdot v^2 + \tfrac{1}{2} \cdot (2/5) \cdot m \cdot r^2 \cdot \omega^2$$

Put in numbers: $KE = \tfrac{1}{2} \cdot (0.125 \ \text{kg}) \cdot (6.00 \ \text{m/s})^2 + \tfrac{1}{2} \cdot (2/5) \cdot (0.125 \ \text{kg}) \cdot (0.100 \ \text{m})^2 \cdot (6.00 \ \text{rad/s})^2 = \underline{2.26 \ \text{Joules}}$

Note: v and ω are equal in this example, but they usually are not.

Problem to Try Yourself

A ball with mass 4.00 kg and radius 0.0700 m is moving with a linear velocity of 2.00 m/s and rotating with angular velocity 28.6 rad/s. What is the kinetic energy?

Solution:

This situation has linear motion and rotational motion.

$$KE = \tfrac{1}{2} \cdot m \cdot v^2 + \tfrac{1}{2} \cdot I \cdot \omega^2 \tag{10.2.2}$$

A ball is a sphere, so $I = (2/5) \cdot m \cdot r^2$ Put this into the equation:

$$KE = \tfrac{1}{2} \cdot m \cdot v^2 + \tfrac{1}{2} \cdot (2/5) \cdot m \cdot r^2 \cdot \omega^2$$

Put in numbers: $KE = \tfrac{1}{2} \cdot (4.00 \text{ kg}) \cdot (2.00 \text{ m/s})^2 + \tfrac{1}{2} \cdot (2/5) \cdot (4.00 \text{ kg}) \cdot (0.0700 \text{ m})^2 \cdot (28.6 \text{ rad/s})^2 = \underline{11.2 \text{ Joules}}$

Note: these numbers are very roughly based on a rolling Bocce ball; we'll cover more theory related to rolling in the next section.

10.3 ROLLING (WITHOUT SLIPPING)

When an object rolls, it is moving linearly and with rotation. That means that its kinetic energy is given by an equation from the last section:

$$KE = \tfrac{1}{2} \cdot m \cdot v^2 + \tfrac{1}{2} \cdot I \cdot \omega^2 \tag{10.2.2}$$

If the object is not slipping, we can say two additional things about the motion:

$$\Delta x = \Delta\theta \cdot r \tag{10.3.1}$$

$$v = \omega \cdot r \tag{10.3.2}$$

Δx is linear displacement, $\Delta\theta$ is the angular displacement, r is the radius of the rotation, v is linear velocity, ω is angular velocity. Equations 10.3.1 and 10.3.2 are basically the same as Equations 9.8.1 and 9.8.2, when we related linear motion and angular motion.

Rolling problems often combine Equations 10.3.1 and 10.3.2 with 10.2.2 (kinetic energy) and sometimes an additional equation for moment of inertia. In addition to this, it can use the calculation of kinetic energy as part of a conservation of mechanical energy problem.

Example 10.3.1 – Problem with Kinetic Energy

A wheel is moving. It is shaped like a disk with mass 0.500 kg and radius 0.100 m. If it has kinetic energy 5.00 Joules, what is the linear velocity and the angular velocity of the wheel?

Solution:

$$KE = \tfrac{1}{2}{\cdot}m{\cdot}v^2 + \tfrac{1}{2}{\cdot}I{\cdot}\omega^2 \qquad\qquad (10.2.2)$$

For a disk, $I = \tfrac{1}{2}{\cdot}m{\cdot}r^2$
For rolling, $v = \omega{\cdot}r$
Put the equations for I and for v into the equation for KE:

$$KE = \tfrac{1}{2}{\cdot}m{\cdot}(\omega{\cdot}r)^2 + \tfrac{1}{2}{\cdot}(\tfrac{1}{2}{\cdot}m{\cdot}r^2){\cdot}\omega^2$$

We now have an equation where ω (angular velocity) is the only unknown.

Multiply out the terms on the right side:

$$KE = \tfrac{1}{2}{\cdot}m{\cdot}r^2{\cdot}\omega^2 + (1/4){\cdot}m{\cdot}r^2{\cdot}\omega^2$$

Since both terms are something times ω^2, you can rewrite this as:

$$KE = (\tfrac{1}{2}{\cdot}m{\cdot}r^2 + (1/4){\cdot}m{\cdot}r^2){\cdot}\omega^2$$

Divide both sides by everything next to the ω^2:

$$\omega^2 = KE/(\tfrac{1}{2}{\cdot}m{\cdot}r^2 + (1/4){\cdot}m{\cdot}r^2)$$

Take the square root. Here we will assume that the direction of the disk's rotation is the positive direction.

$$KE = \sqrt{\frac{KE}{\tfrac{1}{2}{\cdot}m{\cdot}r^2 + (1/4){\cdot}m{\cdot}r^2}}$$

Put in numbers:

$$KE = \sqrt{\frac{5.00\,J}{\frac{1}{2}\cdot 0.500\,kg\cdot\left(0.100\,m\right)^2 +\left(1/4\right)\cdot 0.500\,kg\cdot\left(0.100\,m\right)^2}} = \underline{36.5\,rad/s}$$

To find v, use $v = \omega\cdot r$

Put in numbers $= (36.5\ rad/s)\cdot(0.100\ m) = \underline{3.65\ m/s}$

Note: the term $(\frac{1}{2}\cdot m\cdot r^2 + (1/4)\cdot m\cdot r^2)$ that appeared in the math could have been simplified to $(3/4)\cdot m\cdot r^2$

Problem to Try Yourself

You take a ball with mass 4.00 kg and radius 0.0700 kg, and roll it with a linear velocity of 4.50 m/s. How much kinetic energy did you put into the ball? Hint: use v to calculate ω.

Solution:

$$KE = \tfrac{1}{2}\cdot m\cdot v^2 + \tfrac{1}{2}\cdot I\cdot\omega^2$$

For a ball (with a sphere shape), $I = (2/5)\cdot m\cdot r^2$
For rolling, $v = \omega\cdot r$
Divide both sides by r: $\omega = v/r$
Put the equation for I into the equation for KE:

$$KE = \tfrac{1}{2}\cdot m\cdot v^2 + \tfrac{1}{2}\cdot((2/5)\cdot m\cdot r^2)\cdot\omega^2$$

We know m, v, and r. We need ω:

$$\omega = v/r$$

We could calculate ω and then just add it in, but to make a specific point we will substitute in v/r for ω in the equation above:

$$KE = \tfrac{1}{2}\cdot m\cdot v^2 + \tfrac{1}{2}\cdot((2/5)\cdot m\cdot r^2)\cdot(v/r)^2$$

The second term becomes $(1/5)\cdot m\cdot v^2$. This is because you get r^2/r^2, which cancels the r terms, and the $\frac{1}{2}\cdot(2/5) = (1/5)$.

$$KE = \tfrac{1}{2}\cdot m\cdot v^2 + (1/5)\cdot m\cdot v^2$$

Put in numbers: ½·(4.00 kg)·(4.50 m/s)² + (1/5)·(4.00 kg)·(4.50 m/s)² = 56.7 Joules

The reason for doing the $\omega = v/r$ substitution was to show that you actually didn't need to know the radius in this problem because it cancels out.

10.4 ROTATION EQUIVALENT OF WORK

Work is another quantity that we can write for rotation. The work is given by:

$$W = \pm \tau \cdot \Delta\theta \tag{10.4.1}$$

W is work, τ is magnitude (amount) of torque, $\Delta\theta$ is the angular displacement. The ± sign is a + when τ and $\Delta\theta$ are in the same direction (clockwise or counterclockwise), and the sign is a – sign when τ and $\Delta\theta$ are in opposite directions.

Where does this come from? Remember from Chapter 6 that the equation for work was like (force magnitude)·(linear displacement magnitude)·cos(angle between force and displacement). Torque replaces force, angular displacement replaces linear displacement, and the cosine term can only be equal to + 1 or –1 in this case because the torque and angular displacement can only be 0 degrees apart or 180 degrees apart.

Example 10.4.1

Playing a board game, you flick a spinner to see what number you land on. You put a torque of $4.00 \cdot 10^{-2}$ N·m on the spinner, which is an OK amount because the spinner is small. If your finger is in contact with the spinner for 0.600 radians (about 1/10 of a one time around the circle), and both the direction of the torque and the direction of the rotation are counterclockwise, what is the work done?

$$W = \pm \tau \cdot \Delta\theta \tag{10.4.1}$$

We choose the positive sign because the torques and displacement are in the same direction.

$$W = (4.00 \cdot 10^{-2} \text{ N·m}) \cdot (0.600 \text{ rad}) = \underline{2.40 \cdot 10^{-2} \text{ N·m}}$$

Problem to Try Yourself

Imagine swinging a hammer to hit a nail as doing a rotation of 1.50 radians (about one quarter of a circle). If you do 12.0 N·m of work, how much torque did you put on the hammer?

Solution:

$$W = \pm \tau \cdot \Delta\theta$$

We choose the positive sign because the torques and displacement are in the same direction.

Divide both sides by $\Delta\theta$:

$$\tau = W/\Delta\theta$$

Put in numbers: $\tau = (12.0 \text{ N·m})/(1.50 \text{ rad}) = \underline{8.00 \text{ N·m}}$

10.5 ROTATION EQUIVALENT OF POWER

The rotation equivalent of power can be written as:

$$P = \pm \tau \cdot \omega \tag{10.5.1}$$

P is power, τ is magnitude (amount) of torque, and ω is magnitude (amount) of angular velocity. Just like the work equation above, the $\pm$ sign is $+$ when torque and angular velocity are in the same direction, and $-$ when they are not. This is similar to the linear equation $P = F \cdot v \cdot \cos(\theta)$ from Section 6.10.

Example 10.5.1

A steering wheel turns, from a torque of 0.300 N·m. If the wheel is turning at 1.20 rad/s in the direction of the torque, what is the power?

Solution:

$$P = \pm \tau \cdot \omega = (0.300 \text{ N·m}) \cdot (1.20 \text{ rad/s}) = \underline{0.360 \text{ Watts}}$$

Note 1: this type of problem could also be added onto a longer example – for example, the "problem to try yourself" in Section 9.7 that involved calculating a torque using r·F·sin(φ) and involved calculating ω using kinematics (Section 9.4).

Note 2: Watts are the same as kg·m²/s³.

Problem to Try Yourself

The "problem to try yourself" in the last section involved swinging a hammer with a torque of 8.00 N·m. If the power is 40.0 Watts at one point, how fast is the hammer rotating?

Solution:

$$P = \pm \tau \cdot \omega \qquad (10.5.1)$$

Divide both sides by τ: ω = P/τ

Put in numbers: ω = 40.0 W/8.00 N·m = <u>5.00 rad/s</u>

Note: remember that radians appear in angular quantities. 1 Watt is kg·m²/s³ and 1 N·m is kg·m²/s².

10.6 ANGULAR MOMENTUM

Angular momentum is the rotation equivalent of linear momentum. The magnitude of linear momentum is equal to:

$$p = m \cdot v$$

p is magnitude momentum, m is mass, and v is magnitude linear velocity

Putting in moment of inertia (I) for mass (m) and angular velocity (ω) for linear velocity (v), we get:

$$L = I \cdot \omega \qquad (10.6.1)$$

L is angular momentum, I is moment of inertia, and ω is angular velocity. The units of angular momentum are kg·m²/s, with the 1/s coming from the units of ω and the units of kg·m² coming from the units of I.

Side note: you can also write angular momentum as:

$$L = m \cdot v \cdot r \cdot \sin(\varphi)$$

L is angular momentum, m is mass, and v is the amount (magnitude) of linear velocity. After choosing an axis (like a center of rotation), r is the distance of the object from the axis and φ (Greek letter phi) is the angle between the direction of a line from the axis to the object and the direction of the velocity of the object. In this section, we will focus on things that are moving in a circle or rotating; in those cases this equation simplifies to become Equation 10.6.1. To get to $L = I \cdot \omega$ for a small mass moving in a circle, use $v = \omega \cdot r$, $I = m \cdot r^2$ assuming a point mass, and $φ = 90$ degrees.

Side note 2: angular momentum can be expressed as a vector. Here, we deal only with the magnitude.

Example 10.6.1

Inside of a motor, a fan spins. The fan is made of three rods, each of length 0.150 m and mass 0.200 kg. It spins at a constant rate of 4.00 rad/s. What is its angular momentum?

Solution:

$$L = I \cdot \omega \qquad (10.6.1)$$

For three rods that are the same mass and length, $I = 3 \cdot (1/3) \cdot m \cdot r^2 = m \cdot r^2$ (note: here we are using r for the length of the rod, so as not to confuse it with the L for angular momentum)

Put $m \cdot r^2$ in for I in the equation for angular momentum:

$$L = m \cdot r^2 \cdot \omega$$

Put in numbers: $L = (0.200 \text{ kg}) \cdot (0.150 \text{ m})^2 \cdot (4.00 \text{ rad/s}) = \underline{1.80 \cdot 10^{-2}}$ __kg·m²/s__

Problem to Try Yourself

A different fan in a different motor has four rods that are exactly the same, each with mass 0.170 kg. It is spinning at 3.00 rad/s, and has angular momentum $1.35 \cdot 10^{-2}$ kg·m²/s. How long are the rods?

Solution:

$$L = I \cdot \omega \qquad (10.6.1)$$

With four rods (that are the same), $I = 4 \cdot (1/3) \cdot m \cdot r^2 = (4/3) \cdot m \cdot r^2$

Put this in for I: $L = (4/3) \cdot m \cdot r^2 \cdot \omega$

r is now the only unknown variable in this equation, and we need to solve for it.

Divide by everything next to r^2: $r^2 = L / ((4/3) \cdot m \cdot \omega)$

Take the square root: $r = \sqrt{\dfrac{L}{\dfrac{4}{3} \cdot m \cdot \omega}}$

(Take the positive sign on the square root since lengths are positive.)

Put in numbers: $r = \sqrt{\dfrac{1.35 \cdot 10^{-2}\,kg \cdot m^2/s}{\dfrac{4}{3} \cdot (0.170\,kg) \cdot (3.00\,rad/s)}} = \underline{0.141\ m}$

10.7 CONSERVATION OF ANGULAR MOMENTUM

In Chapter 7, we covered the conservation of linear momentum, with a particular focus on collision applications. Here, we will look at the conservation of angular momentum, with focus on more than just collisions. We will look just at things that are rotating or moving in a circle.

When is the angular momentum conserved? (Reminder: conserved means that the total amount stays the same for a group.) We mentioned that for a group of objects, the total linear momentum is conserved if the total force on them from objects outside of the group is zero. Conservation of angular momentum is almost the same, but total torque is zero instead of force.

Looking at some initial and final time point, for two objects:

$$I_{1i} \cdot \omega_{1i} + I_{2i} \cdot \omega_{2i} = I_{1f} \cdot \omega_{1f} + I_{2f} \cdot \omega_{2f} \qquad (10.7.1)$$

I_{1i} is the moment of inertia of the first object (the subscript 1) at the initial time point (the subscript i). ω_{1i} is the angular velocity for the first object at the initial time point. I_{2i} and ω_{2i} are the same but for the second object. The right side of the equation is the same, but for the final time point (so f for final instead of i for initial).

If there is only one object, this equation becomes:

$$I_i \cdot \omega_i = I_f \cdot \omega_f \qquad (10.7.2)$$

If there are more than two objects, you just add more terms. For example, if you have three objects, the left and right side of Equation 10.7.1 each get one more term – something like $I_{3i} \cdot \omega_{3i}$ on the left side and $I_{3f} \cdot \omega_{3f}$ on the right side.

We'll do an example shortly with collisions. If the objects stick together after the collision, their final angular velocities are equal:

$$\omega_{1f} = \omega_{2f} \tag{10.7.3}$$

This is just like the condition from perfectly inelastic linear collisions when two objects stuck together and had final linear velocities be the same (Equation 7.5.1).

Example 10.7.1 – Basic Problem with Conservation of Angular Momentum

A very common physics demonstration is the following: have a person spin around on a stool, while holding two weights very close to them. Suddenly the person holds the weights out 'away from their body, and the spinning slows down. With the weights in, the moment of inertia of the person is 1.10 kg·m^2, and they are rotating at 4.00 rad/s. If they slow down to 2.35 rad/s after holding the weights outward, what is the moment of inertia at that point? Assume that angular momentum is conserved.

Solution:

The initial angular momentum is equal to the final angular momentum. In each case, it's one (large) rotating object, so we use Equation 10.7.2:

$$I_i \cdot \omega_i = I_f \omega_f \tag{10.7.2}$$

This is just like Equation 10.7.1, but with only one object instead of two. (We're considering the person and the weights like one big object; we could also use Equation 10.7.1 but this is simpler.)

We want to solve for the moment of inertia at the final time point: I_f. Divide both sides by ω_f:

$$I_f = I_i \cdot \omega_i / \omega_f$$

Put in numbers: $I_f = (1.10$ kg·m$^2) \cdot (4.00$ rad/s$)/(2.35$ rad/s$) = \underline{1.87 \text{ kg·m}^2}$

Note 1: the point of this demonstration is to show that the angular velocity decreases when the moment of inertia increases, which happens when the weights are moved further away. Moment of inertia is always some number times mass times radius squared, and moving the weights away increases the value of radius for the weights.

Note 2: problems with conservation of angular momentum can certainly involve more than one object.

Example 10.7.2

A string trimmer is like a rod spinning around its end, with a moment of inertia of $3.33 \cdot 10^{-5}$ kg·m². It is spinning at 30.0 rad/s initially, without any help from a motor (maybe just after someone turns the motor off). Then, it collides with a clump of mud, which sticks to the string. You can say that the mud is approximately a point mass with mass 0.0250 kg and a distance of 0.0800 m from the center of rotation. Assume that angular momentum is conserved. What is the new angular velocity?

Solution:

This is the rotational equivalent of collisions, which we need Equations 10.7.1 and 10.7.3 for:

$$I_{1i} \cdot \omega_{1i} + I_{2i} \cdot \omega_{2i} = I_{1f} \cdot \omega_{1f} + I_{2f} \cdot \omega_{2f} \qquad (10.7.1)$$

$$\omega_{1f} = \omega_{2f} \qquad (10.7.3)$$

We'll call the string object 1, and the mud object 2. For the trimmer, we're given moment of inertia. For the point mass, the moment of inertia is equal to m·r² (Equation 9.5.1).

Since the objects stick together, $\omega_{1f} = \omega_{2f}$. Put in ω_{1f} everywhere that ω_{2f} appears in the first equation:

$$I_{1i} \cdot \omega_{1i} + I_{2i} \cdot \omega_{2i} = I_{1f} \cdot \omega_{1f} + I_{2f} \cdot \omega_{1f}$$

The right side can be rewritten since both terms are something times ω_{1f}:

$$I_{1i} \cdot \omega_{1i} + I_{2i} \cdot \omega_{2i} = (I_{1f} + I_{2f}) \cdot \omega_{1f}$$

Divide both sides by $(I_{1f} + I_{2f})$:

$$\omega_{1f} = (I_{1i} \cdot \omega_{1i} + I_{2i} \cdot \omega_{2i})/(I_{1f} + I_{2f})$$

This math might feel familiar – it's very similar to the perfectly inelastic collisions in Chapter 7.

Before putting in numbers, remember that I_{2i} and I_{2f} (the mud) are equal to $m \cdot r^2$:

$$\omega_{1f} = (I_{1i} \cdot \omega_{1i} + m \cdot r^2 \cdot \omega_{2i})/(I_{1f} + m \cdot r^2)$$

Put in numbers. Remember that the moments of inertia don't change, and that the mud is not moving initially.

$\omega_{1f} = (3.33 \cdot 10^{-5}\,\text{kg} \cdot \text{m}^2 \cdot 30.0\,\text{rad/s} + 0.0250\,\text{kg} \cdot (0.0800\,\text{m})^2 \cdot 0)/(3.33 \cdot 10^{-5}\,\text{kg} \cdot \text{m}^2 + 0.0250\,\text{kg} \cdot (0.0800\,\text{m})^2) = \underline{5.17\,\text{rad/s}}$

Note: if the mud had been moving initially, you could use $L = I \cdot \omega$, $\omega = v/r$, and $I = m \cdot r^2$ to find the initial angular momentum for it.

Problem to Try Yourself

Another very common demonstration related to the conservation of angular momentum is the following: a person sits on a spinnable stool, at rest, holding a wheel that is spinning. In this case, the wheel has angular momentum of $0.00600\,\text{kg} \cdot \text{m}^2/\text{s}$. The person flips the wheel over, so that the angular momentum of the wheel is reversed and is now $-0.00600\,\text{kg} \cdot \text{m}^2/\text{s}$. The person suddenly begins to spin slowly, keeping the total angular momentum the same as before the wheel was flipped. If the person has a moment of inertia of $0.800\,\text{kg} \cdot \text{m}^2$, how fast does the person spin? Assume that angular momentum is conserved. Hint: now there are two objects – the person, and the wheel. Any values given for angular momentum are equivalent to $I \cdot \omega$ for that object, at that time point.

Solution:

With two objects, we start with Equation 10.7.1:

$$I_{1i} \cdot \omega_{1i} + I_{2i} \cdot \omega_{2i} = I_{1f} \cdot \omega_{1f} + I_{2f} \cdot \omega_{2f}$$

We'll call the wheel object 1, and the person object 2. For the wheel, instead of being given I and ω, we are just given the angular momentum at each time point, which is equal to $I\cdot\omega$. The person's moment of inertia stays the same, meaning that $I_{2i} = I_{2f}$. We want to find ω_{2f}, which is the angular velocity of the person (object 2) at the final time point.

Subtract $I_{1f} \cdot \omega_{1f}$ from both sides: $I_{1i} \cdot \omega_{1i} + I_{2i} \cdot \omega_{2i} - I_{1f} \cdot \omega_{1f} = I_{2f} \cdot \omega_{2f}$

Divide both sides by I_{2f}: $\omega_{2f} = (I_{1i} \cdot \omega_{1i} + I_{2i} \cdot \omega_{2i} - I_{1f} \cdot \omega_{1f})/I_{2f}$

Now put in numbers. $I_{1i} \cdot \omega_{1i}$ is the wheel's angular momentum at the initial time point (0.00600 kg·m²/s) and $I_{1f} \cdot \omega_{1f}$ is the wheel's angular momentum at the final time point. (–0.00600 kg·m²/s). $\omega_{2i} = 0$ because the person isn't rotating at the start.

$\omega_{2f} = (0.00600$ kg·m²/s $+ 0.800$ kg·m²·0 $-$ -0.00600 kg·m²/s$)/(0.800$ kg·m²$) = \underline{0.015 \text{ rad/s}}$

10.8 CHAPTER 10 SUMMARY

Kinetic energy is:

½·m·v² for something moving linearly

$$\tfrac{1}{2}\cdot I\cdot\omega^2 \text{ for something that is rotating} \qquad (10.2.1)$$

$$\tfrac{1}{2}\cdot m\cdot v^2 + \tfrac{1}{2}\cdot I\cdot\omega^2 \text{ for something that is rotating and}$$
$$\text{moving linearly (like rolling)} \qquad (10.2.2)$$

When an object rolls without slipping, we can relate the linear and angular quantities with:

$$\Delta x = \Delta\theta\cdot r \qquad (10.3.1)$$

$$v = \omega\cdot r \qquad (10.3.2)$$

Work and power have rotation equivalents:

$$W = \pm\tau\cdot\Delta\theta \qquad (10.4.1)$$

$$P = \pm\tau\cdot\omega \qquad (10.5.1)$$

In Equations 10.4.1 and 10.5.1 the + sign is for when τ and $\Delta\theta$ are in the same direction.

The rotation equivalent of momentum is called angular momentum:

$$L = I\cdot\omega \qquad (10.6.1)$$

The total amount of angular momentum can be conserved. If there are two objects:

$$I_{1i} \cdot \omega_{1i} + I_{2i} \cdot \omega_{2i} = I_{1f} \cdot \omega_{1f} + I_{2f} \cdot \omega_{2f} \qquad (10.7.1)$$

If there is only one object:

$$I_i \cdot \omega_i = I_f \cdot \omega_f \qquad (10.7.2)$$

If two objects are rotating and collide, and if they stick together, their velocities are the same:

$$\omega_{1f} = \omega_{2f} \qquad (10.7.3)$$

Chapter 10 problem types:

Problems using the kinetic energy, either for rotational motion or rotational and linear motion (Section 10.2)

Solve problems using rolling without slipping, including using the kinetic energy, angular velocity, and linear velocity (Section 10.3)

Problems using the rotation equivalent of work (Section 10.4)

Problems using the rotation equivalent of power (Section 10.5)

Problems using angular momentum (Section 10.6)

Conservation of angular momentum problems, including rotation collisions (Section 10.7)

Practice Problems
Section 10.2 (Rotation and Kinetic Energy)

1. A sprinkler in a dishwasher has moment of inertia $8.00 \cdot 10^{-2}$ kg·m². If it rotates at 4.00 rad/s, what is the kinetic energy?

   Solution:
   Since the object is only rotating,

   $$KE = \tfrac{1}{2} \cdot I \cdot \omega^2 \qquad (10.2.1)$$

Put in numbers: $KE = \frac{1}{2} \cdot (8.00 \cdot 10^{-2} \text{ kg} \cdot \text{m}^2) \cdot (4.00 \text{ rad/s})^2 = \underline{0.640 \text{ Joules}}$

2. A rolling pin rolls across a piece of pie dough. It has mass 4.00 kg, moment of inertia $8.00 \cdot 10^{-4}$ kg·m², angular velocity 4.50 rad/s, and linear velocity $9.00 \cdot 10^{-2}$ m/s. What is the kinetic energy?

Solution:

This object is rotating and moving linearly.

$$KE = \frac{1}{2} \cdot m \cdot v^2 + \frac{1}{2} \cdot I \cdot \omega^2 \qquad (10.2.2)$$

Put in numbers: $KE = \frac{1}{2} \cdot (4.00 \text{ kg}) \cdot (9.00 \cdot 10^{-2} \text{ m/s})^2 + \frac{1}{2} \cdot (8.00 \cdot 10^{-4}$ kg·m²)·(4.50 rad/s)² $= \underline{2.43 \cdot 10^{-2} \text{ J}}$

3. Your water bottle gets away from you, and rolls away. It has mass 0.700 kg, moment of inertia $1.40 \cdot 10^{-4}$ kg·m², angular velocity 2.50 rad/s, and linear velocity $5.00 \cdot 10^{-2}$ m/s. What is the kinetic energy?

Solution:

This object is rotating and moving linearly.

$$KE = \frac{1}{2} \cdot m \cdot v^2 + \frac{1}{2} \cdot I \cdot \omega^2 \qquad (10.2.2)$$

Put in numbers: $KE = \frac{1}{2} \cdot (0.700 \text{ kg}) \cdot (5.00 \cdot 10^{-2} \text{ m/s})^2 + \frac{1}{2} \cdot (1.40 \cdot 10^{-4}$ kg·m²)·(2.50 rad/s)² $= \underline{1.31 \cdot 10^{-3} \text{ J}}$

4. The blade of a food processor is something like a disk (0.100 kg, $1.50 \cdot 10^{-2}$ m radius) plus two rods (the blades; $2.00 \cdot 10^{-2}$ kg mass each, $3.00 \cdot 10^{-2}$ m length each). If it rotates at 15.0 rad/s, what is the kinetic energy? Hint: find the moment of inertia as a first step.

Solution:

This object is only rotating.

$$KE = \frac{1}{2} \cdot I \cdot \omega^2 \qquad (10.2.1)$$

For moment of inertia (I), add the moment of inertia for the disk and two blades. Remember from Section 9.5 that the moment of inertia of a disk is $(1/2) \cdot m \cdot r^2$, and the moment of inertia for a rod spinning around its end is $(1/3) \cdot m \cdot L^2$.

$I = (1/2) \cdot (0.100 \text{ kg}) \cdot (1.50 \cdot 10^{-2} \text{ m})^2 + 2 \cdot (1/3) \cdot (2.00 \cdot 10^{-2} \text{ kg}) \cdot (3.00 \cdot 10^{-2} \text{ m})^2 = 2.33 \cdot 10^{-5} \text{ kg} \cdot \text{m}^2$

Put numbers into Equation 10.2.1: $KE = \frac{1}{2} \cdot (2.33 \cdot 10^{-5} \text{ kg} \cdot \text{m}^2) \cdot (15.0 \text{ rad/s})^2 = \underline{2.62 \cdot 10^{-3} \text{ J}}$

Section 10.3 (Rolling, without Slipping)

5. A rolling pin has radius 0.0200 m, mass 4.00 kg and moment of inertia $8.00 \cdot 10^{-4} \text{ kg} \cdot \text{m}^2$. It has a linear velocity of 0.150 m/s as it rolls over a piece of dough. What is the kinetic energy?

   Solution:

$$KE = \frac{1}{2} \cdot m \cdot v^2 + \frac{1}{2} \cdot I \cdot \omega^2 \qquad (10.2.2)$$

$$v = \omega \cdot r \qquad (10.3.2)$$

We know r = 0.0200 m, m = 4.00 kg, I = $8.00 \cdot 10^{-4} \text{ kg} \cdot \text{m}^2$, and v = 0.150 m/s. We want to find KE. We need to use Equation 10.3.2 to find ω, and then we can use Equation 10.2.2 to find KE.

Take Equation 10.3.2, and divide both sides by r: $\omega = v/r$

Put v/r into Equation 10.2.2 for ω: $KE = \frac{1}{2} \cdot m \cdot v^2 + \frac{1}{2} \cdot I \cdot (v/r)^2$

Put in numbers: $KE = \frac{1}{2} \cdot (4.00 \text{ kg}) \cdot (0.150 \text{ m/s})^2 + \frac{1}{2} \cdot (8.00 \cdot 10^{-4} \text{ kg} \cdot \text{m}^2) \cdot (0.150 \text{ m/s}/0.0200 \text{ m})^2 = \underline{6.75 \cdot 10^{-2} \text{ Joules}}$

Note: Section 13.6 reviews combining two equations.

6. A coffee can falls on the floor, and starts rolling away from you! It has mass 1.00 kg, radius 0.0400 m, and moment of inertia $8.00 \cdot 10^{-4}$ kg·m². If it has kinetic energy 2.25 Joules, what are the magnitudes of the linear and angular velocities?

   Solution:

$$KE = \frac{1}{2} \cdot m \cdot v^2 + \frac{1}{2} \cdot I \cdot \omega^2 \qquad (10.2.2)$$

$$v = \omega \cdot r \qquad (10.3.2)$$

Put $\omega \cdot r$ in for v in Equation 10.2.2, since $v = \omega \cdot r$ in Equation 10.3.2:

$$KE = \frac{1}{2} \cdot m \cdot (\omega \cdot r)^2 + \frac{1}{2} \cdot I \cdot \omega^2$$

In this equation, ω is now the only variable that we don't know. Multiply out the square in the first term:

$$KE = \tfrac{1}{2} \cdot m \cdot \omega^2 \cdot r^2 + \tfrac{1}{2} \cdot I \cdot \omega^2$$

Since both terms on the right side have ω^2 in them, we can rewrite the right side as:

$$KE = \omega^2 \cdot (\tfrac{1}{2} \cdot m \cdot r^2 + \tfrac{1}{2} \cdot I)$$

Divide both sides by $(\tfrac{1}{2} \cdot m \cdot r^2 + \tfrac{1}{2} \cdot I)$: $\omega^2 = KE/(\tfrac{1}{2} \cdot m \cdot r^2 + \tfrac{1}{2} \cdot I)$

Take the square root: $\omega = \pm \sqrt{KE / \left(\tfrac{1}{2} \cdot m \cdot r^2 + \tfrac{1}{2} \cdot I \right)}$

Take the positive root, since we are finding the magnitude. Put in numbers:

$$\omega = \pm \sqrt{(2.25\,\text{J}) / \left(\tfrac{1}{2} \cdot 1.00\,\text{kg} \cdot (0.0400\,\text{m})^2 + \tfrac{1}{2} \cdot 8.00 \cdot 10^{-4}\,\text{kg} \cdot \text{m}^2 \right)}$$

$$= \underline{43.3\ \text{rad/s}}$$

To find v, use Equation 10.3.2: $v = \omega \cdot r$

Put in numbers: $v = (43.3\ \text{rad/s}) \cdot (0.0400\ \text{m}) = \underline{1.73\ \text{m/s}}$

Note: Section 13.6 reviews combining two equations.

7. The roller inside of a vacuum cleaner has radius 0.0100 m, mass 0.750 kg and moment of inertia $3.75 \cdot 10^{-5}$ kg·m². If the kinetic energy is 1.20 Joules, what are the linear velocity and the angular velocity?

Solution:

$$KE = \tfrac{1}{2} \cdot m \cdot v^2 + \tfrac{1}{2} \cdot I \cdot \omega^2 \tag{10.2.2}$$

$$v = \omega \cdot r \tag{10.3.2}$$

Put $\omega \cdot r$ in for v in Equation 10.2.2, since $v = \omega \cdot r$ in Equation 10.3.2:

$$KE = \tfrac{1}{2} \cdot m \cdot (\omega \cdot r)^2 + \tfrac{1}{2} \cdot I \cdot \omega^2$$

In this equation, ω is now the only variable that we don't know. Multiply out the square in the first term:

$$KE = \tfrac{1}{2} \cdot m \cdot \omega^2 \cdot r^2 + \tfrac{1}{2} \cdot I \cdot \omega^2$$

Since both terms on the right side have ω^2 in them, we can rewrite the right side as:

$$KE = \omega^2 \cdot (\tfrac{1}{2} \cdot m \cdot r^2 + \tfrac{1}{2} \cdot I)$$

Divide both sides by $(\tfrac{1}{2} \cdot m \cdot r^2 + \tfrac{1}{2} \cdot I)$: $\omega^2 = KE / (\tfrac{1}{2} \cdot m \cdot r^2 + \tfrac{1}{2} \cdot I)$

Take the square root: $\omega = \pm \sqrt{KE / \left(\tfrac{1}{2} \cdot m \cdot r^2 + \tfrac{1}{2} \cdot I \right)}$

Take the positive root, since we are finding the magnitude. Put in numbers:

$$\omega = \pm \sqrt{(1.20\,J) / \left(\tfrac{1}{2} \cdot 0.750\,kg \cdot (0.0100\,m)^2 + \tfrac{1}{2} \cdot 3.75 \cdot 10^{-5}\,kg \cdot m^2 \right)}$$

$\underline{= 146\ rad/s}$

To find v, use Equation 10.3.2: $v = \omega \cdot r$

Put in numbers: $v = (146\ rad/s) \cdot (0.0100\ m) = \underline{1.46\ m/s}$

Note: Section 13.6 reviews combining two equations.

8. A water glass topples over and rolls. It has radius 0.0300 m, mass 0.500 kg, and moment of inertia $1.75 \cdot 10^{-5}$ kg·m². It rolls with linear velocity 1.10 m/s. What is the kinetic energy?

Solution:

$$KE = \tfrac{1}{2} \cdot m \cdot v^2 + \tfrac{1}{2} \cdot I \cdot \omega^2 \qquad (10.2.2)$$

$$v = \omega \cdot r \qquad (10.3.2)$$

We know $r = 0.0300$ m, $m = 0.500$ kg, $I = 1.75 \cdot 10^{-5}$ kg·m², and $v = 1.10$ m/s. We want to find KE. We need to use Equation 10.3.2 to find ω, and then we can use Equation 10.2.2 to find KE.

Take Equation 10.3.2, and divide both sides by r: $\omega = v/r$

Put v/r into Equation 10.2.2 for ω: $KE = \tfrac{1}{2} \cdot m \cdot v^2 + \tfrac{1}{2} \cdot I \cdot (v/r)^2$

Put in numbers: $KE = \tfrac{1}{2} \cdot (0.500\ kg) \cdot (1.10\ m/s)^2 + \tfrac{1}{2} \cdot (1.75 \cdot 10^{-5}\ kg \cdot m^2) \cdot (1.10\ m/s / 0.0300\ m)^2 = \underline{0.314\ Joules}$

Note: Section 13.6 reviews combining two equations.

Section 10.4 (Rotation and Work)

9. You put torque on a ball using a force of 14.0 N, in the tangential direction and counterclockwise, at a distance of $2.00 \cdot 10^{-2}$ m from the center of the ball. If the ball turns by 1.20 rad counterclockwise as you do this, how much work did you do?

Solution:

$$W = \pm \tau \cdot \Delta \theta \qquad (10.4.1)$$

To calculate τ, we'll use Equation 9.6.1:

$$\tau = r \cdot F \cdot \sin(\varphi) \qquad (9.6.1)$$

Put Equation 9.6.1 in for τ in Equation 10.4.1:

$$W = \pm r \cdot F \cdot \sin(\varphi) \cdot \Delta \theta$$

The $\pm$ sign is $+$ here, since the torque and angular displacement are in the same direction (counterclockwise). As in Section 9.5, a tangential direction for the force means $\varphi = 90°$.

Put in numbers: $W = + (2.00 \cdot 10^{-2}$ m$) \cdot (14.0$ N$) \cdot \sin(90°) \cdot (1.20$ rad$) = \underline{0.336 \text{ N} \cdot \text{m}}$

10. You turn a screwdriver, using two fingers. Each finger puts $1.80 \cdot 10^{-3}$ Newtons of force, at a distance of 0.0100 m from the center, in the tangential direction, causing a clockwise torque. If the screwdriver goes through 60.0 radians in the clockwise direction, what is the work done?

Solution:

$$W = \pm \tau \cdot \Delta \theta \qquad (10.4.1)$$

To calculate τ, we'll use Equation 9.6.1:

$$\tau = r \cdot F \cdot \sin(\varphi) \qquad (9.6.1)$$

Put Equation 9.6.1 in for τ in Equation 10.4.1:

$$W = \pm 2 \cdot r \cdot F \cdot \sin(\varphi) \cdot \Delta \theta$$

The $\pm$ sign is $+$ here, since the torque and angular displacement are in the same direction (clockwise). The 2 is because there are two fingers each giving the same torque. As in Section 9.5, a tangential direction for the force means $\varphi = 90°$.

Put in numbers: $W = +2 \cdot (0.0100 \text{ m}) \cdot (1.80 \cdot 10^{-3} \text{ N}) \cdot \sin(90°) \cdot (60.0 \text{ rad}) = \underline{2.16 \cdot 10^{-3} \text{ N} \cdot \text{m}}$

11. You open the top of a pickle jar, by spinning it. You open it using two fingers. Each finger puts $1.60 \cdot 10^{-3}$ Newtons of force, at a distance of 0.0300 m from the center, in the tangential direction, causing a counterclockwise torque. If the top goes through 12.0 radians in the counterclockwise direction, what is the work done?

Solution:

$$W = \pm \tau \cdot \Delta\theta \qquad (10.4.1)$$

To calculate τ, we'll use Equation 9.6.1:

$$\tau = r \cdot F \cdot \sin(\varphi) \qquad (9.6.1)$$

Put Equation 9.6.1 in for τ in Equation 10.4.1:

$$W = \pm 2 \cdot r \cdot F \cdot \sin(\varphi) \cdot \Delta\theta$$

The $\pm$ sign is $+$ here, since the torque and angular displacement are in the same direction (counterclockwise). The 2 is because there are two fingers each giving the same torque. As in Section 9.5, a tangential direction for the force means $\varphi = 90°$.

Put in numbers: $W = +2 \cdot (0.0300 \text{ m}) \cdot (1.60 \cdot 10^{-3} \text{ N}) \cdot \sin(90°) \cdot (12.0 \text{ rad}) = \underline{1.15 \cdot 10^{-3} \text{ N} \cdot \text{m}}$

Section 10.5 (Rotation and Power)

12. For the situation in problem 9, what is the power if the ball turns at 1.30 rad/s?

Solution:

$$P = \pm \tau \cdot \omega \qquad (10.5.1)$$

To calculate τ, we'll use Equation 9.6.1:

$$\tau = r \cdot F \cdot \sin(\varphi) \quad (9.6.1)$$

Put Equation 9.6.1 in for τ in Equation 10.5.1, with the ± sign set to +:

$$P = r \cdot F \cdot \sin(\varphi) \cdot \omega$$

As in Section 9.5, a tangential direction for the force means $\varphi = 90°$.
Put in numbers: $P = (2.00 \cdot 10^{-2}\,m) \cdot (14.0\,N) \cdot \sin(90°) \cdot (1.30\,rad/s) = \underline{0.364}$
<u>Watts</u>

13. For the situation in problem 10, what is the power if the screwdriver turns at 3.00 rad/s?

    Solution:

$$P = \pm \tau \cdot \omega \quad\quad\quad (10.5.1)$$

To calculate τ, we'll use Equation 9.6.1:

$$\tau = r \cdot F \cdot \sin(\varphi) \quad\quad\quad (9.6.1)$$

Put Equation 9.6.1 in for τ in Equation 10.5.1, with the ± sign set to +:

$$P = 2 \cdot r \cdot F \cdot \sin(\varphi) \cdot \omega$$

The 2 is because there are two fingers each giving the same torque. As in Section 9.5, a tangential direction for the force means $\varphi = 90°$.
Put in numbers: $P = 2 \cdot (0.0100\,m) \cdot (1.80 \cdot 10^{-3}\,N) \cdot \sin(90°) \cdot (3.00$
$rad/s) = \underline{1.08 \cdot 10^{-4}\,W}$

14. For the situation in problem 11, what is the power if the jar top turns at 2.50 rad/s?

    Solution:

$$P = \pm \tau \cdot \omega \quad\quad\quad (10.5.1)$$

To calculate τ, we'll use Equation 9.6.1:

$$\tau = r \cdot F \cdot \sin(\varphi) \quad\quad\quad (9.6.1)$$

Put Equation 9.6.1 in for τ in Equation 10.5.1, with the $\pm$ sign set to $+$:

$$P = 2 \cdot r \cdot F \cdot \sin(\varphi) \cdot \omega$$

The 2 is because there are two fingers each giving the same torque. As in Section 9.5, a tangential direction for the force means $\varphi = 90°$.

Put in numbers: $P = 2 \cdot (0.0300 \text{ m}) \cdot (1.60 \cdot 10^{-3} \text{ N}) \cdot \sin(90°) \cdot (2.50 \text{ rad/s}) = \underline{2.40 \cdot 10^{-4} \text{ W}}$

Section 10.6 (Angular Momentum)

15. A sprinkler in a dishwasher has moment of inertia $8.00 \cdot 10^{-2} \text{ kg} \cdot \text{m}^2$. If it rotates at 4.00 rad/s, what is the angular momentum?

    Solution:

    $$L = I \cdot \omega \qquad (10.6.1)$$

    Put in numbers: $L = (8.00 \cdot 10^{-2} \text{ kg} \cdot \text{m}^2) \cdot (4.00 \text{ rad/s}) = \underline{0.320 \text{ kg} \cdot \text{m}^2/\text{s}}$

16. The blade of a food processor is something like a disk (0.100 kg, $1.50 \cdot 10^{-2}$ m radius) plus two rods (the blades; $2.00 \cdot 10^{-2}$ kg mass each, $3.00 \cdot 10^{-2}$ m length each). If it is rotating at 20.0 rad/s, what is the angular momentum?

    Solution:

    $$L = I \cdot \omega \qquad (10.6.1)$$

    We need to calculate I, by adding the angular momentum for the disk ($I = (1/2) \cdot m \cdot r^2$, Equation 9.5.3) and the two rods (each $I = (1/3) \cdot m \cdot L^2$ for a rod spinning around its end, Equation 9.5.4).

    $I = (1/2) \cdot (0.100 \text{ kg}) \cdot (1.50 \cdot 10^{-2} \text{ m})^2 + 2 \cdot (1/3) \cdot (2.00 \cdot 10^{-2} \text{ kg}) \cdot (3.00 \cdot 10^{-2} \text{ m})^2 = 2.33 \cdot 10^{-5} \text{ kg} \cdot \text{m}^2$

    Put in numbers for Equation 10.6.1: $L = (2.33 \cdot 10^{-5} \text{ kg} \cdot \text{m}^2) \cdot (20.0 \text{ rad/s}) = \underline{2.40 \cdot 10^{-4} \text{ kg} \cdot \text{m}^2/\text{s}}$

17. Think of the arrow at one end of a weathervane as a point mass with mass 0.100 kg and distance 0.300 m from the center. If it spins at 1.50 rad/s, what is its angular momentum?

Solution:

$$L = I \cdot \omega \tag{10.6.1}$$

We need to calculate I, using $I = m \cdot r^2$ for a point mass (Equation 9.5.1).

$I = (0.100 \text{ kg}) \cdot (0.300 \text{ m})^2 = 9.00 \cdot 10^{-3} \text{ kg} \cdot \text{m}^2$

Put in numbers for Equation 10.6.1: $L = (9.00 \cdot 10^{-3} \text{ kg} \cdot \text{m}^2) \cdot (1.50 \text{ rad/s}) = \underline{1.35 \cdot 10^{-2} \text{ kg} \cdot \text{m}^2/\text{s}}$

Section 10.7 (Conservation of Angular Momentum and Rotation Collisions)

18. A (fictional) clothes washer is invented that can speed up and slow down by changing size. At the start of the spin cycle, the moment of inertia is 1.20 kg·m² and the washer is spinning at 3.00 rad/s. Then, the walls close in, to half of the distance they were, which changes the moment of inertia to 0.300 kg·m². Assume that there are no external torques. What is the new angular velocity?

Solution:

No external torques means that angular momentum is conserved. Since there is only one object, we use Equation 10.7.2:

$$I_i \cdot \omega_i = I_f \cdot \omega_f \tag{10.7.2}$$

We know $I_i = 1.20$ kg·m², $\omega_i = 3.00$ rad/s, and $I_f = 0.300$ kg·m². We want to find ω_f.

Divide both sides by I_f: $\omega_f = (I_i \cdot \omega_i)/(I_f)$

Put in numbers: $\omega_f = (1.20 \text{ kg} \cdot \text{m}^2 \cdot 3.00 \text{ rad/s})/(0.300 \text{ kg} \cdot \text{m}^2) = \underline{12.0 \text{ rad/s}}$

19. A person has tripped and is about to fall down. The person is spinning at 1.00 rad/s, and has a moment of inertia of 0.675 kg·m². As a reflex, the person puts their arms out to steady themselves. This changes their moment of inertia to be 1.90 kg·m². Assume that there are no external torques. What is the new rate of spinning? Note: the additional moment of inertia is based on two rods (arms) of length 0.800 m each and mass 3.00 kg each.

Solution:

No external torques means that angular momentum is conserved. Since there is only one object, we use Equation 10.7.2:

$$I_i \cdot \omega_i = I_f \cdot \omega_f \qquad (10.7.2)$$

We know $I_i = 0.675$ kg·m², $\omega_i = 1.00$ rad/s, and $I_f = 1.90$ kg·m². We want to find ω_f.

Divide both sides by I_f: $\omega_f = (I_i \cdot \omega_i)/(I_f)$

Put in numbers: $\omega_f = (0.675$ kg·m²·1.00 rad/s$)/(1.90$ kg·m²$) = \underline{0.355}$ $\underline{\text{rad/s}}$

20. A sprinkler inside of a dishwasher is shaped like two rods, each with length 0.300 m. Right before the washing stops, there is water running through the sprinkler; each rod has mass 2.50 kg and is spinning at 2.00 rad/s. Then, the water is turned off. Without the water inside, each rod now only has mass 0.850 kg. Assume that there are no external torques. What is the new velocity of the sprinkler as it spins? Hint: because the two rods are stuck together, treat them as one object with a moment of inertia that is equal to adding the moment of inertia for the two rods. (Note: this problem ignores friction and possibly forces from the water moving; eventually friction would make the spinning stop.)

Solution:

No external torques means that angular momentum is conserved. As mentioned in the hint, we'll treat the two rods as one object here, and with one object we use Equation 10.7.2:

$$I_i \cdot \omega_i = I_f \cdot \omega_f \qquad (10.7.2)$$

We need to calculate I_i and I_f. In both cases, the moment of inertia for two rods that are exactly the same is:

$$I = 2 \cdot (1/3) \cdot m \cdot L^2$$

This is Equation 9.5.4 for a rod spun around its end, with a 2 at the start because there are two rods.

At the initial time point, m = 2.50 kg and L = 0.300 m. This gives:

$$I_i = 2 \cdot (1/3) \cdot (2.50 \text{ kg}) \cdot (0.300 \text{ m})^2 = 0.150 \text{ kg·m}^2$$

At the final time point, m has changed to 0.850 kg. This gives:

$$I_f = 2 \cdot (1/3) \cdot (0.750 \text{ kg}) \cdot (0.300 \text{ m})^2 = 5.10 \cdot 10^{-2} \text{ kg·m}^2$$

We now know every variable in Equation 10.7.2 except for ω_f.

Divide both sides of Equation 10.7.2 by I_f: $\omega_f = (I_i \cdot \omega_i)/(I_f)$

Put in numbers: $\omega_f = (0.150 \text{ kg·m}^2 \cdot 2.00 \text{ rad/s})/(5.10 \cdot 10^{-2} \text{ kg·m}^2) = \underline{5.88}$
<u>rad/s</u>

21. A "high five" is given! One person's hand and forearm (like a rod with mass 1.50 kg and length 0.300 m) is spinning at 0.800 m/s, and it intentionally collides with a second person's hand (a point mass with mass 0.300 kg and distance 0.300 m from the center of the first person's arm's rotation). Before the collision, the first person's arm and hand are rotating at 1.20 rad/s and the second person's hand is not moving. The two objects stick together right after the collision, with no external torques. What is the angular velocity of both objects (together) right after the collision?

Solution:

This is the rotational equivalent of collisions, which we need Equations 10.7.1 and 10.7.3 for:

$$I_{1i} \cdot \omega_{1i} + I_{2i} \cdot \omega_{2i} = I_{1f} \cdot \omega_{1f} + I_{2f} \cdot \omega_{2f} \tag{10.7.1}$$

$$\omega_{1f} = \omega_{2f} \tag{10.7.3}$$

We'll call the arm/hand that is initially moving as number 1, and the hand that is not moving initially as number 2. Number 1 will be treated like a rod spun around its end, with $I = (1/3) \cdot m \cdot L^2$ (Equation 9.5.4), and number 2 will be a point mass with $I = m \cdot r^2$ (Equation 9.5.1). Nothing changes the moment of inertia between the initial and final time points.

I_{1i} and I_{1f} are both equal to $(1/3)\cdot(1.50 \text{ kg})\cdot(0.300 \text{ m})^2 = 4.50\cdot10^{-2} \text{ kg}\cdot\text{m}^2$

I_{2i} and I_{2f} are both equal to $(0.300 \text{ kg})\cdot(0.300 \text{ m})^2 = 2.70\cdot10^{-2} \text{ kg}\cdot\text{m}^2$

Since the objects stick together,

$$\omega_{1f} = \omega_{2f} \tag{10.7.3}$$

Put in ω_{1f} everywhere that ω_{2f} appears in the first equation:

$$I_{1i}\cdot\omega_{1i} + I_{2i}\cdot\omega_{2i} = I_{1f}\cdot\omega_{1f} + I_{2f}\cdot\omega_{1f}$$

The right side can be rewritten since both terms are something times ω_{1f}:

$$I_{1i}\cdot\omega_{1i} + I_{2i}\cdot\omega_{2i} = (I_{1f} + I_{2f})\cdot\omega_{1f}$$

Divide both sides by $(I_{1f} + I_{2f})$:

$$\omega_{1f} = (I_{1i}\cdot\omega_{1i} + I_{2i}\cdot\omega_{2i})/(I_{1f} + I_{2f})$$

Put in numbers:

$\omega_{1f} = (4.50\cdot10^{-2} \text{ kg}\cdot\text{m}^2\cdot1.20 \text{ rad/s} + 2.70\cdot10^{-2} \text{ kg}\cdot\text{m}^2\cdot0)/(4.50\cdot10^{-2} \text{ kg}\cdot\text{m}^2 + 2.70\cdot10^{-2} \text{ kg}\cdot\text{m}^2)$

$\omega_{1f} = \underline{0.750 \text{ rad/s}}$

22. A ball becomes stuck in a tree, and a person swings a stick at it in order to try and get it down. The stick has length 2.50 m and mass 2.00 kg, and is initially rotating at 0.800 rad/s. The ball has mass 0.500 kg, isn't moving, and can be thought of as a point mass. The two objects stick together right after they collide. The ball ended up only being 2.00 m from the center of the stick's rotation. (The stick was a little longer than needed.) Assume there are no external torques. What is the angular velocity of both objects (together) right after the collision?

Solution:

This is the rotational equivalent of collisions, which we need Equations 10.7.1 and 10.7.3 for:

$$I_{1i} \cdot \omega_{1i} + I_{2i} \cdot \omega_{2i} = I_{1f} \cdot \omega_{1f} + I_{2f} \cdot \omega_{2f} \qquad (10.7.1)$$

$$\omega_{1f} = \omega_{2f} \qquad (10.7.3)$$

We'll call the stick object 1, and we'll call the ball object 2. Object 1 will be treated like a rod spun around its end, with $I = (1/3) \cdot m \cdot L^2$ (Equation 9.5.4), and object 2 will be a point mass with $I = m \cdot r^2$ (Equation 9.5.1). Nothing changes the moment of inertia between the initial and final time points.

I_{1i} and I_{1f} are both equal to $(1/3) \cdot (2.00 \text{ kg}) \cdot (2.50 \text{ m})^2 = 4.17 \text{ kg} \cdot \text{m}^2$

I_{2i} and I_{2f} are both equal to $(0.500 \text{ kg}) \cdot (2.00 \text{ m})^2 = 2.00 \text{ kg} \cdot \text{m}^2$

Since the objects stick together,

$$\omega_{1f} = \omega_{2f} \qquad (10.7.3)$$

Put in ω_{1f} everywhere that ω_{2f} appears in the first equation:

$$I_{1i} \cdot \omega_{1i} + I_{2i} \cdot \omega_{2i} = I_{1f} \cdot \omega_{1f} + I_{2f} \cdot \omega_{1f}$$

The right side can be rewritten since both terms are something times ω_{1f}:

$$I_{1i} \cdot \omega_{1i} + I_{2i} \cdot \omega_{2i} = (I_{1f} + I_{2f}) \cdot \omega_{1f}$$

Divide both sides by $(I_{1f} + I_{2f})$:

$$\omega_{1f} = (I_{1i} \cdot \omega_{1i} + I_{2i} \cdot \omega_{2i})/(I_{1f} + I_{2f})$$

Put in numbers:

$$\omega_{1f} = (4.17 \text{ kg} \cdot \text{m}^2 \cdot 0.800 \text{ rad/s} + 2.00 \text{ kg} \cdot \text{m}^2 \cdot 0)/(4.17 \text{ kg} \cdot \text{m}^2 + 2.00 \text{ kg} \cdot \text{m}^2)$$

$$\omega_{1f} = \underline{0.541 \text{ rad/s}}$$

Newton's More General Law of Gravity

11.1 INTRODUCTION: GRAVITY IS DIFFERENT AWAY FROM THE GROUND

Throughout this book, we have presented gravity as being equal to m·g, where g is 9.80 m/s². It turns out that this is a special case, and only for locations near the "ground" on Earth. In this chapter, we will talk about uses of a more general equation of gravity.

Note: this chapter will use a few pieces of information about gravity and space. The value of the physics constant G (the gravitational constant) is from the NASA Jet Propulsion Laboratory, astrodynamic parameters: https://ssd.jpl.nasa.gov/astro_par.html. The value of Earth's mass and radius are from the NASA Jet Propulsion Laboratory, Planetary Physical Parameters: https://ssd.jpl.nasa.gov/planets/phys_par.html. The value of the distance between Earth and the moon and the sideral period (time to go around one circle of Earth) is from the NASA Jet Propulsion Laboratory, Planetary Satellite Mean Elements, using "semi-major axis": https://ssd.jpl .nasa.gov/sats/elem/.

11.2 NEWTON'S LAW OF GRAVITY

Gravity is described in more general terms by Newton's law of gravity. A force happens between two objects:

$$F = \frac{G \cdot m_1 \cdot m_2}{r^2} \tag{11.2.1}$$

DOI: 10.1201/9781003005049-11

F is the force of gravity, G is a number equal to $6.67 \cdot 10^{-11}$ N·m²/kg², m_1 and m_2 are the masses of two objects, and r is how far apart the centers of the objects are. Notice that the units of G make the units of the force come out to Newtons, just like any other force.

Example 11.2.1: Gravity with No Planet-Sized Objects

Two large trucks are next to each other on the highway. Each has a mass of 950 kg, and the centers of the trucks are 2.00 m apart. What is the gravitational force?

Solution:

$$F = \frac{G \cdot m_1 \cdot m_2}{r^2} \qquad (11.2.1)$$

Put in numbers: $F = \dfrac{\left(6.67 \cdot 10^{-11} \, N \cdot m^2 / kg^2\right) \cdot \left(950\,kg\right) \cdot \left(950\,kg\right)}{\left(2.00\,m\right)^2} =$

$\underline{1.50 \cdot 10^{-5}\,N}$

Note: even with two of the larger everyday objects, the gravitational force between them is pretty small.

Example 11.2.2: One Very Large Object

A person in a flying airplane (roughly 5 mi or 8 km in the air) is about $6.38 \cdot 10^6$ m from Earth. If the person has mass 100 kg, and Earth has mass $5.97 \cdot 10^{24}$ kg, what is the gravitational force?

Solution:

$$F = \frac{G \cdot m_1 \cdot m_2}{r^2} \qquad (11.2.1)$$

Put in numbers: $F = \dfrac{\left(6.67 \cdot 10^{-11} \, N \cdot m^2 / kg^2\right) \cdot \left(100\,kg\right) \cdot \left(5.97 \cdot 10^{24}\,kg\right)}{\left(6.38 \cdot 10^6 \, m\right)^2}$

$= \underline{978\,N}$

Note: this is slightly less than what you'd expect with m·g, although not by much: $(100\,kg) \cdot (9.80 \, m/s^2) = \underline{980\,N}$.

Problem to Try Yourself

Two rocks are 5.60 m apart. If one rock has mass 405 kg, and the gravitational force between the rocks is $2.85 \cdot 10^{-7}$ N, what is the mass of the second rock?

Solution:

$$F = \frac{G \cdot m_1 \cdot m_2}{r^2}$$

Here we are solving for m_2. Multiply both sides by r^2: $F \cdot r^2 = G \cdot m_1 \cdot m_2$

Divide both sides by $G \cdot m_1$: $m_2 = \dfrac{F \cdot r^2}{G \cdot m_1}$

Put in numbers: $m_2 = \dfrac{\left(2.85 \cdot 10^{-7} \, N\right) \cdot \left(5.60 \, m\right)^2}{\left(6.67 \cdot 10^{-11} \, N \cdot m^2 / kg^2\right) \cdot \left(405 \, kg\right)} = \underline{331 \, kg}$

11.3 CONNECTING M·G WITH NEWTON'S LAW OF GRAVITY

In the introduction, it was stated that m·g is a special case of Newton's law of gravity. Here is the connection:

Let the two objects be some object on Earth's surface, with mass m_{object}.

Let the second object be Earth, with mass $5.97 \cdot 10^{24}$ kg.

The distance between the two objects is the distance between their centers. If the first object is on Earth, their centers are one (average) Earth radius apart: $6.37 \cdot 10^6$ m.

If we put these values into Equation 11.2.1, we get:

$$F = \frac{G \cdot m_{Earth} \cdot m_{object}}{r^2}$$

$$= \frac{\left(6.67 \cdot 10^{-11} \, N \cdot m^2 / kg^2\right) \cdot \left(5.97 \cdot 10^{24} \, kg\right) \cdot m_{object}}{\left(6.37 \cdot 10^6 \, m\right)^2}$$

$$= m_{object} \cdot \left(9.80 \, m / s^2\right)$$

(11.3.1)

This shows the relationship between m·g and Newton's law of gravity. If one of the objects is Earth, and the second object is on the ground, on Earth, then the equation happens to look like m·g (with g = 9.80 m/s²).

Side note 1: you'll notice that we used m·g when objects were in the air as well – this is because the value of r in Equation 11.3.1 is so big that changing it by an amount that is much smaller than r doesn't affect the value. For example, if the object went 100 m into the air, it would only change the value of r by about 0.0016%.

Side note 2: if you calculate out Equation 11.3.1, you'll see that it comes to 9.81 m/s². This is still very close to the value of 9.80 m/s² that is often used in physics textbooks.

11.4 POTENTIAL ENERGY FROM GRAVITY USING NEWTON'S LAW OF GRAVITY

When we used m·g for gravity, we used the equation m·g·(change in height) to find the change in potential energy. We won't do the derivation, but this comes from calculus, more specifically taking the area under the curve (integral) of the force. The same methods would give us an equation for the potential energy change from gravity with Newton's law of gravity:

$$\Delta PE = \frac{G \cdot m_1 \cdot m_2}{r_i} - \frac{G \cdot m_1 \cdot m_2}{r_f} \tag{11.4.1}$$

ΔPE is the change in potential energy between a final and initial time point, G is $6.67 \cdot 10^{-11}$ N·m²/kg², m_1 and m_2 are the masses of two objects, r_i is the distance between the two object's centers at the initial time point, and r_f is the distance at the final time point.

Side note: in this book, potential energy is always expressed as a change between two points. Some books permanently set the initial distance to be infinity, which makes the first term on the right side of Equation 11.4.1 equal to zero.

Example 11.4.1

A 100 kg person flies in an airplane. Their distance from Earth's center increases from $6.37 \cdot 10^6$ m to $6.38 \cdot 10^6$ m. What is the change in gravitational potential energy?

Solution:

$$\Delta PE = \frac{G \cdot m_1 \cdot m_2}{r_i} - \frac{G \cdot m_1 \cdot m_2}{r_f} \tag{11.4.1}$$

Put in numbers:

$$\Delta PE = \frac{\left(6.67 \cdot 10^{-11} \, N \cdot m^2 / kg^2\right) \cdot \left(5.97 \cdot 10^{24} \, kg\right) \cdot 100 \, kg}{6.37 \cdot 10^6 \, m}$$

$$-\frac{\left(6.67 \cdot 10^{-11} \, N \cdot \frac{m^2}{kg^2}\right) \cdot \left(5.97 \cdot 10^{24} \, kg\right) \cdot 100 \, kg}{6.38 \cdot 10^6 \, m}$$

$$= \underline{9.80 \cdot 10^6 \, J}$$

Note 1: this amount is the same as what you would expect from $m \cdot g \cdot \Delta h = (100 \, kg) \cdot (9.80 \, m/s^2) \cdot (0.01 \cdot 10^6 \, m)$. It turns out that, in general, you have to go even further away before $m \cdot g \cdot \Delta h$ is not really usable.

Note 2: The initial distance is between the center of the person and the center of Earth. If the person is on Earth's surface, and we approximate Earth as a sphere, that is the radius of Earth.

Problem to Try Yourself

In the example above, if the change in potential energy is $2.93 \cdot 10^7$, and the initial distance from Earth's center is the same, what is the final distance?

Solution:

$$\Delta PE = \frac{G \cdot m_1 \cdot m_2}{r_i} - \frac{G \cdot m_1 \cdot m_2}{r_f} \quad (11.4.1)$$

We now solve for r_f. Subtract $G \cdot m_1 m_2 / r_i$ from both sides:

$$\Delta PE - \frac{G \cdot m_1 \cdot m_2}{r_i} = -\frac{G \cdot m_1 \cdot m_2}{r_f}$$

Multiply both sides by r_f: $r_f \cdot \left(\Delta PE - \frac{G \cdot m_1 \cdot m_2}{r_i}\right) = -G \cdot m_1 \cdot m_2$

Divide both sides by $\left(\Delta PE - \frac{G \cdot m_1 \cdot m_2}{r_i}\right)$: $r_f = \frac{-G \cdot m_1 \cdot m_2}{\left(\Delta PE - \frac{G \cdot m_1 \cdot m_2}{r_i}\right)}$

Put in numbers:

$$r_f = \cfrac{-\left(6.67 \cdot 10^{-11}\,\text{N} \cdot \cfrac{\text{m}^2}{\text{kg}^2}\right) \cdot \left(5.97 \cdot 10^{24}\,\text{kg}\right) \cdot 100\,\text{kg}}{\left(2.93 \cdot 10^7\,\text{J} - \cfrac{\left(6.67 \cdot 10^{-11}\,\text{N} \cdot \cfrac{\text{m}^2}{\text{kg}^2}\right) \cdot \left(5.97 \cdot 10^{24}\,\text{kg}\right) \cdot 100\,\text{kg}}{6.37 \cdot 10^6\,\text{m}}\right)}$$

$$= 6.40 \cdot 10^6\,\text{m}$$

Note: this is going three times further up from Earth's surface than the previous example – a difference of $0.03 \cdot 10^6$ m instead of $0.01 \cdot 10^6$ m. In miles, it is something like 18 instead of 6.

11.5 NEWTON'S LAW OF GRAVITY WITH CONSERVATION OF MECHANICAL ENERGY

Section 11.4 gave us an equation for the change in potential energy, ΔU. In Chapter 6, we had two equations related to the conservation of mechanical energy:

$$\Delta KE + \Delta PE = W_{nc} \tag{6.9.1}$$

$$\Delta KE = \tfrac{1}{2} \cdot m \cdot v_f^2 - \tfrac{1}{2} \cdot m \cdot v_i^2 \tag{6.6.2}$$

ΔKE is the change in kinetic energy (assuming linear motion), ΔPE is the change in potential energy (Equation 11.4.1 gives this for Newton's law of gravity), W_{nc} is the work done by non-conservative forces, m is mass, and v_f and v_i are the final time point (f) and initial time point (i) values of linear velocity. In a case where gravity is the only force, $W_{nc} = 0$ since gravity is conservative.

We can combine the two equations above with Equation 11.4.1 to get the following equation:

$$\frac{1}{2} \cdot m_1 \cdot v_f^2 - \frac{1}{2} \cdot m_1 \cdot v_i^2 + \frac{G \cdot m_1 \cdot m_2}{r_i} - \frac{G \cdot m_1 \cdot m_2}{r_f} = W_{nc} \tag{11.5.1}$$

In this equation, m_1 is a moving object, and m_2 is not moving. In our airplane examples before, m_1 would be the person and m_2 would be Earth.

If gravity is the only force, then W_{nc} (work done by non-conservative forces) is zero, and Equation 11.5.1 becomes the **conservation of mechanical energy**. As a reminder from Section 6.9, this means that the total amount of mechanical energy (kinetic energy plus potential energy) stays the same, and that the amount of change is zero.

Example 11.5.1

A piece of "space junk" is in space, near Earth. (Space junk is left-over pieces from previous objects sent into space.) It starts at $7.00 \cdot 10^6$ m from Earth's surface, and comes closer, to $6.80 \cdot 10^6$ m. If it starts from rest, what is the magnitude of the velocity when it is closer? Assume that gravity is the only force. Note: we are staying away from Earth's surface in this example, so that there will be less air and less air resistance.

Solution:

$$\frac{1}{2} \cdot m_1 \cdot v_f^2 - \frac{1}{2} \cdot m_1 \cdot v_i^2 + \frac{G \cdot m_1 \cdot m_2}{r_i} - \frac{G \cdot m_1 \cdot m_2}{r_f} = W_{nc} \qquad (11.5.1)$$

Here, $v_i = 0$, $m_2 = 5.97 \cdot 10^{24}$ kg (Earth), $G = 6.67 \cdot 10^{-11}$ N·m²/kg², $r_i = 7.00 \cdot 10^6$ m, and $r_f = 6.80 \cdot 10^6$ m. $W_{nc} = 0$ because gravity is the only force. We actually don't need to know m_1 because it is in every term exactly once and will cancel out.

First, take out the term that becomes zero since $v_i = 0$:

$$\frac{1}{2} \cdot m_1 \cdot v_f^2 + \frac{G \cdot m_1 \cdot m_2}{r_i} - \frac{G \cdot m_1 \cdot m_2}{r_f} = 0$$

Next, divide both sides by m_1. It will cancel out m_1 in every term on the left side.

$$\frac{1}{2} \cdot v_f^2 + \frac{G \cdot m_2}{r_i} - \frac{G \cdot m_2}{r_f} = 0$$

Add $G \cdot m_2 / r_f$ to both sides, and subtract $G \cdot m_2 / r_i$ from both sides:

$$\frac{1}{2} \cdot v_f^2 = -\frac{G \cdot m_2}{r_i} + \frac{G \cdot m_2}{r_f}$$

Multiply both sides by 2, to cancel the ½ on the left side:

$$v_f^2 = -\frac{2 \cdot G \cdot m_2}{r_i} + \frac{2 \cdot G \cdot m_2}{r_f}$$

Take the square root:

$$v_f = \pm\sqrt{-\frac{2 \cdot G \cdot m_2}{r_i} + \frac{2 \cdot G \cdot m_2}{r_f}}$$

We'll choose the positive root, since we want the magnitude of velocity.

Put in numbers:

$$v_f = \sqrt{ -\frac{2 \cdot \left(6.67 \cdot 10^{-11}\,N \cdot \frac{m^2}{kg^2}\right) \cdot \left(5.97 \cdot 10^{24}\,kg\right)}{7.00 \cdot 10^6\,m} + \frac{2 \cdot \left(6.67 \cdot 10^{-11}\,N \cdot \frac{m^2}{kg^2}\right) \cdot \left(5.97 \cdot 10^{24}\,kg\right)}{6.80 \cdot 10^6\,m} } = 1.83 \cdot 10^3\,m/s$$

Problem to Try Yourself

An object is traveling away from Earth. At the starting point, it has velocity 806 m/s and is $6.70 \cdot 10^6$ m from Earth's center. As it gets further from Earth, its velocity will decrease. At what distance will the velocity become zero? Assume that gravity is the only force.

Solution:

$$\frac{1}{2} \cdot m_1 \cdot v_f^2 - \frac{1}{2} \cdot m_1 \cdot v_i^2 + \frac{G \cdot m_1 \cdot m_2}{r_i} - \frac{G \cdot m_1 \cdot m_2}{r_f} = W_{nc} \quad (11.5.1)$$

In this problem, $v_f = 0$, $v_i = 800$ m/s, $r_i = 6.70 \cdot 10^6$ m, $m_2 = 5.97 \cdot 10^{24}$ kg (Earth), $G = 6.67 \cdot 10^{-11}$ N·m²/kg², $W_{nc} = 0$ (gravity is the only force), and m_1 will cancel out of every term.

Putting in the zero values gives:

$$-\frac{1}{2} \cdot m_1 \cdot v_i^2 + \frac{G \cdot m_1 \cdot m_2}{r_i} - \frac{G \cdot m_1 \cdot m_2}{r_f} = 0$$

Divide both sides by m_1:

$$-\frac{1}{2} \cdot v_i^2 + \frac{G \cdot m_2}{r_i} - \frac{G \cdot m_2}{r_f} = 0$$

We want to solve for r_f. First, add $G \cdot m_2/r_f$ to both sides:

$$-\frac{1}{2} \cdot v_i^2 + \frac{G \cdot m_2}{r_i} = \frac{G \cdot m_2}{r_f}$$

Multiply both sides by r_f:

$$r_f \cdot \left(-\frac{1}{2} \cdot v_i^2 + \frac{G \cdot m_2}{r_i} \right) = G \cdot m_2$$

Divide both sides by $\left(-\frac{1}{2} \cdot v_i^2 + \frac{G \cdot m_2}{r_i} \right)$:

$$r_f = \frac{G \cdot m_2}{\left(-\frac{1}{2} \cdot v_i^2 + \frac{G \cdot m_2}{r_i} \right)}$$

Put in numbers:

$$r_f = \frac{\left(6.67 \cdot 10^{-11} \, N \cdot \frac{m^2}{kg^2} \right) \cdot \left(5.97 \cdot 10^{24} \, kg \right)}{\left(-\frac{1}{2} \cdot \left(806 \, m/s \right)^2 + \frac{\left(6.67 \cdot 10^{-11} \, N \cdot \frac{m^2}{kg^2} \right) \cdot \left(5.97 \cdot 10^{24} \, kg \right)}{6.70 \cdot 10^6 \, m} \right)}$$

$$= \underline{6.74 \cdot 10^6 \, m}$$

11.6 NEWTON'S LAW OF GRAVITY
WITH CIRCULAR MOTION

In Chapter 8, we said that the total force on an object traveling in a circle is given by:

$$F_{c,\,net} = m \cdot v^2 / r$$

If an object is going in a circle, and gravity is the force causing this, then $F_{c,\,net}$ is equal to Newton's law of gravity (Equation 11.2.1):

$$\frac{G \cdot m_1 \cdot m_2}{r^2} = \frac{m_1 \cdot v^2}{r} \tag{11.6.1}$$

In this equation, m_1 is mass traveling in a circle, and m_2 is a mass that is in the center. You can picture this as a moon going around a planet – the moon would be m_1, and the planet would be m_2. (Side note: per Kepler's first law, orbits in space are actually an ellipse, not a circle.)

Example 11.6.1

The moon is approximately $3.84 \cdot 10^8$ m from Earth. Based on Equation 11.6.1, what is the magnitude of the moon's velocity, traveling around Earth?

Solution:

$$\frac{G \cdot m_1 \cdot m_2}{r^2} = \frac{m_1 \cdot v^2}{r} \tag{11.6.1}$$

Multiply both sides by r, and divide both sides by m_1 (which cancels m_1):

$$\frac{G \cdot m_2}{r} = v^2$$

Take the square root:

$$v = \sqrt{\frac{G \cdot m_2}{r}}$$

Put in numbers:

$$v = \sqrt{\frac{\left(6.67 \cdot 10^{-11}\, N \cdot \frac{m^2}{kg^2}\right) \cdot \left(5.97 \cdot 10^{24}\, kg\right)}{3.84 \cdot 10^8\, m}} = \underline{1.02 \cdot 10^3\, m/s}$$

Note: this is a tangential velocity as the moon goes in a (approximate) circle, whereas you can think of the velocity in Section 11.5 as being direction towards or away from Earth. At this velocity, and making a circle with radius $3.84 \cdot 10^8$ m, you could use $v = (2 \cdot \pi \cdot r)/T$ (Equation 8.3.3) to find $T = 27.4$ days. In other words, the moon takes 27.4 days to go around Earth. The data table used for the radius gives a "sidereal period" of 27.3 days.

Problem to Try Yourself

An object is going in a circle around Earth, at a constant speed of 798 m/s. How far is it from Earth's center?

Solution:

$$\frac{G \cdot m_1 \cdot m_2}{r^2} = \frac{m_1 \cdot v^2}{r} \qquad (11.6.1)$$

This time we solve for r. Multiply both sides by r^2, and divide both sides by m_1:

$$G \cdot m_2 = r \cdot v^2$$

Divide both sides by v^2:

$$\frac{G \cdot m_2}{v^2} = r$$

Put in numbers: $r = \dfrac{\left(6.67 \cdot 10^{-11}\, N \cdot \frac{m^2}{kg^2}\right) \cdot \left(5.97 \cdot 10^{24}\, kg\right)}{\left(798 m/s\right)^2}$

$= \underline{6.25 \cdot 10^8\, m}$

11.7 CHAPTER 11 SUMMARY

Newton's law of gravity is a more general equation for gravity's force than m·g. It is:

$$F = \frac{G \cdot m_1 \cdot m_2}{r^2} \tag{11.2.1}$$

F is the force of gravity, G is a number equal to $6.67 \cdot 10^{-11}$ N·m²/kg², m_1 and m_2 are the masses of two objects, and r is how far apart the objects are.

When one of the masses is Earth, and the other object is on Earth (one Earth radius from its center), you get m·g. This means that m·g is a special case of Newton's law of gravity.

Newton's law of gravity has a potential energy related to it:

$$\Delta PE = \frac{G \cdot m_1 \cdot m_2}{r_i} - \frac{G \cdot m_1 \cdot m_2}{r_f} \tag{11.4.1}$$

This potential energy can be used with the conservation of mechanical energy. If gravity is the only force on an object:

$$\frac{1}{2} \cdot m_1 \cdot v_f^2 - \frac{1}{2} \cdot m_1 \cdot v_i^2 + \frac{G \cdot m_1 \cdot m_2}{r_i} - \frac{G \cdot m_1 \cdot m_2}{r_f} = W_{nc} \tag{11.5.1}$$

If an object (m_1) is traveling in a circle around object m_2, and the circular motion is caused by gravity, then gravity is the net centripetal direction force and:

$$\frac{G \cdot m_1 \cdot m_2}{r^2} = \frac{m_1 \cdot v^2}{r} \tag{11.6.1}$$

Chapter 11 problem types:

Calculate the force of gravity using Newton's law of gravity (Section 11.2)

Calculate the potential energy from gravity using Newton's law of gravity (Section 11.4)

Conservation of mechanical energy problems with potential energy from Newton's law of gravity (Section 11.5)

Circular motion problems with the net force in the centripetal direction from Newton's law of gravity (Section 11.6)

Practice Problems

Note: these problems use the same values for G, mass of Earth, and radius of Earth given earlier in Chapter 11.

Section 11.2 (Newton's Law of Gravity)

1. Calculate the force of gravity between two cars (with masses 902 kg and 904 kg) with centers that are 3.00 m apart in a parking lot.

   Solution:

   The equation for Newton's law of gravity is:

   $$F = \frac{G \cdot m_1 \cdot m_2}{r^2} \qquad (11.2.1)$$

   G is $6.67 \cdot 10^{-11}$ N·m²/kg², m_1 is 902 kg, m_2 is 904 kg, and r = 3.00 m. Put in numbers:

   $$F = \frac{\left(6.67 \cdot 10^{-11} \, N \cdot m^2 / kg^2\right) \cdot (902 \, kg) \cdot (904 \, kg)}{(3.00 \, m)^2} = \underline{6.04 \cdot 10^{-6} \, N}$$

2. The force of gravity between Earth and a 100.0 kg person is 980. Newtons. Using $G = 6.67 \cdot 10^{-11}$ N·m²/kg² and the mass of Earth as $5.97 \cdot 10^{24}$ kg, calculate the distance between Earth's center and the person.

   Solution:

   The equation for Newton's law of gravity is:

   $$F = \frac{G \cdot m_1 \cdot m_2}{r^2} \qquad (11.2.1)$$

   We know F, G, m_1 and m_2, and we want to find r.

   Multiply both sides by r², and divide both sides by F: $r^2 = \dfrac{G \cdot m_1 \cdot m_2}{F}$

   Take the square root: $r = \pm \sqrt{\dfrac{G \cdot m_1 \cdot m_2}{F}}$

Take the positive root, since we are looking for the distance between the two objects

Put in numbers:
$$r = +\sqrt{\dfrac{\left(6.67 \cdot 10^{-11}\,\text{N} \cdot \text{m}^2 / \text{kg}^2\right) \cdot \left(100.0\,\text{kg}\right) \cdot \left(5.97 \cdot 10^{24}\,\text{kg}\right)}{980.\,\text{N}}}$$

$$= \underline{6.37 \cdot 10^6\,\text{m}}$$

3. A person is on an airplane, $6.38 \cdot 10^6$ m from Earth's center (mass $5.97 \cdot 10^{24}$ kg). If the force of gravity on the person is 809 Newtons, what is the person's mass?

Solution:

The equation for Newton's law of gravity is:

$$F = \frac{G \cdot m_1 \cdot m_2}{r^2} \qquad (11.2.1)$$

We know r, m_1, G and F, and we want to find m_2.

Multiply both sides by r^2: $F \cdot r^2 = G \cdot m_1 \cdot m_2$

Divide both sides by ($G \cdot m_1$): $m_2 = \dfrac{F \cdot r^2}{G \cdot m_1}$

Put in numbers: $m_2 = \dfrac{\left(809\,\text{N}\right) \cdot \left(6.38 \cdot 10^6\,\text{m}\right)^2}{\left(6.67 \cdot 10^{-11}\,\text{N} \cdot \text{m}^2 / \text{kg}^2\right) \cdot \left(5.97 \cdot 10^{24}\,\text{kg}\right)} =$ __82.7 kg__

Section 11.4 (Potential Energy from Gravity Using Newton's Law of Gravity)

4. A person with mass 60.0 kg is on airplane that lands. The airplane goes from distance $6.38 \cdot 10^6$ m from Earth's center (mass $5.97 \cdot 10^{24}$ kg) to distance $6.37 \cdot 10^6$ m from Earth's center. What is the change in potential energy?

Solution:

For the change in potential energy with Newton's law of gravity, we use Equation 11.4.1:

$$\Delta PE = \frac{G \cdot m_1 \cdot m_2}{r_i} - \frac{G \cdot m_1 \cdot m_2}{r_f} \qquad (11.4.1)$$

Here, $G = 6.67 \cdot 10^{-11}$ kg·m²/kg², $m_1 = 60.0$ kg, $m_2 = 5.97 \cdot 10^{24}$ kg, $r_i = 6.38 \cdot 10^6$ m, and $r_f = 6.37 \cdot 10^6$ m. Note: choosing m_1 versus m_2 doesn't affect any results.

Put in numbers:

$$\Delta PE = \frac{\left(6.67 \cdot 10^{-11}\, N \cdot m^2 / kg^2\right) \cdot \left(60.0\, kg\right) \cdot \left(5.97 \cdot 10^{24}\, kg\right)}{\left(6.38 \cdot 10^6\, m\right)}$$

$$- \frac{\left(6.67 \cdot 10^{-11}\, N \cdot m^2 / kg^2\right) \cdot \left(60.0\, kg\right) \cdot \left(5.97 \cdot 10^{24}\, kg\right)}{\left(6.37 \cdot 10^6\, m\right)}$$

$\Delta PE = \underline{-5.88 \cdot 10^6}$ J

5. A different person on the same flight as problem 4 had a potential energy change of $\Delta PE = -8.00 \cdot 10^6$ Joules. What was the mass of that person?

Solution:

For the change in potential energy with Newton's law of gravity, we use Equation 11.4.1:

$$\Delta PE = \frac{G \cdot m_1 \cdot m_2}{r_i} - \frac{G \cdot m_1 \cdot m_2}{r_f} \qquad (11.4.1)$$

We know that $\Delta PE = -8.00 \cdot 10^6$ J, $G = 6.67 \cdot 10^{-11}$ kg·m²/kg², $m_2 = 5.97 \cdot 10^{24}$ kg, $r_i = 6.38 \cdot 10^6$ m, and $r_f = 6.37 \cdot 10^6$ m. We want to find m_1.

Since m_1 is in both terms on the right, we can rewrite the right side as:

$$\Delta PE = m_1 \cdot \left(\frac{G \cdot m_2}{r_i} - \frac{G \cdot m_2}{r_f} \right)$$

Divide both sides by everything next to m_1, which is $\left(\dfrac{G \cdot m_2}{r_i} - \dfrac{G \cdot m_2}{r_f} \right)$:

$$m_1 = \frac{\Delta PE}{\left(\dfrac{G \cdot m_2}{r_i} - \dfrac{G \cdot m_2}{r_f} \right)}$$

Put in numbers:

$$m_1 = \frac{-8.00 \cdot 10^6 \, J}{\left(\dfrac{\left(6.67 \cdot 10^{-11} \, N \cdot m^2 \, / \, kg^2\right) \cdot \left(5.97 \cdot 10^{24} \, kg\right)}{\left(6.38 \cdot 10^6 \, m\right)} - \dfrac{\left(6.67 \cdot 10^{-11} \, N \cdot m^2 \, / \, kg^2\right) \cdot \left(5.97 \cdot 10^{24} \, kg\right)}{\left(6.37 \cdot 10^6 \, m\right)} \right)}$$

$\underline{m_1 = 81.6 \, kg}$

Section 11.5 (Newton's Law of Gravity with Conservation of Mechanical Energy)

6. A piece of space junk is moving towards Earth at 5.00 m/s, at a distance of $7.20 \cdot 10^6$ m from Earth's center. What will the magnitude of its velocity be when it gets to $6.80 \cdot 10^6$ m from Earth's center?

Solution:

$$\frac{1}{2} \cdot m_1 \cdot v_f^2 - \frac{1}{2} \cdot m_1 \cdot v_i^2 + \frac{G \cdot m_1 \cdot m_2}{r_i} - \frac{G \cdot m_1 \cdot m_2}{r_f} = W_{nc} \quad (11.5.1)$$

In this problem, $v_i = 5.00$ m/s, $r_i = 7.20 \cdot 10^6$ m, $r_f = 6.80 \cdot 10^6$ m, $m_2 = 5.97 \cdot 10^{24}$ kg (Earth), $G = 6.67 \cdot 10^{-11}$ N·m²/kg², $W_{nc} = 0$ (gravity is the only force), and m_1 will cancel out of every term. We want to find v_f.

Move every term except $\frac{1}{2} \cdot m_1 \cdot v_f^2$ to the right side of the equation, and take out W_{nc} since it is equal to zero. We can do this by adding $\frac{1}{2} \cdot m_1 \cdot v_i^2$ and $\frac{G \cdot m_1 \cdot m_2}{r_f}$ to both sides, and subtracting $\frac{G \cdot m_1 \cdot m_2}{r_i}$ from both sides.

$$\frac{1}{2} \cdot m_1 \cdot v_f^2 = \frac{1}{2} \cdot m_1 \cdot v_i^2 - \frac{G \cdot m_1 \cdot m_2}{r_i} + \frac{G \cdot m_1 \cdot m_2}{r_f}$$

Divide both sides of the equation by m_1:

$$\frac{1}{2}\cdot v_f^{\,2} = \frac{1}{2}\cdot v_i^{\,2} - \frac{G\cdot m_2}{r_i} + \frac{G\cdot m_2}{r_f}$$

Divide both sides of the equation by ½: $v_f^{\,2} = v_i^{\,2} - \dfrac{G\cdot m_2}{\tfrac{1}{2}\cdot r_i} + \dfrac{G\cdot m_2}{\tfrac{1}{2}\cdot r_f}$

Take the square root: $v_f = \pm\sqrt{v_i^{\,2} - \dfrac{G\cdot m_2}{\tfrac{1}{2}\cdot r_i} + \dfrac{G\cdot m_2}{\tfrac{1}{2}\cdot r_f}}$

Take the positive root, since magnitudes are always positive and we are looking for the magnitude of velocity. Put in numbers:

$$v_f = +\sqrt{\left(5.00\,\mathrm{m/s}\right)^2 - \frac{\left(6.67\cdot10^{-11}\,\mathrm{N\cdot m^2/kg^2}\right)\cdot\left(5.97\cdot10^{24}\,\mathrm{kg}\right)}{\frac{1}{2}\cdot\left(7.20\cdot10^6\,\mathrm{m}\right)} + \frac{\left(6.67\cdot10^{-11}\,\mathrm{N\cdot m^2/kg^2}\right)\cdot\left(5.97\cdot10^{24}\,\mathrm{kg}\right)}{\frac{1}{2}\cdot\left(6.80\cdot10^6\,\mathrm{m}\right)}}$$

$v_f = 2.55\cdot10^3\,\mathrm{m/s}$

7. An object is located $6.85\cdot10^6$ m from Earth's center, and it is moving away from Earth at a velocity of 1105 m/s. How far will it get before the velocity becomes zero?

Solution:

$$\frac{1}{2}\cdot m_1\cdot v_f^{\,2} - \frac{1}{2}\cdot m_1\cdot v_i^{\,2} + \frac{G\cdot m_1\cdot m_2}{r_i} - \frac{G\cdot m_1\cdot m_2}{r_f} = W_{nc} \quad (11.5.1)$$

In this problem, $v_i = 1105$ m/s, $r_i = 6.85\cdot10^6$ m, $v_f = 0$, $m_2 = 5.97\cdot10^{24}$ kg (Earth), $G = 6.67\cdot10^{-11}$ N·m²/kg², $W_{nc} = 0$ (gravity is the only force), and m_1 will cancel out of every term. We want to find r_f.

Add $\dfrac{G\cdot m_1\cdot m_2}{r_f}$ to both sides, and remove W_{nc} since it is equal to zero:

$$\frac{1}{2}\cdot m_1\cdot v_f^{\,2} - \frac{1}{2}\cdot m_1\cdot v_i^{\,2} + \frac{G\cdot m_1\cdot m_2}{r_i} = \frac{G\cdot m_1\cdot m_2}{r_f}$$

Divide both sides by m_1, which cancels it from every term

$$\frac{1}{2} \cdot v_f^2 - \frac{1}{2} \cdot v_i^2 + \frac{G \cdot m_2}{r_i} = \frac{G \cdot m_2}{r_f}$$

Multiply both sides by r_f. Keep it separate from the terms:

$$r_f \cdot \left(\frac{1}{2} \cdot v_f^2 - \frac{1}{2} \cdot v_i^2 + \frac{G \cdot m_2}{r_i} \right) = G \cdot m_2$$

Divide both sides by everything next to r_f, which is $\left(\frac{1}{2} \cdot v_f^2 - \frac{1}{2} \cdot v_i^2 + \frac{G \cdot m_2}{r_i} \right)$:

$$r_f = \frac{G \cdot m_2}{\left(\dfrac{1}{2} \cdot v_f^2 - \dfrac{1}{2} \cdot v_i^2 + \dfrac{G \cdot m_2}{r_i} \right)}$$

Put in numbers:

$$r_f = \frac{\left(6.67 \cdot 10^{-11}\, N \cdot m^2 / kg^2 \right) \cdot \left(5.97 \cdot 10^{24}\, kg \right)}{\left(\dfrac{1}{2} \cdot \left(0\,m/s \right)^2 - \dfrac{1}{2} \cdot \left(1105\,m/s \right)^2 + \dfrac{\left(6.67 \cdot 10^{-11}\, N \cdot m^2 / kg^2 \right) \cdot \left(5.97 \cdot 10^{24}\, kg \right)}{6.85 \cdot 10^6\,m} \right)}$$

$$= \underline{6.92 \cdot 10^6\,m}$$

8. A piece of space junk is moving towards Earth at 5.00 m/s, at a distance of $7.20 \cdot 10^6$ m from Earth's center. How far from Earth's center will it be when the velocity is 915 m/s?

Solution:

$$\frac{1}{2} \cdot m_1 \cdot v_f^2 - \frac{1}{2} \cdot m_1 \cdot v_i^2 + \frac{G \cdot m_1 \cdot m_2}{r_i} - \frac{G \cdot m_1 \cdot m_2}{r_f} = W_{nc} \quad (11.5.1)$$

In this problem, $v_i = 5.00\,m/s$, $r_i = 7.20 \cdot 10^6\,m$, $v_f = 915\,m/s$, $m_2 = 5.97 \cdot 10^{24}$ kg (Earth), $G = 6.67 \cdot 10^{-11}$ N·m²/kg², $W_{nc} = 0$ (gravity is the only force), and m_1 will cancel out of every term. We want to find r_f.

Add $\dfrac{G \cdot m_1 \cdot m_2}{r_f}$ to both sides, and remove W_{nc} since it is equal to zero:

$$\frac{1}{2} \cdot m_1 \cdot v_f^2 - \frac{1}{2} \cdot m_1 \cdot v_i^2 + \frac{G \cdot m_1 \cdot m_2}{r_i} = \frac{G \cdot m_1 \cdot m_2}{r_f}$$

Divide both sides by m_1, which cancels it from every term

$$\frac{1}{2} \cdot v_f^2 - \frac{1}{2} \cdot v_i^2 + \frac{G \cdot m_2}{r_i} = \frac{G \cdot m_2}{r_f}$$

Multiply both sides by r_f. Keep it separate from the terms:

$$r_f \cdot \left(\frac{1}{2} \cdot v_f^2 - \frac{1}{2} \cdot v_i^2 + \frac{G \cdot m_2}{r_i} \right) = G \cdot m_2$$

Divide both sides by everything next to r_f, which is $\left(\frac{1}{2} \cdot v_f^2 - \frac{1}{2} \cdot v_i^2 + \frac{G \cdot m_2}{r_i} \right)$:

$$r_f = \frac{G \cdot m_2}{\left(\frac{1}{2} \cdot v_f^2 - \frac{1}{2} \cdot v_i^2 + \frac{G \cdot m_2}{r_i} \right)}$$

Put in numbers:

$$r_f = \frac{\left(6.67 \cdot 10^{-11} \, N \cdot m^2 / kg^2 \right) \cdot \left(5.97 \cdot 10^{24} \, kg \right)}{\left(\begin{array}{c} \frac{1}{2} \cdot \left(915 m/s \right)^2 - \frac{1}{2} \cdot \left(5.00 m/s \right)^2 \\ + \dfrac{\left(6.67 \cdot 10^{-11} \, N \cdot m^2 / kg^2 \right) \cdot \left(5.97 \cdot 10^{24} \, kg \right)}{7.20 \cdot 10^6 \, m} \end{array} \right)}$$

$$= \underline{7.15 \cdot 10^6 \, m}$$

9. An object is located $6.81 \cdot 10^6$ m from Earth's center, and it is moving away from Earth at a velocity of 1105 m/s. What is the velocity when it reaches a distance of $6.84 \cdot 10^6$ m?

Solution:

$$\frac{1}{2} \cdot m_1 \cdot v_f^2 - \frac{1}{2} \cdot m_1 \cdot v_i^2 + \frac{G \cdot m_1 \cdot m_2}{r_i} - \frac{G \cdot m_1 \cdot m_2}{r_f} = W_{nc} \quad (11.5.1)$$

In this problem, $v_i = 1105$ m/s, $r_i = 6.81 \cdot 10^6$ m, $r_f = 6.84 \cdot 10^6$ m, $m_2 = 5.97 \cdot 10^{24}$ kg (Earth), $G = 6.67 \cdot 10^{-11}$ N·m²/kg², $W_{nc} = 0$ (gravity is the only force), and m_1 will cancel out of every term. We want to find v_f.

Move every term except $\frac{1}{2} \cdot m_1 \cdot v_f^2$ to the right side of the equation, and take out W_{nc} since it is equal to zero. We can do this by adding $\frac{1}{2} \cdot m_1 \cdot v_i^2$ and $\frac{G \cdot m_1 \cdot m_2}{r_f}$ to both sides, and subtracting $\frac{G \cdot m_1 \cdot m_2}{r_i}$ from both sides.

$$\frac{1}{2} \cdot m_1 \cdot v_f^2 = \frac{1}{2} \cdot m_1 \cdot v_i^2 - \frac{G \cdot m_1 \cdot m_2}{r_i} + \frac{G \cdot m_1 \cdot m_2}{r_f}$$

Divide both sides of the equation by m_1:

$$\frac{1}{2} \cdot v_f^2 = \frac{1}{2} \cdot v_i^2 - \frac{G \cdot m_2}{r_i} + \frac{G \cdot m_2}{r_f}$$

Divide both sides of the equation by ½: $v_f^2 = v_i^2 - \dfrac{G \cdot m_2}{\frac{1}{2} \cdot r_i} + \dfrac{G \cdot m_2}{\frac{1}{2} \cdot r_f}$

Take the square root: $v_f = \pm \sqrt{v_i^2 - \dfrac{G \cdot m_2}{\frac{1}{2} \cdot r_i} + \dfrac{G \cdot m_2}{\frac{1}{2} \cdot r_f}}$

Take the positive root, since magnitudes are always positive and we are looking for the magnitude of velocity. Put in numbers:

$$v_f = +\sqrt{\begin{array}{c} \left(1105\,\text{m/s}\right)^2 - \dfrac{\left(6.67 \cdot 10^{-11}\,\text{N·m}^2/\text{kg}^2\right) \cdot \left(5.97 \cdot 10^{24}\,\text{kg}\right)}{\frac{1}{2} \cdot \left(6.81 \cdot 10^6\,\text{m}\right)} + \\ \dfrac{\left(6.67 \cdot 10^{-11}\,\text{N·m}^2/\text{kg}^2\right) \cdot \left(5.97 \cdot 10^{24}\,\text{kg}\right)}{\frac{1}{2} \cdot \left(6.84 \cdot 10^6\,\text{m}\right)} \end{array}}$$

$v_f = \underline{841 \text{ m/s}}$

Section 11.6 (Newton's Law of Gravity with Circular Motion)

10. An object is going around Earth, at a distance of $7.00 \cdot 10^6$ m. What is the magnitude of the velocity?

Solution:

If an object is going in a circle, and gravity is the force causing this, then $F_{c,\,net}$ is equal to Newton's law of gravity (Equation 11.2.1):

$$\frac{G \cdot m_1 \cdot m_2}{r^2} = \frac{m_1 \cdot v^2}{r} \qquad\qquad (11.6.1)$$

m_1 is the object going around something else. Here, we want to solve for v.

Multiply both sides by r: $\dfrac{G \cdot m_1 \cdot m_2}{r} = m_1 \cdot v^2$

Note: multiplying by r made the $1/r^2$ on the left side become $1/r$.

Divide both sides by m_1: $\dfrac{G \cdot m_2}{r} = v^2$

Note: the m_1 cancels from both sides.

Take the square root: $v = \pm\sqrt{\dfrac{G \cdot m_2}{r}}$

Take the positive value in the $\pm$ sign, since magnitudes are always positive and we are asked for the magnitude of velocity.

Put in numbers: $v = +\sqrt{\dfrac{\left(6.67 \cdot 10^{-11}\, N \cdot m^2\,/\,kg^2\right) \cdot \left(5.97 \cdot 10^{24}\, kg\right)}{7.00 \cdot 10^6\, m}} =$

$\underline{7.54 \cdot 10^3\, m/s}$

Note on units: remember that 1 N is the same as 1 kg·m/s^2.

11. An object is going around Earth, at a velocity of 999 m/s. How far from Earth's center is it?

Solution:

If an object is going in a circle, and gravity is the force causing this, then $F_{c,\,net}$ is equal to Newton's law of gravity (Equation 11.2.1):

$$\frac{G \cdot m_1 \cdot m_2}{r^2} = \frac{m_1 \cdot v^2}{r} \qquad\qquad (11.6.1)$$

Multiply both sides by r^2. This will cancel the $1/r^2$ on the left side, and turn $1/r$ into r on the right side.

$$G \cdot m_1 \cdot m_2 = m_1 \cdot v^2 \cdot r$$

Divide both sides by $(m_1 \cdot v^2)$, which will cancel the m_1 on both sides:

$$\frac{G \cdot m_2}{v^2} = r$$

Put in numbers: $r = \dfrac{\left(6.67 \cdot 10^{-11} \, N \cdot m^2 / kg^2\right) \cdot \left(5.97 \cdot 10^{24} \, kg\right)}{\left(999\right)^2} =$ $\underline{3.99 \cdot 10^8 \, m}$

Simple Harmonic Motion

12.1 INTRODUCTION: MOTION REPEATS

We've talked about motion in lines and in circles, looking at kinematics, forces, energy, and momentum. We'll finish by looking at motion that repeats itself. Two classic things that repeat motion are an object connected to a spring, and a pendulum (like on an old clock).

12.2 REPETITIVE MOTION QUANTITIES, AND SIMPLE HARMONIC MOTION

When something moves back and forth, we use the following quantities to describe them:

- **Amplitude** (A): how far from the middle of the motion the object. Units are meters (m).

- **Period** (T): how much time the object takes to repeat its motion

- **Frequency** (f): how many times per second the object repeats its motion. Units are 1/seconds (1/s), also called Hertz (Hz).

The period and frequency are related to each other:

$$f = 1/T \qquad (12.2.1)$$

In this chapter, we will talk about a specific type of repeating motion, called **simple harmonic motion**. Simple harmonic motion happens when

DOI: 10.1201/9781003005049-12

341

the acceleration is equal to −1 times a number times position, where position is measured from the middle of a repeating motion.

Side note: in the description of simple harmonic motion above, the "number" is equal to $(2 \cdot \pi \cdot f)^2$. $2 \cdot \pi \cdot f$ is given the letter ω in many books, and called the **angular frequency**.

12.3 POSITION, VELOCITY AND ACCELERATION

For an object moving in simple harmonic motion, the position, velocity and acceleration are given by:

$$x = A \cdot \cos(2 \cdot \pi \cdot f \cdot t) \qquad (12.3.1)$$

$$v = 2 \cdot \pi \cdot f \cdot A \cdot \sin(2 \cdot \pi \cdot f \cdot t) \qquad (12.3.2)$$

$$a = -(2 \cdot \pi \cdot f)^2 \cdot A \cdot \cos(2 \cdot \pi \cdot f \cdot t) \qquad (12.3.3)$$

The position is defined so that the middle of the motion is $x = 0$.

x is linear position, v is linear velocity, a is linear acceleration, A is amplitude, f is frequency, t is time.

An important note! Your calculator should be in radians mode for these equations to work correctly.

Notice that each equation has either a sine or a cosine function in it. This makes the values go back and forth as time changes. Notice too that you could rewrite acceleration as $a = -1 \cdot (2 \cdot \pi \cdot f)^2 \cdot x$, which matches the description of simple harmonic motion given in Section 12.2.

Some textbook problems ask for the largest possible value of an equation with a cosine or sine term in them. For these, remember that sine and cosine functions only give values between −1 and 1, so put in 1 for the value of the sine or cosine function if a largest possible value is requested.

Side note: These equations assume that the objects start at $x = A$. If the objects start somewhere else, you can add a **phase** term inside the cosine or sine, which is sometimes called δ (Greek letter delta). Equation 12.3.1 would look like this: $x = A \cdot \cos(2 \cdot \pi \cdot f \cdot t + \delta)$. Set $\delta = \pi$ to start at $x = -A$, $\delta = \pi/2$ or $3 \cdot \pi/2$ to start at $x = 0$. You would have to add δ to the velocity and acceleration equations as well.

Example 12.3.1

A part goes up and down inside of an engine. Let's model it as going through simple harmonic motion. It goes through 1.60 repeats of its motion every second, and goes a distance of 0.0300 m (3.00 cm) away from the center of the motion each time.

A. What is the period of the motion?
B. What is the position of the part, 2.00 seconds after the motion starts? Assume it starts at x = +A, so that there is no need for a phase change.
C. What is the value of the acceleration, 2.00 seconds after the motion starts?

Solution:

Part A:
The frequency is the number of repeats per second, which is 1.60. The units are 1/s ("per second"), which are usually called Hertz (Hz).
Period (T) and frequency (f) are related by Equation 12.2.1:

$$f = 1/T \qquad (12.2.1)$$

Take the inverse of both sides: T = 1/f
Put in numbers: T = 1/(1.60 1/s) = 0.625 s
Notes: this means that the motion repeats every 0.625 seconds. Instead of taking the inverse of both sides, you could also multiply both sides by T and divide both sides by f.

Part B:

$$x = A \cdot \cos(2 \cdot \pi \cdot f \cdot t) \qquad (12.3.1)$$

A is the amplitude (largest distance from center), which is 0.0300 m, t is 2.00 s, and f is 1.60 1/s.
Put in numbers: x = (0.0300 m)·cos(2·π·(1.60 1/s)·(2.00 s)) = 9.27·10⁻³ m

Note: remember to have your calculator in radians mode.

Part C:

$$a = -(2 \cdot \pi \cdot f)^2 \cdot A \cdot \cos(2 \cdot \pi \cdot f \cdot t) \qquad (12.3.3)$$

Put in numbers: a = -(2·π·(1.60 1/s))²·(0.0300 m)·cos(2·π·(1.60 1/s)·(2.00 s)) = -0.937 m/s²

Note 1: this is equal to $-(2 \cdot \pi \cdot f)^2$ times the position.

Note 2: the negative sign means away from the initial point, which is defined as being in the positive direction. A positive sign would mean towards the point.

Problem to Try Yourself

In the same setup as Example 12.3.1, what is the value of velocity 3.00 s after the motion starts?

Solution:

$$v = 2 \cdot \pi \cdot f \cdot A \cdot \sin(2 \cdot \pi \cdot f \cdot t) \qquad (12.3.2)$$

Put in numbers: $v = 2 \cdot \pi \cdot (1.60 \ 1/s) \cdot (0.0300 \ m) \cdot \sin(2 \cdot \pi \cdot (1.60 \ 1/s) \cdot (3.00 \ s)) = \underline{-0.287 \ m/s}$

Note: the negative sign means away from the initial point, which is defined as being in the positive direction. A positive sign would mean towards the point.

12.4 SIMPLE HARMONIC MOTION AND OBJECTS ON SPRINGS

When an object is connected to the end of a spring, the object can move back and forth, with the spring compressing and then stretching (and then repeating). The position $x = 0$ happens when the spring is not stretched and not compressed (its equilibrium length).

For this situation, the frequency is equal to:

$$f = \frac{1}{2 \cdot \pi} \cdot \sqrt{\frac{k}{m}} \qquad (12.4.1)$$

k is the spring constant of the spring, and m is the mass of the object on the end of the spring.

Where does this come from? Combine $F = m \cdot a$ (Newton's second law) and $F = -kx$ (spring force), to find that $a = -(k/m) \cdot x$. Then remember that in simple harmonic motion, you have $a = -1 \cdot (\text{number}) \cdot x$ and that the number is equal to $(2 \cdot \pi \cdot f)^2$.

Equations 12.2.1, 12.3.1, 12.3.2, 12.3.3, and 12.4.1 contain all of the information needed to solve for all of the various quantities described in this chapter. You'll often use each equation to solve for a specific variable, and then sometimes you'll use the result of one equation to solve for a second variable, and so on.

Example 12.4.1

A 2.00 kg object is going back and forth, with amplitude 0.200 m, while connected to a spring. This could be a model for someone using an elastic band to help them as they move their hand back and forth while stretching.

A. If the frequency is 0.800 repeats per second, what is the spring constant of the spring?
B. What is the position of the object after 1.85 s, assuming that the motion starts at x = +A?

Solution:

Part A:

$$f = \frac{1}{2 \cdot \pi} \cdot \sqrt{\frac{k}{m}} \qquad (12.4.1)$$

We know f and m, and want to find k.

Multiply both sides by (2·π): $2 \cdot \pi \cdot f = \sqrt{\frac{k}{m}}$

Square both sides: $(2 \cdot \pi \cdot f)^2 = \frac{k}{m}$

Multiply both sides by m: $k = m \cdot (2 \cdot \pi \cdot f)^2$

Put in numbers: $k = (2.00\,\text{kg}) \cdot (2 \cdot \pi \cdot 0.8001/\text{s})^2 = \underline{50.5\,\text{N/m}}$

Note on units: 1 N/m = 1 kg/s², since 1 N = 1 kg·m/s².

Part B:

$$x = A \cdot \cos(2 \cdot \pi \cdot f \cdot t) \qquad (12.3.1)$$

Put in numbers: x = (0.200 m)·cos(2·π·(0.800 1/s)·(1.85 s)) = $\underline{-0.198\,\text{m}}$

Note: a negative position means on the opposite side of the middle from the starting point, since the starting point here is defined as + 1·amplitude.

Problem to Try Yourself

An object is moving while connected to a spring, with the same application listed above. The spring has spring constant 20.0 N/m, the object has mass 3.60 kg, and the amplitude of the motion is 0.400 m.

A. How many times does the motion repeat per second?
B. What is the period of the motion?
C. What is the position, 2.67 s after the motion starts, assuming that the motion starts at x = +A?

Solution:

Part A:
   The number of repeats per second is the frequency.

$$f = \frac{1}{2 \cdot \pi} \cdot \sqrt{\frac{k}{m}} \qquad (12.4.1)$$

Put in numbers: $f = \dfrac{1}{2 \cdot \pi} \cdot \sqrt{\dfrac{20.0\,\mathrm{N/m}}{3.60\,\mathrm{kg}}} = 0.375\ 1/s$

Part B:
   Period (T) and frequency (f) are both in the equation:

$$f = 1/T \qquad (12.2.1)$$

Take the inverse of both sides: T = 1/f
Put in numbers: T = 1/(0.375 1/s) = <u>2.67 s</u>
Note: this means that the motion will repeat every 2.67 seconds.

Part C:

$$x = A \cdot \cos(2 \cdot \pi \cdot f \cdot t) \qquad (12.3.1)$$

Put in numbers: x = (0.400 m)·cos(2·π·(0.375 1/s)·(2.67 s)) = <u>0.400 m</u>

Note: this is exactly what you would expect – the position is back at +1·amplitude after one repeat. If you tried 2·2.67 s, or 3·2.67 s, etc., you would also get 0.400 m. If you tried ½·2.67 s, you would get −0.400 m (the other end of the motion).

12.5 SIMPLE HARMONIC MOTION: SIMPLE PENDULUM

A **simple pendulum** is an object on the end of a string, swinging back and forth. It will repeat its motion, with a frequency and period that we can calculate. In this case, the frequency is equal to:

$$f = \frac{1}{2 \cdot \pi} \cdot \sqrt{\frac{g}{L}} \qquad (12.5.1)$$

g is 9.80 m/s², and L is the length of the pendulum.

Getting this equation is more complicated than for the object connected to a spring, but the basic process is the same.

Side note 1: technically, this object is rotating and not moving linearly. For this reason, we will not use equations 12.3.1–12.3.3 for this section.

Side note 2: some textbooks cover a similar but more realistic pendulum setup, called the **physical pendulum**. In this case, the frequency is related to the moment of inertia and the distance between the center of the pendulum and the point where the pendulum rotates from.

Example 12.5.1

What is the period of a simple pendulum with length 0.605 m?

Solution:

We'll use two equations here: one to get frequency, and one to go from frequency to period.

$$f = \frac{1}{2 \cdot \pi} \cdot \sqrt{\frac{g}{L}} \qquad (12.5.1)$$

$$f = 1/T \qquad (12.2.1)$$

First, find the frequency using Equation 12.5.1. Put in numbers:

$$f = \frac{1}{2 \cdot \pi} \cdot \sqrt{\frac{9.80 \, m/s^2}{0.605 \, m}} = 0.641 \, 1/s$$

Now, use $f = 1/T$. Take the inverse of both sides: $T = 1/f$
Put in numbers: $T = 1/(0.641 \ 1/s) = \underline{1.56 \ s}$

Problem to Try Yourself

A simple pendulum has period 1.25 s. How long is it?

Solution:
First find frequency, then use that to find length.

$$f = 1/T \tag{12.2.1}$$

Put in numbers: $f = 1/T = 1/1.25 \ s = 0.800 \ 1/s$
Now use the equation that is specifically for the simple pendulum.

$$f = \frac{1}{2 \cdot \pi} \cdot \sqrt{\frac{g}{L}} \tag{12.5.1}$$

Multiply both sides by $(2 \cdot \pi)$: $2 \cdot \pi \cdot f = \sqrt{\frac{g}{L}}$

Square both sides: $\left(2 \cdot \pi \cdot f\right)^2 = \frac{g}{L}$

Multiply both sides by L, and divide both sides by $(2 \cdot \pi \cdot f)^2$:

$$L = \frac{g}{\left(2 \cdot \pi \cdot f\right)^2}$$

Put in numbers: $L = \dfrac{9.80 \, \text{m}/\text{s}^2}{\left(2 \cdot \pi \cdot \left(0.8001/s\right)\right)^2} = 0.388 \, \text{m}$

12.6 OBJECT CONNECTED TO A SPRING AND MECHANICAL ENERGY

With an object connected to a spring, imagine that the object is moving on a flat surface, with no friction. The mechanical energy of this system is conserved, because the only force doing work (spring) is conservative. The total amount of mechanical energy is equal to:

$$ME = \frac{1}{2} \cdot k \cdot A^2 \tag{12.6.1}$$

k is the spring constant for the spring, and A is the amplitude of the motion.

At the very edges of the motion (at $x = A$ for example), all of the energy is potential energy and the object is not moving. In the middle of the motion, all of the energy is kinetic, and the potential energy is zero. For the middle, this means that:

$$\tfrac{1}{2} \cdot k \cdot A^2 = \tfrac{1}{2} \cdot m \cdot v_{\text{middle}}^2 \qquad (12.6.2)$$

Side note: one way of getting Equation 12.6.1 is to look specifically at the point where the object is moving through the middle of the motion. All of the energy is kinetic at that point, and v is equal to its maximum value: $2 \cdot \pi \cdot f \cdot A$, based on Equation 12.3.2. Using Equation 12.4.1 for f and $KE = \tfrac{1}{2} \cdot m \cdot v^2$ results in Equation 12.6.1.

Example 12.6.1

Example 12.4.1 had an object with mass 2.00 kg, moving while connected to a spring with spring constant 50.5 N/m, and an amplitude of 0.200 m.

A. What is the mechanical energy?
B. What is the magnitude of velocity when in the middle?

Part A:

$$ME = \tfrac{1}{2} \cdot k \cdot A^2 \qquad (12.6.1)$$

Put in numbers: $ME = \tfrac{1}{2} \cdot (50.5 \text{ N/m}) \cdot (0.200 \text{ m})^2 = \underline{1.01 \text{ Joules}}$
Note on units: remember that 1 N·m and 1 Joule are equal.

Part B:

$$\tfrac{1}{2} \cdot k \cdot A^2 = \tfrac{1}{2} \cdot m \cdot v_{\text{middle}}^2 \qquad (12.6.2)$$

Divide both sides by $(\tfrac{1}{2} \cdot m)$: $v_{\text{middle}}^2 = \dfrac{\tfrac{1}{2} \cdot k \cdot A^2}{\tfrac{1}{2} \cdot m}$

Take the square root: $v_{\text{middle}} = \pm \sqrt{\dfrac{\tfrac{1}{2} \cdot k \cdot A^2}{\tfrac{1}{2} \cdot m}}$

Put in numbers: $v_{middle} = \sqrt{\dfrac{\frac{1}{2}\cdot(50.5\,N/m)\cdot(0.200\,m)^2}{\frac{1}{2}\cdot(2.00\,kg)}}$

$= 1.00$ m/s

Note: the + sign was chosen because the magnitude was asked for. The object goes through the middle in both directions during the motion.

Problem to Try Yourself

An object with mass 1.35 kg is moving in simple harmonic motion, with amplitude 0.250 m, connected to a spring with spring constant 75.0 N/m.

A. How fast is the object moving when in the middle?
B. How much mechanical energy is in the system?

Solution:

Part A:

$$\frac{1}{2}\cdot k\cdot A^2 = \frac{1}{2}\cdot m\cdot v_{middle}^2 \qquad (12.6.2)$$

Divide both sides by $(\frac{1}{2}\cdot m)$: $v_{middle}^2 = \dfrac{\frac{1}{2}\cdot k\cdot A^2}{\frac{1}{2}\cdot m}$

Take the square root: $v_{middle} = \pm\sqrt{\dfrac{\frac{1}{2}\cdot k\cdot A^2}{\frac{1}{2}\cdot m}}$

Put in numbers: $v_{middle} = \sqrt{\dfrac{\frac{1}{2}\cdot(75.0\,N/m)\cdot(0.250\,m)^2}{\frac{1}{2}\cdot(1.35\,kg)}}$ = $\underline{1.86\ m/s}$

Note: the + sign was chosen because the magnitude was asked for. The object goes through the middle in both directions during the motion.

Part B:

$$ME = \frac{1}{2}\cdot k\cdot A^2 \qquad (12.6.1)$$

Put in numbers: $ME = \frac{1}{2}\cdot(75.0\ N/m)\cdot(0.250\ m)^2 = \underline{2.34\ Joules}$

12.7 CHAPTER 12 SUMMARY

Some objects move back and forth. Chapter 12 describes one specific type of motion like this, called **simple harmonic motion**.

In addition to having a position, velocity, and acceleration, we introduced three quantities related to an object moving in simple harmonic motion:

- **Amplitude** (A): how far from the middle of the motion the object gets before repeating

- **Period** (T): how much time the object takes to repeat its motion

- **Frequency** (f): how many times per second the object repeats its motion

The variables are related using the following equations:

$$f = 1/T \qquad\qquad (12.2.1)$$

$$x = A \cdot \cos(2 \cdot \pi \cdot f \cdot t) \qquad\qquad (12.3.1)$$

$$v = 2 \cdot \pi \cdot f \cdot A \cdot \sin(2 \cdot \pi \cdot f \cdot t) \qquad\qquad (12.3.2)$$

$$a = -(2 \cdot \pi \cdot f)^2 \cdot A \cdot \cos(2 \cdot \pi \cdot f \cdot t) \qquad\qquad (12.3.3)$$

(Note: equations 12.3.1–3 have no phase term in them.)

We then looked at specific situations, which add equations that only apply for that situation

Object moving on a mass:

$$f = \frac{1}{2 \cdot \pi} \cdot \sqrt{\frac{k}{m}} \qquad\qquad (12.4.1)$$

$$ME = \frac{1}{2} \cdot k \cdot A^2 \qquad\qquad (12.6.1)$$

$$\frac{1}{2} \cdot k \cdot A^2 = \frac{1}{2} \cdot m \cdot v_{middle}^2 \qquad\qquad (12.6.2)$$

Note: the last two equations were used when the object is moving horizontally moving left and right

Simple pendulum (a mass at the end of a string):

$$f = \frac{1}{2 \cdot \pi} \cdot \sqrt{\frac{g}{L}} \qquad (12.5.1)$$

Note: we did not use equations 12.3.1–12.3.3 for the pendulum, since those equations are linear motion and the pendulum is rotating.

Chapter 12 problem types:

Two situations: object on spring, simple pendulum (Sections 12.4, 12.5)

For each of these, you could be asked to compute any of amplitude, period, frequency, position, velocity, acceleration, or some situation-specific variable like mass or spring constant (Sections 12.2 and 12.3, plus one equation for each situation from Sections 12.4 or 12.5)

Energy problems for an object connected to a spring (Section 12.6)

Practice Problems

Sections 12.2, 12.3, and 12.4 (Repetitive Motion Quantities; Position, Velocity, and Acceleration; Simple Harmonic Motion and Objects on Springs)

1. A 1.50 kg object is moving back and forth in simple harmonic motion while connected to a spring – for example, maybe someone stretching or a part inside of a machine. The spring constant is 54.0 N/m, and the amplitude of the motion is 0.500 m.

    A. What is the frequency of the motion?

    B. What is the period of the motion?

    C. What are the values of position, velocity, and acceleration after 1.00 seconds?

    Solution:

    Part A:

    We know m = 1.50 kg, k = 54.0 N/m, and A = 0.500 m. To find f, we use Equation 12.4.1:

$$f = \frac{1}{2 \cdot \pi} \cdot \sqrt{\frac{k}{m}} \qquad (12.4.1)$$

Put in numbers: $f = \dfrac{1}{2 \cdot \pi} \cdot \sqrt{\dfrac{(54.0\,\text{N}/\text{m})}{(1.50\,\text{kg})}} = 0.955\ 1/\text{s}\,(\text{Hz})$

Note on units: remember that $1\,\text{Hz} = 1/\text{s}$, and $1\,\text{N} = 1\,\text{kg·m/s}^2$.

Part B:

To get period (T), use Equation 12.2.1:

$$f = 1/T \qquad\qquad (12.2.1)$$

Multiply both sides by T, and divide both sides by f:

$$T = 1/f$$

Put in numbers: $T = 1/(0.955\ 1/\text{s}) = \underline{1.05\ \text{s}}$

Part C:

$$x = A \cdot \cos(2 \cdot \pi \cdot f \cdot t) \qquad\qquad (12.3.1)$$

$$v = 2 \cdot \pi \cdot f \cdot A \cdot \sin(2 \cdot \pi \cdot f \cdot t) \qquad\qquad (12.3.2)$$

$$a = -(2 \cdot \pi \cdot f)^2 \cdot A \cdot \cos(2 \cdot \pi \cdot f \cdot t) \qquad\qquad (12.3.3)$$

Put in numbers for each:

$x = (0.500\,\text{m}) \cdot \cos(2 \cdot \pi \cdot 0.955\ 1/\text{s} \cdot 1.00\ \text{s}) = \underline{0.480\ \text{m}}$

$v = 2 \cdot \pi \cdot (0.955\ 1/\text{s}) \cdot (0.500\,\text{m}) \cdot \sin(2 \cdot \pi \cdot 0.955\ 1/\text{s} \cdot 1.00\ \text{s}) = \underline{-0.838\ \text{m/s}}$

$a = -(2 \cdot \pi \cdot (0.955\ 1/\text{s}))^2 \cdot (0.500\,\text{m}) \cdot \cos(2 \cdot \pi \cdot 0.955\ 1/\text{s} \cdot 1.00\ \text{s}) = \underline{-17.3\ \text{m/s}^2}$

2. A 5.00 kg object is moving back and forth in simple harmonic motion while connected to a spring. The motion gets 0.800 m from the center of the motion, and the frequency is 1.35 1/s (Hz).

A. What is the period of the motion?

B. What is the spring constant?

C. What is the position after 2.00 seconds?

Solution:

Part A:

To get period (T), use Equation 12.2.1:

$$f = 1/T \qquad\qquad (12.2.1)$$

Multiply both sides by T, and divide both sides by f:

$$T = 1/f$$

Put in numbers: $T = 1/(1.35\ 1/s) = \underline{0.741\ s}$

Part B:

We know $m = 5.00$, $A = 0.800$ m, $f = 1.35$ 1/s, and $T = 0.741$ s. We want to find k. The equation that has k and things that we know is Equation 12.4.1:

$$f = \frac{1}{2 \cdot \pi} \cdot \sqrt{\frac{k}{m}} \qquad (12.4.1)$$

Multiply both sides by $(2 \cdot \pi)$: $2 \cdot \pi \cdot f = \sqrt{\frac{k}{m}}$

Square both sides: $\left(2 \cdot \pi \cdot f\right)^2 = \frac{k}{m}$

Multiply both sides by m: $k = m \cdot \left(2 \cdot \pi \cdot f\right)^2$

Put in numbers: $k = \left(5.00\,\text{kg}\right) \cdot \left(2 \cdot \pi \cdot \left(1.35\,1/s\right)\right)^2 = 360\,\text{N/m}.$

Note on units: remember that $1\ N = 1\ kg \cdot m/s^2$, so $1\ N/m = 1\ kg/s^2$.

Part C:

$$x = A \cdot \cos(2 \cdot \pi \cdot f \cdot t) \qquad (12.3.1)$$

Put in numbers: $x = (0.800\ m) \cdot \cos(2 \cdot \pi \cdot 1.35\ 1/s \cdot 2.00\ s) = \underline{-0.247\ m}$

3. An object is moving back and forth in simple harmonic motion while connected to a spring. The spring constant is 60.0 N/m, and the frequency is 3.00 1/s (Hz). 5.20 seconds after the motion ends, the position is −0.600 m.

   A. What is the mass of the object?

   B. What is the period?

   C. What is the amplitude?

Solution:

Part A:

We know k=60.0 N/m, f=3.00 1/s, and that x=−0.600 m when t=5.20. To find m, we'll use Equation 12.4.1 because it has m and only variables that we know.

$$f = \frac{1}{2 \cdot \pi} \cdot \sqrt{\frac{k}{m}}$$

(12.4.1)

Multiply both sides by (2·π): $2 \cdot \pi \cdot f = \sqrt{\frac{k}{m}}$

Square both sides: $\left(2 \cdot \pi \cdot f\right)^2 = \frac{k}{m}$

Multiply both sides by m: $m \cdot \left(2 \cdot \pi \cdot f\right)^2 = k$

Divide both sides by (2·π·f)²: $m = \dfrac{k}{\left(2 \cdot \pi \cdot f\right)^2}$

Put in numbers: $m = \dfrac{60.0 \, \text{N}/\text{m}}{\left(2 \cdot \pi \cdot 3.00 \, 1/s\right)^2} = 3.18 \, \text{kg}$

Part B:

To get period (T), use Equation 12.2.1:

$$f = 1/T$$

(12.2.1)

Multiply both sides by T, and divide both sides by f:

$$T = 1/f$$

Put in numbers: T = 1/(3.00 1/s) = 0.333 s

Part C:

Since we have information about the position at a specific time, and the position equation has amplitude, let's use Equation 12.3.1:

$$x = A \cdot \cos(2 \cdot \pi \cdot f \cdot t)$$

(12.3.1)

Divide both sides by cos(2·π·f·t): A = x/cos(2·π·f·t)

Put in numbers: A = (−0.600 m)/cos(2·π·3.00 1/s·5.20 s) = <u>0.742 m</u>

Section 12.5 (Simple Pendulum)

4. A simple pendulum has a frequency of 1.70 1/s (Hz). What is the length of it?

Solution:

$$f = \frac{1}{2 \cdot \pi} \cdot \sqrt{\frac{g}{L}} \tag{12.5.1}$$

Multiply both sides by (2·π): $2 \cdot \pi \cdot f = \sqrt{\frac{g}{L}}$

Square both sides: $\left(2 \cdot \pi \cdot f\right)^2 = \frac{g}{L}$

Multiply both sides by L: $L \cdot \left(2 \cdot \pi \cdot f\right)^2 = g$

Divide both sides by (2·π·f)²: $L = g / \left(2 \cdot \pi \cdot f\right)^2$

Put in numbers: $L = \left(9.80 \, \text{m}/\text{s}^2\right) / \left(2 \cdot \pi \cdot 1.70 \, 1/\text{s}\right)^2 = 8.59 \cdot 10^{-2} \, \text{m}$

5. A simple pendulum has a length of 0.300 m. What is the frequency?

Solution:

$$f = \frac{1}{2 \cdot \pi} \cdot \sqrt{\frac{g}{L}} \tag{12.5.1}$$

Put in numbers: $f = \frac{1}{2 \cdot \pi} \cdot \sqrt{\frac{\left(9.80 \, \text{m}/\text{s}^2\right)}{0.300 \, \text{m}}} = 0.910 \, 1/\text{s} \, (\text{Hz})$

6. A simple pendulum has a frequency of 3.20 1/s (Hz). What is the length of it?

Solution:

$$f = \frac{1}{2 \cdot \pi} \cdot \sqrt{\frac{g}{L}} \tag{12.5.1}$$

Multiply both sides by (2·π): $2 \cdot \pi \cdot f = \sqrt{\dfrac{g}{L}}$

Square both sides: $\left(2 \cdot \pi \cdot f\right)^2 = \dfrac{g}{L}$

Multiply both sides by L: $L \cdot \left(2 \cdot \pi \cdot f\right)^2 = g$

Divide both sides by (2·π·f)²: $L = g / \left(2 \cdot \pi \cdot f\right)^2$

Put in numbers: $L = \left(9.80\,\text{m}/\text{s}^2\right) / \left(2 \cdot \pi \cdot 3.20\,1/\text{s}\right)^2 = 2.42 \cdot 10^{-2}\,\text{m}$

7. A simple pendulum has a length of 3.00 m. What is the frequency?

   Solution:

$$f = \frac{1}{2 \cdot \pi} \cdot \sqrt{\frac{g}{L}} \qquad (12.5.1)$$

Put in numbers: $f = \dfrac{1}{2 \cdot \pi} \cdot \sqrt{\dfrac{\left(9.80\,\text{m}/\text{s}^2\right)}{3.00\,\text{m}}} = 0.288\,1/\text{s}\,(\text{Hz})$

Section 12.6 (Object Connected to a Spring and Mechanical Energy)

8. An object is moving back and forth in simple harmonic motion while connected to a spring. The spring constant is 50.0 N/m, the amplitude is 0.400 m, and the mass of the object is 1.20 kg.

   A. What is the mechanical energy?

   B. What is the magnitude of the velocity at the middle of the motion?

   Solution:
   Part A:

$$ME = \tfrac{1}{2} \cdot k \cdot A^2 \qquad (12.6.1)$$

Put in numbers: $ME = \tfrac{1}{2} \cdot (50.0\ \text{N/m}) \cdot (0.400\ \text{m})^2 = \underline{4.00\ \text{J}}$

Note on units: 1 Joule is the same as 1 N·m in terms of units. Each one is 1 kg·m²/s².

Part B:

$$\frac{1}{2}\cdot k\cdot A^2 = \frac{1}{2}\cdot m\cdot v_{middle}^2 \tag{12.6.2}$$

• Divide each side by ($\frac{1}{2}\cdot m$): $k\cdot A^2/m = v_{middle}^2$

Take the square root: $v_{middle} = \pm\sqrt{k\cdot A^2/m}$

Take the positive value, since we are asked for a magnitude and magnitudes are always positive.

Put in numbers: $v_{middle} = +\sqrt{(50.0\,N/m)\cdot(0.400\,m)^2/(1.20\,kg)} =$ 2.58 m/s

9. An object is moving back and forth in simple harmonic motion while connected to a spring. The spring constant is 40.0 N/m, the amplitude is 0.350 m, and the velocity in the middle of the motion is 1.00 m/s.

A. What is the mechanical energy?

B. What is the mass of the object?

Solution:

Part A:

$$ME = \frac{1}{2}\cdot k\cdot A^2 \tag{12.6.1}$$

Put in numbers: ME = $\frac{1}{2}\cdot$(40.0 N/m)$\cdot$(0.350 m)^2 = 2.45 J

Note on units: 1 Joule is the same as 1 N·m in terms of units. Each one is 1 kg·m^2/s^2.

Part B:

$$\frac{1}{2}\cdot k\cdot A^2 = \frac{1}{2}\cdot m\cdot v_{middle}^2 \tag{12.6.2}$$

Divide both sides by ($\frac{1}{2}\cdot v_{middle}^2$): m = ($\frac{1}{2}\cdot k\cdot A^2$)/($\frac{1}{2}\cdot v_{middle}^2$)

Put in numbers: m = ($\frac{1}{2}\cdot$40.0 N/m·(0.350 m)^2)/($\frac{1}{2}\cdot$(1.00 m/s)^2) = 4.90 kg

10. An object is moving back and forth in simple harmonic motion while connected to a spring. The mass is 2.00 kg, the amplitude is 0.350 m, and the velocity in the middle of the motion is 1.50 m/s.

A. What is the mechanical energy? Hint: you can relate ME and $\frac{1}{2} \cdot m \cdot v_{middle}^2$ to each other, because both of them are equal to $\frac{1}{2} \cdot k \cdot A^2$.

B. What is the spring constant?

Solution:

Part A:

$$ME = \frac{1}{2} \cdot k \cdot A^2 \qquad (12.6.1)$$

$$\frac{1}{2} \cdot k \cdot A^2 = \frac{1}{2} \cdot m \cdot v_{middle}^2 \qquad (12.6.2)$$

If you look at these two equations together, you can see that $ME = \frac{1}{2} \cdot m \cdot v_{middle}^2$. This is because both sides of this equation (ME and $\frac{1}{2} \cdot m \cdot v_{middle}^2$) are both equal to $\frac{1}{2} \cdot k \cdot A^2$.

Putting numbers into this: $ME = \frac{1}{2} \cdot (2.00 \text{ kg}) \cdot (1.50 \text{ m/s})^2 = \underline{2.25 \text{ J}}$

Part B:

$$ME = \frac{1}{2} \cdot k \cdot A^2 \qquad (12.6.1)$$

Divide both sides by $(\frac{1}{2} \cdot A^2)$: $k = ME/(\frac{1}{2} \cdot A^2)$

Put in numbers: $k = (2.25 \text{ J})/(\frac{1}{2} \cdot (0.350 \text{ m})^2) = \underline{36.7 \text{ N/m}}$

11. An object is moving back and forth in simple harmonic motion while connected to a spring. The spring constant is 65.0 N/m, the mass is 2.50 kg, and the velocity in the middle of the motion is 1.00 m/s.

A. What is the mechanical energy? Hint: you can relate ME and $\frac{1}{2} \cdot m \cdot v_{middle}^2$ to each other, because both of them are equal to $\frac{1}{2} \cdot k \cdot A^2$.

B. What is the amplitude? Hint: take the positive value if you get two values.

Solution:

Part A:

$$ME = \frac{1}{2} \cdot k \cdot A^2 \qquad (12.6.1)$$

$$\frac{1}{2} \cdot k \cdot A^2 = \frac{1}{2} \cdot m \cdot v_{middle}^2 \qquad (12.6.2)$$

If you look at these two equations together, you can see that $ME = \frac{1}{2} \cdot m \cdot v_{middle}^2$. This is because both sides of this equation (ME and $\frac{1}{2} \cdot m \cdot v_{middle}^2$) are both equal to $\frac{1}{2} \cdot k \cdot A^2$.

Putting numbers into this: $ME = \frac{1}{2} \cdot (2.50 \text{ kg}) \cdot (1.00 \text{ m/s})^2 = \underline{1.25 \text{ J}}$

Part B:

$$ME = \frac{1}{2} \cdot k \cdot A^2 \qquad\qquad (12.6.1)$$

Divide both sides by $(\frac{1}{2} \cdot k)$: $A^2 = ME/(\frac{1}{2} \cdot k)$

Take the square root: $A = \pm\sqrt{ME/\left(\frac{1}{2} \cdot k\right)}$

We will take the positive value, since amplitudes are positive.

Put in numbers: $A = +\sqrt{(1.25 \text{J})/\left(\frac{1}{2} \cdot 65.0 \text{N/m}\right)} = 0.196 \text{m}$

Math Review

13.1 INTRODUCTION: PHYSICS USES MATH

Although physics has many concepts, many physics problems require math. This chapter covers much of the math that is needed to do the problems in this textbook. Each of the sections in this chapter use specific equations from other chapters in this textbook.

13.2 ALGEBRA PROBLEMS – SOLVING FOR A VARIABLE

One of the most common math problems in physics is solving for one variable, given an equation and given values for all variables except for the one that you are solving for. For this type of problem, you do a series of steps to both sides of the equation until the variable you are solving for is on one side of the equation by itself. You then put in the numbers that you are given.

Example 13.2.1

Solve for the variable a in

$$v_f^2 = v_i^2 + 2 \cdot a \cdot \Delta x$$

Given that $v_f = 2$, $v_i = 5$, and $\Delta x = 50$
(This is Equation 2.6.4, from Section 2.6.)

DOI: 10.1201/9781003005049-13

Solution:

We want to get a by itself. First, we get (a times something) by itself.

Subtract v_i^2 from both sides: $v_f^2 - v_i^2 = 2 \cdot a \cdot \Delta x$

Divide by everything that is multiplied by a, which is $2 \cdot \Delta x$: $(v_f^2 - v_i^2)/(2 \cdot \Delta x) = a$

Put in numbers from the problem: $a = (2^2 - 5^2)/(2 \cdot 50) = \underline{-0.21}$

Note: we switched the left and right sides, so that the equation was a = something instead of something = a. Since the sides are equal, this is OK to do.

Problem to Try Yourself

Equation 2.6.1 is:

$$v_f = v_i + a \cdot \Delta t$$

Find the value of Δt, given $v_f = 3$, $v_i = -2$, and $a = 6$

Solution:

Get $a \cdot \Delta t$ by itself, by subtracting v_i from both sides: $v_f - v_i = a \cdot \Delta t$

Get Δt by itself, by dividing both sides by a: $\Delta t = (v_f - v_i)/a$

Put in numbers: $\Delta t = (3 - (-2))/6 = \underline{0.833}$

13.3 EXPONENTIAL NUMBERS

With some topics in physics, the numbers get really large and really small. When this happens, the numbers are often written as exponentials. For example, 3.15 million (3,150,000) can be written as $3.15 \cdot 10^6$. The number that goes with the 10 is called an **exponent**. Notice that there are 6 numbers after the 3; this is where the 6 in the exponential comes from.

When you multiply two numbers that are written as exponentials, you multiply the numbers in front and then add the numbers in the exponential:

$$(4 \cdot 10^5) \cdot (2 \cdot 10^6) = (4 \cdot 2) \cdot 10^{(5+6)} = \underline{8 \cdot 10^{11}}$$

In the very last step, had the number in front been more than 10, we could have added one more to the exponent, for example: $12 \cdot 10^{11} = 1.2 \cdot 10^{12}$

When you divide two numbers that are written as exponentials, you divide the numbers in front and then subtract the exponents:

$$(4 \cdot 10^5)/(2 \cdot 10^6) = (4/2) \cdot 10^{(5-6)} = \underline{2 \cdot 10^{-1}}$$

Note: $2 \cdot 10^{-1}$ is the same number as 0.2

Example 13.3.1

Solve for m in the equation: $I = m \cdot r^2$, given that $I = 5.00 \cdot 10^{-5}$ and $r = 1.20 \cdot 10^{-2}$

Note: this is Equation 9.5.1.

Solution:

Divide both sides by r^2: $m = I/r^2$

Put in numbers: $m = (5.00 \cdot 10^{-5})/(1.20 \cdot 10^{-2})^2 = \underline{0.347}$

Note: if you wanted to, you could square the $1.20 \cdot 10^{-2}$ first, which comes to $1.44 \cdot 10^{-4}$, which is like $(1.2)^2 \cdot 10^{-2 + -2}$. Then, $(5.00 \cdot 10^{-5})/(1.44 \cdot 10^{-4}) = (5/1.44) \cdot 10^{-5 - (-4)} = 3.47 \cdot 10^{-1} = 0.347$

Problem to Try Yourself

Solve $L = I \cdot \omega$ (Equation 10.6.1) for ω, given $L = 1.6 \cdot 10^{-4}$ and $I = 3.45 \cdot 10^{-6}$

Solution:

Divide both sides by I: $\omega = L/I$

Put in numbers: $\omega = (1.6 \cdot 10^{-4})/(3.45 \cdot 10^{-6}) = \underline{46.4}$

Note: this is the same as $(1.6/3.45) \cdot 10^{-4 - (-6)}$

13.4 SOLVING WHEN SOMETHING IS SQUARED, PART 1

When an equation has a variable that is **squared** (exponent of 2), and the equation doesn't also have the same variable with exponent 1, you can solve for that variable by taking the square root. (Note: this means that $v^2 = 16$ could be solved by a square root, but $v^2 + v = 16$ could not.)

There's one complication, however. When you take the square root, the answer can be positive or negative. You will have to choose which answer is correct, based on the problem. For example, if you are solving for velocity and you know that the object is going downward (and that upward is positive), then you will choose negative. If you are just looking for the amount, you will probably want to choose positive. In the math, this shows up as a ±, which is called a "plus/minus sign".

Example 13.4.1

You throw a ball downward, and the behavior is described by the equation $v_f^2 = v_i^2 + 2 \cdot a \cdot \Delta x$. Solve for v_i, with $v_f = -10.5$, $a = -9.80$, and $\Delta x = -2.00$. Note that v_f and v_i are two different variables.

Note: this is Equation 2.6.4, from Section 2.6.

Solution:

Subtract $2 \cdot a \cdot \Delta x$ from both sides: $v_f^2 - 2 \cdot a \cdot \Delta x = v_i^2$

Take the square root: $v_i = \pm \sqrt{v_f^2 - 2 \cdot a \cdot \Delta x}$

Put in numbers: $v_i = \pm \sqrt{(-10.5)^2 - 2 \cdot (-9.80) \cdot (-2.00)}$

$v_i = \pm 8.43$

Note: the ± sign means that the math would let the value be either positive or negative. You have to make the choice yourself, based on the word problem this would have come from. In this case, the ball is thrown downward, which we will assume makes the initial velocity negative:

$\underline{v_i = -8.43}$

Problem to Try Yourself

Solve $W_{net} = \frac{1}{2} \cdot m \cdot v_f^2 - \frac{1}{2} \cdot m \cdot v_i^2$ (Equation 6.6.2) for v_f, given $W_{net} = 2.5$, $m = 3.2$, and $v_i = 7.6$. Assume that v_f has a direction of "to the right", and that the positive direction is "to the right".

Solution:

Add $\frac{1}{2} \cdot m \cdot v_i^2$ to both sides: $\frac{1}{2} \cdot m \cdot v_f^2 = W_{net} + \frac{1}{2} \cdot m \cdot v_i^2$

Divide both sides by ($\frac{1}{2} \cdot m$): $v_f^2 = (W_{net} + \frac{1}{2} \cdot m \cdot v_i^2)/(\frac{1}{2} \cdot m)$

Take the square root: $v_f = \pm \sqrt{\dfrac{W_{net} + 1/2 \cdot m \cdot v_i^2}{1/2 \cdot m}}$

Here, you choose the positive of the ± sign, since the object is going in the positive direction.

$$v_f = \sqrt{\dfrac{W_{net} + 1/2 \cdot m \cdot v_i^2}{1/2 \cdot m}}$$

Put in numbers: $v_f = \sqrt{\dfrac{2.5 + 1/2 \cdot 3.2 \cdot (7.6)^2}{1/2 \cdot 3.2}} = 7.70$

13.5 SOLVING WHEN SOMETHING IS SQUARED, PART 2

When you are solving for a variable that is squared and also has a linear term (something like $x^2 + x + \text{something else} = 0$), you need to use the **quadratic formula**:

If $a \cdot x^2 + b \cdot x + c = 0$, where a, b, and c are numbers and x is the variable you want to solve for,

$$x = \frac{-b \pm \sqrt{b^2 - 4 \cdot a \cdot c}}{2 \cdot a}$$

Notice that there is a ± sign in this equation. This means that there are two possible answers: one where + is used in the equation and one where − is used in the equation. You will need to use the problem to figure out which of the answers is correct. In physics, this equation is often used to solve for time with projectile motion (an object flying in the air), and one common result is that one of the two values for time comes out negative. When this happens, the positive time is the one to choose. (Occasionally both times are positive; that is often when the object is at a certain height twice, once on the way up and once on the way down.)

Example 13.5.1

Solve the equation $\Delta x = v_i \cdot t + \frac{1}{2} \cdot a \cdot (\Delta t)^2$ for Δt, with $\Delta x = -0.500$, $v_i = 1.00$, and $a = -9.80$.

Note: this is Equation 2.6.2

Solution:

First, arrange the equation so that it looks like (first number)· $(\Delta t)^2 +$ (second number)·$\Delta t +$ (third number) $= 0$.

Subtract Δx from both sides: $\frac{1}{2} \cdot a \cdot (\Delta t)^2 + v_i \cdot \Delta t - \Delta x = 0$

The number in front of $(\Delta t)^2$ ("a" in the quadratic formula equation, different from the variable a) is $\frac{1}{2} \cdot -9.80 = -4.90$. The number in front of Δt ("b" in the quadratic equation) is 1.00. The number without Δt ("c" in the quadratic equation) is 0.500. (Note: "c" is $-\Delta x$.)

Put these numbers into the quadratic equation:

$$\Delta t = \frac{-b \pm \sqrt{b^2 - 4 \cdot a \cdot c}}{2 \cdot a} = \frac{-1.00 \pm \sqrt{1.00^2 - 4 \cdot -4.90 \cdot 0.500}}{2 \cdot -4.90}$$

$$= -0.233 \, \text{and} + 0.437$$

In projectile motion problems (which this is based on), if one answer is negative, you choose the positive answer.

$t = 0.437$

Problem to Try Yourself

Solve $\Delta x = v_i t + \frac{1}{2} \cdot a \cdot (\Delta t)^2$ (Equation 2.6.2) for Δt, with $\Delta x = 1.75$, $v_i = 15$, and $a = -9.80$. Take the larger of the two times as the answer.

Solution:

First, arrange the equation so that it looks like (first number)· $(\Delta t)^2$ + (second number)· Δt + (third number) = 0.

Subtract Δx from both sides: $\frac{1}{2} \cdot a \cdot (\Delta t)^2 + v_i \cdot \Delta t - \Delta x = 0$

The number in front of t^2 ("a" in the quadratic formula equation, different from the variable a) is $\frac{1}{2} \cdot -9.80 = -4.90$. The number in front of t ("b" in the quadratic equation) is 15. The number without t ("c" in the quadratic equation) is -1.75. (Note: "c" is $-\Delta x$.)

Put these numbers into the quadratic equation:

$$\Delta t = \frac{-b \pm \sqrt{b^2 - 4 \cdot a \cdot c}}{2 \cdot a} = \frac{-15 \pm \sqrt{15^2 - 4 \cdot -4.90 \cdot -1.75}}{2 \cdot -4.90} = 0.121 \, \text{and} \, 2.94$$

We were asked to choose the larger value of Δt.

$t = 2.94$

13.6 TWO EQUATIONS AT ONCE

Some solutions in physics require the use of two equations at once. Usually the equations will have more than one variable in common. One example of this is the equations for two objects that collide in an elastic collision:

$$m_1 \cdot v_{1i} + m_2 \cdot v_{2i} = m_1 \cdot v_{1f} + m_2 \cdot v_{2f}$$

$$v_{1i} + v_{1f} = v_{2i} + v_{2f}$$

Note: these are Equations 7.4.1 and 7.6.1; Equation 7.6.1 is specifically from *Physics for Scientists and Engineers* by Tipler and Mosca (which is cited in the acknowledgements section).

In this setup, you will usually have two variables that are unknown. The strategy is to use one of the two equations to solve for one unknown variable in terms of the other, and then put in this result into the second equation wherever the first variable is located.

Example 13.6.1

Using the two equations above, solve for v_{1f} and v_{2f} if $m_1 = 2.90$, $m_2 = 3.60$, $v_{1i} = 1.20$ and $v_{2i} = -0.600$.

Solution:

Use the second equation to relate v_{1f} and v_{2f}. It doesn't matter which one we solve for to start; we'll choose v_{1f}.

Take the second equation, and subtract v_{1i} from both sides:

$$v_{1f} = v_{2i} + v_{2f} - v_{1i}$$

We have now solved one equation for one of the unknown variables (v_{1f}), in terms of the other unknown variable (v_{2f}) and two variables that we know the values of.

Next, put the equation for v_{1f} in for v_{1f} in the first equation:

$$m_1 \cdot v_{1i} + m_2 \cdot v_{2i} = m_1 \cdot (v_{2i} + v_{2f} - v_{1i}) + m_2 \cdot v_{2f}$$

Multiply out m_1 times the three velocities in parentheses:

$$m_1 \cdot v_{1i} + m_2 \cdot v_{2i} = m_1 \cdot v_{2i} + m_1 \cdot v_{2f} - m_1 \cdot v_{1i} + m_2 \cdot v_{2f}$$

v_{2f} is now the only unknown variable in this equation. We want to get it by itself. Start by moving all terms on the right that do not have v_{2f} in it.

Add $m_1 \cdot v_{1i}$ to both sides, and subtract $m_1 \cdot v_{2i}$ from both sides:

$$m_1 \cdot v_{1i} + m_2 \cdot v_{2i} + m_1 \cdot v_{1i} - m_1 \cdot v_{2i} = m_1 \cdot v_{2f} + m_2 \cdot v_{2f}$$

We can rewrite the right side as $(m_1 + m_2) \cdot v_{2f}$, since both sides have v_{2f} multiplied in it

$$m_1 \cdot v_{1i} + m_2 \cdot v_{2i} + m_1 \cdot v_{1i} - m_1 \cdot v_{2i} = (m_1 + m_2) \cdot v_{2f}$$

Divide both sides by $(m_1 + m_2)$:

$$(m_1 \cdot v_{1i} + m_2 \cdot v_{2i} + m_1 \cdot v_{1i} - m_1 \cdot v_{2i})/(m_1 + m_2) = v_{2f}$$

Put in numbers:
$v_{2f} = (2.90 \cdot 1.20 + 3.60 \cdot -0.600 + 2.90 \cdot 1.20 - 2.90 \cdot -0.600)/(2.90 + 3.60)$
$= \underline{1.01}$

We now have v_{2f}. To find v_{1f}, use the equation for v_{1f} that we solved for earlier in this problem:

$$v_{1f} = v_{2i} + v_{2f} - v_{1i} = -0.600 + 1.01 - 1.20 = \underline{-0.79}$$

Problem to Try Yourself

Solve $KE = \frac{1}{2} \cdot m \cdot v^2 + \frac{1}{2} \cdot I \cdot \omega^2$ and $v = \omega \cdot r$ (Equation 10.3.2) for v, given $r = 0.5$, $KE = 1.2$, $m = 1.5$, and $I = 0.25$. Take the positive answer for v.

Solution:

Solve $v = \omega \cdot r$ for ω, and then put that answer in for ω in $KE = \frac{1}{2} \cdot m \cdot v^2 + \frac{1}{2} \cdot I \cdot \omega^2$

Divide both sides of $v = \omega \cdot r$ by r: $\omega = v/r$

Put this into $KE = \frac{1}{2} \cdot m \cdot v^2 + \frac{1}{2} \cdot I \cdot \omega^2$:

$$KE = \frac{1}{2} \cdot m \cdot v^2 + \frac{1}{2} \cdot I \cdot (v/r)^2$$

Both of the terms on the right are equal to something times v^2:

$$KE = (\frac{1}{2} \cdot m) \cdot v^2 + (\frac{1}{2} \cdot I/r^2) \cdot v^2$$

Since both are multiplied by v^2, you can combine them:

$$KE = (\frac{1}{2} \cdot m + \frac{1}{2} \cdot I/r^2) \cdot v^2$$

Divide both sides by $(\frac{1}{2} \cdot m + \frac{1}{2} \cdot I/r^2)$:

$$v^2 = KE/(\frac{1}{2} \cdot m + \frac{1}{2} \cdot I/r^2)$$

Take the square root: $v = \pm \sqrt{\dfrac{KE}{\frac{1}{2} \cdot m + 1/2 \cdot I/r^2}}$

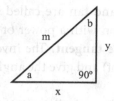

FIGURE 13.1 A typical right triangle trigonometry setup, with three angles (a, b, and 90°), a horizontal side with length x, a vertical side with length y, and a hypotenuse of length m (m for magnitude, like with a vector).

Take the positive answer, since the problem says to:

$$v = \sqrt{\dfrac{KE}{\dfrac{1}{2} \cdot m + 1/2 \cdot I/r^2}}$$

Put in numbers: $v = \sqrt{\dfrac{1.2}{\dfrac{1}{2} \cdot 1.5 + \dfrac{1}{2} \cdot 0.25/0.5^2}} = 0.980$

Note: you could now go back and solve for ω using ω = v/r, which comes to 1.96.

13.7 TRIGONOMETRY

Figure 13.1 shows a typical setup of right triangle trigonometry. The triangle has 3 sides, of length x, y, and m, and 3 angles, a (across from y), b (across from x), and 90 degrees (across from m). The side labeled m is often called the **hypotenuse**.

The names of x, y, and m are chosen because trigonometry is often used with vectors in this textbook. The side labeled x is usually the x component of a vector. The side labeled y is usually the y component of a vector. The side labeled m is usually the magnitude.

The lengths and angles relate in the following ways:

$m = \sqrt{x^2 + y^2}$

$a + b + 90° = 180°$

$\sin(a) = y/m$ and $a = \sin^{-1}(y/m)$

$\cos(a) = x/m$ and $a = \cos^{-1}(x/m)$

$\tan(a) = y/x$ and $a = \tan^{-1}(y/x)$

The functions sin, cos, and tan are called **sine, cosine, and tangent**. The versions of this function with a power of −1 are called **inverse sine, inverse cosine, and inverse tangent.** The inverse functions take a value for sine (or cosine or tangent) and give the angle with that value of sine (or cosine or tangent).

Side note: you could write a similar set of equations for angle b. One example would be tan(b) = x/y. In this textbook, we use angle a to prevent confusion. This is because the angles for vectors in this textbook are generally given with respect to the horizontal direction, which is angle a in figure 13.7.1, and not angle b.

Example 13.7.1

Looking at figure 13.7.1, if x = 12 and y = 15, what are:

A. m
B. cosine of angle a
C. sine of angle a
D. tangent of angle a
E. Angle a
F. Angle b

Solutions

A. $m = \sqrt{x^2 + y^2} = \sqrt{(12)^2 + (15)^2} = 19.2$

B. cos(a) = x/m = 12/19.2 = 0.625
C. sin(a) = y/m = 15/19.2 = 0.781
D. tan(a) = y/x = 15/12 = 1.25
E. a = cos⁻¹(x/m) or sin⁻¹(y/m) or tan⁻¹(y/x). Any of these come to
51.3°
F. a + b + 90° = 180°

Subtract 90° and a from both sides: b = 180° − 90° − a = 180° − 90° − 51.3° = 38.7°

Problem To Try Yourself

If x = 7 and y = 8, solve for:

A. m
B. angle a

C. cos(a)

D. angle b

Solution:

A. $m = \sqrt{x^2 + y^2} = \sqrt{(7)^2 + (8)^2} = 10.6$

B. $a = \tan^{-1}(y/x) = \tan^{-1}(8/7) = \underline{48.8°}$

C. $\cos(a) = \cos(48.8°) = \underline{0.659}$. Note: this is the same as x/m

D. $a + b + 90° = 180°$. Subtract 90° and a from both sides: $b = 180° - 90° - a = 180° - 90° - 48.8° = \underline{41.2°}$

13.8 CIRCLE GEOMETRY REVIEW

Chapters 8–10 in the textbook deal with moving in circles. A few equations related to circles are occasionally useful.

Number of degrees in a circle: 360

1 revolution = 360 degrees

Circumference of a circle: $2 \cdot \pi \cdot r$ (r is radius; circumference is the distance around the circle)

One example of where this is useful is the equation for the velocity of an object going through uniform circular motion (Chapter 8, Equation 8.3.3):

$v = \dfrac{2 \cdot \pi \cdot r}{T}$, where r is radius and T is the amount of time it takes to

go around the circle once. The $(2 \cdot \pi \cdot r)$ is the circumference, which is the

distance needed to go around the circle once.

13.9 CHAPTER 13 SUMMARY

This chapter reviews math that is necessary for the physics problems in this textbook.

When solving for a variable, do the same math to both sides of an equation until that variable is by itself.

If you are multiplying two exponential numbers, add the exponents. If you are dividing them, subtract the exponents.

If you are solving for a variable that is squared in an equation, take a square root and use a $\pm$ sign if there is no term with exponent 1 (like $v^2 = 16$), and use the quadratic equation if there is a term with exponent 1 (like $v^2 + v = 16$).

When solving two equations at once, you will likely be solving for two variables. Solve for one variable in terms of the other using one of the two equations, then use that in the second equation.

Based on figure 13.7.1, which is similar to a vector setup, the equations relating the side lengths and angles are:

$$m = \sqrt{x^2 + y^2}$$

$a + b + 90° = 180°$

$\sin(a) = y/m$ and $a = \sin^{-1}(y/m)$

$\cos(a) = x/m$ and $a = \cos^{-1}(x/m)$

$\tan(a) = y/x$ and $a = \tan^{-1}(y/x)$

Circles have 360 degrees, one revolution, and circumference equation $2 \cdot \pi \cdot r$.

Chapter 13 problem types:

Solving for a variable using algebra (Section 13.2)

Problems involving numbers with exponents (Section 13.3)

Problems where the variable you are solving for is squared (Sections 13.4 and 13.5), including using the quadratic formula (Section 13.5)

Problems where two equations are used together (Section 13.6)

Problems using trigonometry: sine, cosine, tangent, magnitude, and size of angle (Section 13.7)

Practice Problems

Note: many of the values here have zeroes on the end – for example, 4.00 instead of 4. This is keeping with significant figures, as discussed in Section 1.4. For the math practice in this chapter, you can ignore these extra zeros if it is helpful.

Section 13.2 (Algebra Problems – Solving for a Variable)
Note: Equations 2.6.1, 2.6.2, and 2.6.4 are used as examples here

1. Solve $v_f = v_i + a \cdot \Delta t$ for Δt, given that $v_f = -15.0$, $v_i = -1.00$ and $a = -9.80$.

Solution:

Subtract v_i from both sides: $v_f - v_i = a \cdot \Delta t$

Divide both sides by a: $\dfrac{v_f - v_i}{a} = \Delta t$

Put in numbers: $\Delta t = \dfrac{-15.0 - (-1.00)}{-9.80} = \underline{1.43}$

2. Solve $v_f = v_i + a \cdot \Delta t$ for v_i, given that $v_f = 10.0$, $\Delta t = 1.50$ and $a = 5.00$.

Solution:

Subtract a·Δt from both sides: $v_f - a \cdot \Delta t = v_i$

Put in numbers: $v_i = 10.0 - (5.00) \cdot (1.50) = \underline{2.50}$

3. Solve $v_f = v_i + a \cdot \Delta t$ for a, given that $v_f = 12.0$, $\Delta t = 3.00$ and $v_i = 8.00$.

Solution:

Subtract v_i from both sides: $v_f - v_i = a \cdot \Delta t$

Divide both sides by Δt: $\dfrac{v_f - v_i}{\Delta t} = a$

Put in numbers: $a = \dfrac{12.0 - (8.00)}{3.00} = \underline{1.33}$

4. Solve $v_f^2 = v_i^2 + 2 \cdot a \cdot \Delta x$ for Δx, given $v_f = -9.00$, $v_i = 0$, and $a = -9.80$.

Solution:

Subtract v_i^2 from both sides: $v_f^2 - v_i^2 = 2 \cdot a \cdot \Delta x$

Divide both sides by (2·a): $\dfrac{v_f^2 - v_i^2}{2 \cdot a} = \Delta x$

Put in numbers: $\Delta x = \dfrac{(-9.00)^2 - (0)^2}{2 \cdot (-9.8)} = \underline{-4.13}$

5. Solve $v_f^2 = v_i^2 + 2 \cdot a \cdot \Delta x$ for a, given $v_f = 15.0$, $v_i = 9.00$ and $\Delta x = 30.0$.

Solution:

Subtract v_i^2 from both sides: $v_f^2 - v_i^2 = 2 \cdot a \cdot \Delta x$

Divide both sides by $(2 \cdot \Delta x)$: $\dfrac{v_f^2 - v_i^2}{2 \cdot \Delta x} = a$

Put in numbers: $a = \dfrac{(15.0)^2 - (9.00)^2}{2 \cdot (30.0)} = \underline{2.40}$

6. Solve $\Delta x = v_i \cdot \Delta t + \dfrac{1}{2} \cdot a \cdot (\Delta t)^2$ for v_i, given $\Delta x = 0$, $\Delta t = 5.00$, and $a = -9.80$.

Solution:

Subtract $\dfrac{1}{2} \cdot a \cdot (\Delta t)^2$ from both sides: $\Delta x - \dfrac{1}{2} \cdot a \cdot (\Delta t)^2 = v_i \cdot \Delta t$

Divide both sides by Δt: $\dfrac{\Delta x - \dfrac{1}{2} \cdot a \cdot (\Delta t)^2}{\Delta t} = v_i$

Put in numbers: $v_i = \dfrac{0 - \dfrac{1}{2} \cdot (-9.80) \cdot (5.00)^2}{5.00} = \underline{24.5}$

7. Solve $\Delta x = v_i \cdot \Delta t + \dfrac{1}{2} \cdot a \cdot (\Delta t)^2$ for a, given $\Delta x = 50.0$, $\Delta t = 12.0$, and $v_i = 4.00$.

Solution:

Subtract $v_i \cdot \Delta t$ from both sides: $\Delta x - v_i \cdot \Delta t = \dfrac{1}{2} \cdot a \cdot (\Delta t)^2$

Divide both sides by $\dfrac{1}{2} \cdot (\Delta t)^2$: $\dfrac{\Delta x - v_i \cdot \Delta t}{\dfrac{1}{2} \cdot (\Delta t)^2} = a$

Put in numbers: $a = \dfrac{50.0 - (4.00) \cdot (12.0)}{\dfrac{1}{2} \cdot (12.0)^2} = \underline{0.0278}$

Section 13.3 (Exponential Numbers)
Note: Equations 9.5.1 and 10.6.1 will be used in this section

8. Solve $I = m \cdot r^2$ for m, given $I = 1.50 \cdot 10^{-5}$ and $r = 0.080$.

Solution:

Divide both sides by r^2: $I/r^2 = m$

Put in numbers: $m = (1.50 \cdot 10^{-5})/(0.080)^2 = \underline{2.34 \cdot 10^{-3}}$

9. Solve $I = m \cdot r^2$ for m, given $I = 7.50 \cdot 10^{-5}$ and $r = 0.0200$.

Solution:

Divide both sides by r^2: $I/r^2 = m$

Put in numbers: $m = (7.50 \cdot 10^{-5})/(0.020)^2 = \underline{0.188}$

10. Solve $L = I \cdot \omega$ for ω, given $L = 2.50 \cdot 10^{-4}$ and $I = \underline{3.50 \cdot 10^{-5}}$.

Solution:

Divide both sides by I: $L/I = \omega$

Put in numbers: $\omega = (2.50 \cdot 10^{-4})/(3.50 \cdot 10^{-5}) = \underline{7.14}$

11. Solve $L = I \cdot \omega$ for I, given $L = 2.50 \cdot 10^{-3}$ and $\omega = 2.70$.

Solution:

Divide both sides by ω: $L/\omega = I$

Put in numbers: $I = (2.50 \cdot 10^{-3})/(2.70) = \underline{9.26 \cdot 10^{-4}}$

Section 13.4 (Solving When Something is Squared, Part 1)
Note: Equations 2.6.4 and 6.6.2 will be used in this section

12. Solve $v_f^2 = v_i^2 + 2 \cdot a \cdot \Delta x$ for v_f, given $v_i = 8.00$, $a = -9.80$, and $\Delta x = 1.50$. Assume that v_f is in the negative direction.

Solution:

Take the square root of both sides: $v_f = \pm\sqrt{v_i^2 + 2 \cdot a \cdot \Delta x}$

Take the negative value, since v_f is in the negative direction:
$v_f = -\sqrt{v_i^2 + 2 \cdot a \cdot \Delta x}$

Put in numbers: $v_f = -\sqrt{(8.00)^2 + 2 \cdot (-9.80) \cdot (1.50)} = \underline{-5.88}$

13. Solve $v_f^2 = v_i^2 + 2 \cdot a \cdot \Delta x$ for v_f, given $v_i = -14.0$, $a = -9.80$, and $\Delta x = 5.00$. Assume that v_f is in the positive direction.

Solution:

Take the square root of both sides: $v_f = \pm\sqrt{v_i^2 + 2\cdot a \cdot \Delta x}$

Take the positive value, since v_f is in the positive direction: $v_f = +\sqrt{v_i^2 + 2\cdot a \cdot \Delta x}$

Put in numbers: $v_f = +\sqrt{(-14.0)^2 + 2\cdot(-9.80)\cdot(5.00)} = \underline{9.90}$

14. Solve $W_{net} = \frac{1}{2}\cdot m\cdot v_f^2 - \frac{1}{2}\cdot m\cdot v_i^2$ for v_f, given $W_{net} = 1.5$, $m = 0.8$, and $v_i = 2.0$. Assume that v_f is in the positive direction.

Solution:

Add $\frac{1}{2}\cdot m\cdot v_i^2$ to both sides: $W_{net} + \frac{1}{2}\cdot m\cdot v_i^2 = \frac{1}{2}\cdot m\cdot v_f^2$

Divide both sides by $(\frac{1}{2}\cdot m)$: $(W_{net} + \frac{1}{2}\cdot m\cdot v_i^2)/(\frac{1}{2}\cdot m) = v_f^2$
  Take the square root of both sides:
  $v_f = \pm\sqrt{\left(W_{net} + \frac{1}{2}\cdot m\cdot v_i^2\right)/\left(\frac{1}{2}\cdot m\right)}$

Take the positive value, since v_f is in the positive direction:

$v_f = +\sqrt{\left(W_{net} + \frac{1}{2}\cdot m\cdot v_i^2\right)/\left(\frac{1}{2}\cdot m\right)}$

Put in numbers: $v_f = +\sqrt{\left(1.5 + \frac{1}{2}\cdot 0.8\cdot(2.0)^2\right)/\left(\frac{1}{2}\cdot 0.8\right)} = \underline{2.78}$

15. Solve $W_{net} = \frac{1}{2}\cdot m\cdot v_f^2 - \frac{1}{2}\cdot m\cdot v_i^2$ for v_i, given $W_{net} = 8.5$, $m = 1.9$, and $v_f = 14.0$. Assume that v_i is in the negative direction.

Solution:

Subtract $\frac{1}{2}\cdot m\cdot v_f^2$ from both sides: $W_{net} - \frac{1}{2}\cdot m\cdot v_f^2 = -\frac{1}{2}\cdot m\cdot v_i^2$

Divide both sides by $(-\frac{1}{2}\cdot m)$: $(W_{net} - \frac{1}{2}\cdot m\cdot v_f^2)/(-\frac{1}{2}\cdot m) = v_i^2$
  Take the square root of both sides:
  $v_i = \pm\sqrt{\left(W_{net} - \frac{1}{2}\cdot m\cdot v_f^2\right)/\left(-\frac{1}{2}\cdot m\right)}$
Take the negative value, since v_i is in the negative direction:

$v_i = -\sqrt{\left(W_{net} - \frac{1}{2}\cdot m\cdot v_f^2\right)/\left(-\frac{1}{2}\cdot m\right)}$

Put in numbers: $v_i = -\sqrt{\left(8.5 - \frac{1}{2}\cdot1.9\cdot(14.0)^2\right)/\left(-\frac{1}{2}\cdot1.9\right)} = \underline{-13.7}$

*Section 13.5 (Solving When Something Is
Squared, Part 2 – Quadratic Formula)*

16. Solve $\Delta x = v_i \cdot \Delta t + \dfrac{1}{2}\cdot a \cdot (\Delta t)^2$ for Δt, given $\Delta x = -2.00$, $a = -9.80$, and

   $v_i = 2.00$. Choose the positive value.

Solution:

First, arrange the equation so that it looks like (first number)·$(\Delta t)^2$ + (second number)·Δt + (third number) = 0.

Subtract Δx from both sides: $\frac{1}{2}\cdot a\cdot(\Delta t)^2 + v_i\cdot\Delta t - \Delta x = 0$

The number in front of $(\Delta t)^2$ ("a" in the quadratic formula equation, different from the variable a) is $\frac{1}{2}\cdot-9.80 = -4.90$. The number in front of Δt ("b" in the quadratic equation) is 2.00. The number without Δt ("c" in the quadratic equation) is 2.00. (Note: "c" is $-\Delta x$.)

   Put these numbers into the quadratic equation:

$$\Delta t = \frac{-b \pm \sqrt{b^2 - 4\cdot a\cdot c}}{2\cdot a} = \frac{-2.00 \pm \sqrt{2.00^2 - 4\cdot-4.90\cdot2.00}}{2\cdot-4.90}$$

$$= \underline{-0.467 \text{ and} +0.875}$$

We'll choose the positive value, as the problem asks.

$\underline{\Delta t = 0.875}$

17. Solve $\Delta x = v_i \cdot \Delta t + \dfrac{1}{2}\cdot a \cdot (\Delta t)^2$ for Δt, given $\Delta x = -0.750$, $a = -9.80$,

   and $v_i = 5.00$. Choose the positive value.

Solution:

First, arrange the equation so that it looks like (first number)·$(\Delta t)^2$ + (second number)·Δt + (third number) = 0.

Subtract Δx from both sides: $\frac{1}{2}\cdot a\cdot(\Delta t)^2 + v_i\cdot\Delta t - \Delta x = 0$

The number in front of $(\Delta t)^2$ ("a" in the quadratic formula equation, different from the variable a) is $\frac{1}{2} \cdot -9.80 = -4.90$. The number in front of Δt ("b" in the quadratic equation) is 5.00. The number without Δt ("c" in the quadratic equation) is 0.750. (Note: "c" is $-\Delta x$.)

Put these numbers into the quadratic equation:

$$\Delta t = \frac{-b \pm \sqrt{b^2 - 4 \cdot a \cdot c}}{2 \cdot a} = \frac{-5.00 \pm \sqrt{5.00^2 - 4 \cdot -4.90 \cdot 0.750}}{2 \cdot -4.90}$$

$$= -0.133 \text{ and } +1.15$$

We'll choose the positive value, as the problem asks.

$\underline{\Delta t = 1.15}$

18. Solve $\Delta x = v_i \cdot \Delta t + \dfrac{1}{2} \cdot a \cdot (\Delta t)^2$ for Δt, given $\Delta x = -5.00$, a $= -9.80$, and $v_i = 7.50$. Choose the positive value.

Solution:

First, arrange the equation so that it looks like (first number)$\cdot(\Delta t)^2$ + (second number)$\cdot \Delta t$ + (third number) $= 0$.

Subtract Δx from both sides: $\frac{1}{2} \cdot a \cdot (\Delta t)^2 + v_i \cdot \Delta t - \Delta x = 0$

The number in front of $(\Delta t)^2$ ("a" in the quadratic formula equation, different from the variable a) is $\frac{1}{2} \cdot -9.80 = -4.90$. The number in front of Δt ("b" in the quadratic equation) is 7.50. The number without Δt ("c" in the quadratic equation) is 5.00. (Note: "c" is $-\Delta x$.)

Put these numbers into the quadratic equation:

$$\Delta t = \frac{-b \pm \sqrt{b^2 - 4 \cdot a \cdot c}}{2 \cdot a} = \frac{-7.50 \pm \sqrt{7.50^2 - 4 \cdot -4.90 \cdot 5.00}}{2 \cdot -4.90}$$

$$= -0.502 \text{ and } +2.03$$

We'll choose the positive value, as the problem asks.

$\underline{\Delta t = 2.03}$

Section 13.6 (Two Equations at Once)
Note: equations 7.4.1, 7.5.1, 7.6.1, 10.2.2 and 10.3.2 will be used in this section. Equation 7.6.1 is a form of an equation seen in *Physics for Scientists and Engineers* by Tipler and Mosca.

19. Using $m_1 \cdot v_{1i} + m_2 \cdot v_{2i} = m_1 \cdot v_{1f} + m_2 \cdot v_{2f}$ and $v_{1f} = v_{2f}$ solve for v_{1f} and v_{2f} given $m_1 = 80.0$, $m_2 = 110$, $v_{1i} = 2.00$ and $v_{2i} = -1.00$.

Solution:

Since $v_{1f} = v_{2f}$, put v_{1f} in for v_{2f} in $m_1 \cdot v_{1i} + m_2 \cdot v_{2i} = m_1 \cdot v_{1f} + m_2 \cdot v_{2f}$:

$$m_1 \cdot v_{1i} + m_2 \cdot v_{2i} = m_1 \cdot v_{1f} + m_2 \cdot v_{1f}$$

Since both terms on the right side now have v_{1f} in them, we can rewrite (factor) the right side: $m_1 \cdot v_{1i} + m_2 \cdot v_{2i} = v_{1f}(m_1 + m_2)$

Divide both sides by $(m_1 + m_2)$: $v_{1f} = (m_1 \cdot v_{1i} + m_2 \cdot v_{2i})/(m_1 + m_2)$

Put in numbers: $v_{1f} = (80.0 \cdot 2.00 + 110 \cdot -1.00)/(80.0 + 110) = \underline{0.263}$

$v_{2f} = \underline{0.263}$ as well, since $v_{1f} = v_{2f}$

20. Using $m_1 \cdot v_{1i} + m_2 \cdot v_{2i} = m_1 \cdot v_{1f} + m_2 \cdot v_{2f}$ and $v_{1i} + v_{1f} = v_{2i} + v_{2f}$ solve for v_{1f} and v_{2f} given $m_1 = 80.0$, $m_2 = 110$, $v_{1i} = 2.00$ and $v_{2i} = -1.00$.

Solution:

Use the second equation to relate v_{1f} and v_{2f}. It doesn't matter which one we solve for to start; we'll choose v_{1f}.

Take the second equation, and subtract v_{1i} from both sides:

$$v_{1f} = v_{2i} + v_{2f} - v_{1i}$$

We have now solved one equation for one of the unknown variables (v_{1f}), in terms of the other unknown variable (v_{2f}) and two variables that we know the values of.

Next, put the equation for v_{1f} in for v_{1f} in the first equation:

$$m_1 \cdot v_{1i} + m_2 \cdot v_{2i} = m_1 \cdot (v_{2i} + v_{2f} - v_{1i}) + m_2 \cdot v_{2f}$$

Multiply out m_1 times the three velocities in parentheses:

$$m_1 \cdot v_{1i} + m_2 \cdot v_{2i} = m_1 \cdot v_{2i} + m_1 \cdot v_{2f} - m_1 \cdot v_{1i} + m_2 \cdot v_{2f}$$

v_{2f} is now the only unknown variable in this equation. We want to get it by itself. Start by moving all terms on the right that do not have v_{2f} in it.

Add $m_1 \cdot v_{1i}$ to both sides, and subtract $m_1 \cdot v_{2i}$ from both sides:

$$m_1 \cdot v_{1i} + m_2 \cdot v_{2i} + m_1 \cdot v_{1i} - m_1 \cdot v_{2i} = m_1 \cdot v_{2f} + m_2 \cdot v_{2f}$$

We can rewrite the right side as $(m_1 + m_2) \cdot v_{2f}$, since both sides have v_{2f} multiplied in it

$$m_1 \cdot v_{1i} + m_2 \cdot v_{2i} + m_1 \cdot v_{1i} - m_1 \cdot v_{2i} = (m_1 + m_2) \cdot v_{2f}$$

Divide both sides by $(m_1 + m_2)$:

$$(m_1 \cdot v_{1i} + m_2 \cdot v_{2i} + m_1 \cdot v_{1i} - m_1 \cdot v_{2i})/(m_1 + m_2) = v_{2f}$$

Put in numbers:

$$v_{2f} = (80.0 \cdot 2.00 + 110 \cdot -1.00 + 80.0 \cdot 2.00 - 80.0 \cdot -1.00)/(80.0 + 110) = \underline{1.53}$$

We now have v_{2f}. To find v_{1f}, use the equation for v_{1f} that we solved for earlier in this problem:

$$v_{1f} = v_{2i} + v_{2f} - v_{1i} = -1.00 + 1.53 - 2.00 = \underline{-1.47}$$

21. Solve $KE = \frac{1}{2} \cdot m \cdot v^2 + \frac{1}{2} \cdot I \cdot \omega^2$ and $v = \omega \cdot r$ for KE, given $r = 0.300$, $v = 2.50$, $m = 1.5$, and $I = 0.150$.

Solution:

Take $v = \omega \cdot r$ and divide both sides by r: $\omega = v/r$

Put v/r in for ω in $KE = \frac{1}{2} \cdot m \cdot v^2 + \frac{1}{2} \cdot I \cdot \omega^2$: $KE = \frac{1}{2} \cdot m \cdot v^2 + \frac{1}{2} \cdot I \cdot (v/r)^2$

Put in numbers: $KE = \frac{1}{2} \cdot (1.5) \cdot (2.50)^2 + \frac{1}{2} \cdot (0.150) \cdot (2.50/0.300)^2 = \underline{9.90}$

22. Using $m_1 \cdot v_{1i} + m_2 \cdot v_{2i} = m_1 \cdot v_{1f} + m_2 \cdot v_{2f}$ and $v_{1f} = v_{2f}$ solve for v_{1f} and v_{2f} given $m_1 = 50.0$, $m_2 = 90.0$, $v_{1i} = -1.75$ and $v_{2i} = 1.80$.

Solution:

Since $v_{1f} = v_{2f}$, put v_{1f} in for v_{2f} in $m_1 \cdot v_{1i} + m_2 \cdot v_{2i} = m_1 \cdot v_{1f} + m_2 \cdot v_{2f}$:

$$m_1 \cdot v_{1i} + m_2 \cdot v_{2i} = m_1 \cdot v_{1f} + m_2 \cdot v_{1f}$$

Since both terms on the right side now have v_{1f} in them, we can rewrite (factor) the right side: $m_1 \cdot v_{1i} + m_2 \cdot v_{2i} = v_{1f}(m_1 + m_2)$

Divide both sides by $(m_1 + m_2)$: $v_{1f} = (m_1 \cdot v_{1i} + m_2 \cdot v_{2i})/(m_1 + m_2)$

Put in numbers: $v_{1f} = (50.0 \cdot -1.75 + 90.0 \cdot 1.80)/(50.0 + 90.0) = \underline{0.532}$

$v_{2f} = \underline{0.532}$ as well, since $v_{1f} = v_{2f}$

23. Using $m_1 \cdot v_{1i} + m_2 \cdot v_{2i} = m_1 \cdot v_{1f} + m_2 \cdot v_{2f}$ and $v_{1i} + v_{1f} = v_{2i} + v_{2f}$ solve for v_{1f} and v_{2f} given $m_1 = 50.0$, $m_2 = 90.0$, $v_{1i} = -1.75$ and $v_{2i} = 1.80$.

Solution:

Use the second equation to relate v_{1f} and v_{2f}. It doesn't matter which one we solve for to start; we'll choose v_{1f}.

Take the second equation, and subtract v_{1i} from both sides:

$$v_{1f} = v_{2i} + v_{2f} - v_{1i}$$

We have now solved one equation for one of the unknown variables (v_{1f}), in terms of the other unknown variable (v_{2f}) and two variables that we know the values of.

Next, put the equation for v_{1f} in for v_{1f} in the first equation:

$$m_1 \cdot v_{1i} + m_2 \cdot v_{2i} = m_1 \cdot (v_{2i} + v_{2f} - v_{1i}) + m_2 \cdot v_{2f}$$

Multiply out m_1 times the three velocities in parentheses:

$$m_1 \cdot v_{1i} + m_2 \cdot v_{2i} = m_1 \cdot v_{2i} + m_1 \cdot v_{2f} - m_1 \cdot v_{1i} + m_2 \cdot v_{2f}$$

v_{2f} is now the only unknown variable in this equation. We want to get it by itself. Start by moving all terms on the right that do not have v_{2f} in it.

Add $m_1 \cdot v_{1i}$ to both sides, and subtract $m_1 \cdot v_{2i}$ from both sides:

$$m_1 \cdot v_{1i} + m_2 \cdot v_{2i} + m_1 \cdot v_{1i} - m_1 \cdot v_{2i} = m_1 \cdot v_{2f} + m_2 \cdot v_{2f}$$

We can rewrite the right side as $(m_1 + m_2) \cdot v_{2f}$, since both sides have v_{2f} multiplied in it

$$m_1 \cdot v_{1i} + m_2 \cdot v_{2i} + m_1 \cdot v_{1i} - m_1 \cdot v_{2i} = (m_1 + m_2) \cdot v_{2f}$$

Divide both sides by $(m_1 + m_2)$:

$$(m_1 \cdot v_{1i} + m_2 \cdot v_{2i} + m_1 \cdot v_{1i} - m_1 \cdot v_{2i})/(m_1 + m_2) = v_{2f}$$

Put in numbers:

$v_{2f} = (50.0 \cdot -1.75 + 90.0 \cdot 1.80 + 50.0 \cdot -1.75 - 50.0 \cdot 1.80)/$
$(50.0 + 90.0) = \underline{-0.736}$

We now have v_{2f}. To find v_{1f}, use the equation for v_{1f} that we solved for earlier in this problem:

$$v_{1f} = v_{2i} + v_{2f} - v_{1i} = 1.80 + -0.736 - (-1.75) = \underline{2.81}$$

24. Solve $KE = \frac{1}{2} \cdot m \cdot v^2 + \frac{1}{2} \cdot I \cdot \omega^2$ and $v = \omega \cdot r$ (Equation 10.3.2) for KE, given $r = 0.500$, $\omega = 7.00$, $m = 0.750$, and $I = 0.0500$.

Solution:

Since $v = \omega \cdot r$, put $\omega \cdot r$ in for v in $KE = \frac{1}{2} \cdot m \cdot v^2 + \frac{1}{2} \cdot I \cdot \omega^2$:
$KE = \frac{1}{2} \cdot m \cdot (\omega \cdot r)^2 + \frac{1}{2} \cdot I \cdot \omega^2$

Put in numbers: $KE = \frac{1}{2} \cdot (0.750) \cdot (7.00 \cdot 0.500)^2 + \frac{1}{2} \cdot (0.0500) \cdot (7.00)^2 = \underline{5.82}$

Section 13.7 (Trigonometry)

25. Using the figure (figure 13.2), if $x = 3.50$ and $y = 4.75$, what are m, a, b, cos(a), sin(a) and tan(a)?

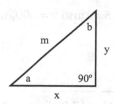

FIGURE 13.2 A typical right triangle trigonometry setup, with three angles (a, b, and 90°), a horizontal side with length x, a vertical side with length y, and a hypotenuse of length m (m for magnitude, like with a vector).

Solution:

$$m = \sqrt{x^2 + y^2} = \sqrt{(3.50)^2 + (4.75)^2} = \underline{5.90}$$

$$a = \tan^{-1}(y/x) = \tan^{-1}(4.75/3.50) = \underline{53.6°}$$

$$a + b + 90° = 180°$$

Subtract a and 90° from both sides: $b = 180° - 90° - a = 180° - 90° - 53.6° = \underline{36.4°}$

$\cos(a) = \cos(53.6°) = \underline{0.593}$ (note: this is equal to x/m = 3.50/5.90)

$\sin(a) = \sin(53.6°) = \underline{0.805}$ (note: this is equal to y/m = 4.75/5.90)

$\tan(a) = \tan(53.6°) = \underline{1.36}$ (note: this is equal to y/x = 4.75/3.50)

26. Using the figure, if $a = 32.0°$ and $m = 5.00$, what are x, y, b, and tan(a)?

Solution:

x: $\cos(a) = x/m$ multiply both sides by m: $x = m \cdot \cos(a) = 5.00 \cdot \cos(32.0°) = \underline{4.24}$

y: $\sin(a) = y/m$ multiply both sides by m: $y = m \cdot \sin(a) = 5.00 \cdot \sin(32.0°) = \underline{2.65}$

b: $a + b + 90° = 180°$ subtract a and 90° from both sides: $b = 180° - 90° - a$

$b = 180° - 90° - 32.0° = \underline{58.0°}$

$\tan(a) = \tan(32.0°) = \underline{0.625}$ (note: this is equal to y/x = 2.65/4.24)

27. Using the figure, if $x = 8.00$ and $y = 7.00$, what are m, a, b, cos(a), sin(a) and tan(a)?

Solution:

$m = \sqrt{x^2 + y^2} = \sqrt{(8.00)^2 + (7.00)^2} = \underline{10.6}$

$a = \tan^{-1}(y/x) = \tan^{-1}(7.00/8.00) = \underline{41.2°}$

$a + b + 90° = 180°$ Subtract a and 90° from both sides: $b = 180° - 90° - a$

$b = 180° - 90° - 41.2° = \underline{48.8°}$

$\cos(a) = \cos(41.2°) = \underline{0.753}$ (note: this is equal to $x/m = 8.00/10.6$)

$\sin(a) = \sin(41.2°) = \underline{0.659}$ (note: this is equal to $y/m = 7.00/10.6$)

$\tan(a) = \tan(41.2°) = \underline{0.875}$ (note: this is equal to $y/x = 7.00/8.00$)

28. Using the figure, if $a = 50.0°$ and $m = 2.30$, what are x, y, b, and tan(a)?

Solution:

x: $\cos(a) = x/m$ multiply both sides by m: $x = m \cdot \cos(a) = 2.30 \cdot \cos(50.0°) = \underline{1.48}$

y: $\sin(a) = y/m$ multiply both sides by m: $y = m \cdot \sin(a) = 2.30 \cdot \sin(50.0°) = \underline{1.76}$

b: $a + b + 90° = 180°$ subtract a and 90° from both sides: $b = 180° - 90° - a$

$b = 180° - 90° - 50.0° = \underline{40.0°}$

$\tan(a) = \tan(50.0°) = \underline{1.19}$ (note: this is equal to $y/x = 1.76/1.48$)

Index

Printed in the United States
by Baker & Taylor Publisher Services